W0235136

Advances in
Nephrourology

ETTORE MAJORANA INTERNATIONAL SCIENCE SERIES

Series Editor:
Antonino Zichichi
European Physical Society
Geneva, Switzerland

(LIFE SCIENCES)

Advances in Nephrourology

Edited by

Michele Pavone-Macaluso

University Polyclinic Hospital
Palermo, Italy

and

Philip H. Smith

St. James's University Hospital
Leeds, England

Associate Editors

Antonio Vercellone
University of Turin
Turin, Italy

Rosario Maiorca
United Hospitals
Brescia, Italy

Ugo Rotolo
University Polyclinic Hospital
Palermo, Italy

Plenum Press · New York and London

Library of Congress Cataloging in Publication Data

International Nephrourological Course (3rd : 1980 : Erice, Italy)
  Advances in nephrourology.

  (Ettore Majorana international science series. Life sciences; v. 9)
  "Proceedings of the Third International Nephrourological Course, held May 12-18, 1980, in Erice, Sicily" — T.p. verso.
  Bibliography: p.
  Includes index.
  1. Kidneys—Calculi—Congresses. 2. Kidneys—Diseases—Immunological aspects—Congresses. I. Pavone—Macaluso, M. II. Smith, Philip Henry. III. Title. IV. Series. [DNLM: 1. Kidney diseases—Immunology—Congresses. 2. Kidney calculi—Congresses. W1 ET712M v. 9/WJ 300 I598 1980a]
  RC916.I56 1980   616.6′1   81-13934
  ISBN 978-1-4684-8946-0   ISBN 978-1-4684-8944-6 (eBook)   AACR2
  DOI 10.1007/978-1-4684-8944-6

Proceedings of the Third International Nephrourological Course, held May 12–18, 1980, in Erice, Sicily

© 1981 Plenum Press, New York
Softcover reprint of the hardcover 1st edition 1981
A Division of Plenum Publishing Corporation
233 Spring Street, New York, N.Y. 10013

All rights reserved

No part of this book may be reproduced, stored in a retrieval system, or transmitted, in any form or by any means, electronic, mechanical, photocopying, microfilming, recording, or otherwise, without written permission from the publisher

PREFACE

This volume is a report of the proceedings of the Third
International Nephro-Urological Course held in the Ettore Majorana
Centre for Scientific Culture, Erice, Sicily, from 12th to 18th
May 1980.

Contributions were accepted on the understanding that the
editors could make certain changes leading towards a uniformity of
style but accepting as a priority the importance of early publication, if necessary at the expense of stylistic perfection.

The meeting, directed by A. Vercellone (Torino), R. Maiorca
(Brescia) and M. Pavone-Macaluso (Palermo), was sponsored by: the
Italian Associations of Nephrology, Immunology and Urology; the
Ministries of Scientific Research and Public Education of the
Republic of Italy; the Sicilian Regional Government; the National
Research Council and the University of Palermo.

Immunologic problems in renal disease and metabolic and medical
aspects of urolithiasis were the two subjects of the Congress,
which was attended by numerous invited speakers and participants.

The first part was introduced by A. Vercellone (Torino), who
discussed the major steps in the development and the present per-
spectives of nephrology, a relatively newly born science, recognized
as such only in 1960. He called attention to the great significance
of our present knowledge of the immunologic mechanisms (circulating
immune complexes or in situ mounting, cellular immunity, activation
of complement) which are involved in the pathogenesis of glomerulonephritis. The routine use of assays capable of elucidating the
abnormalities of such mechanisms will increase the accuracy of our
diagnostic assessment and allow new rational approaches and treatments in the fields of glomerulonephritis and kidney transplantation.

The second part, devoted to medical and metabolic aspects of urolithiasis, was introduced by S. Maiorca (Brescia). It was pointed out that stone disease is very frequent in the Mediterranean area, especially in the southern regions of the Italian peninsula. The increased incidence of calcium-phosphate and oxalate stones may depend on the high calorie content and protein intake of the Western industrialized countries. A decrease of uric acid lithiasis can be observed, particularly in Europe, although it is still endemic in some parts of Asia. A more detailed investigation of certain high-risk groups will undoubtedly elucidate some obscure points in the genesis of urolithiasis.

Another important point of interest, which led to a stimulating discussion, was the problem of the pathophysiology and diagnosis of calcium nephrolithiasis. The pathogenesis of absorptive hypercalciuria seems to be due to enhanced renal synthesis and to high serum levels of 1-25 $(OH)_2D3$. However, some patients with absorptive hypercalciuria show a vitamin-D independent calcium hyperabsorption.

The concept of a primary renal disease, the so called "renal calcium leak theory", as responsible for idiopathic hypercalciuria, was an important and much debated issue of discussion.

A significant step forward is the prognostic value of crystalluria in the follow-up of stone patients. By this easy, quick and simple method it is possible to observe microscopically the type, the number, the size and also the aggregation of crystals in the urinary sediment. Another development of great potential is the concept of the treatment of urinary tract infection by urease inhibitors.

We would like to thank Miss Francesca Montano for collating the material and typing the manuscript and gratefully acknowledge the cooperation and the generous help provided by Ente Fiuggi, Rome, for the section devoted to urolithiasis.

The next Course of the International School of Urology and Nephrology will deal with tumours of the kidney and of the prostate. It will be held in Erice, 2-12 June 1981.

CONTENTS

PART 2

MEDICAL AND METABOLIC ASPECTS OF UROLITHIASIS

CONTENTS

PART 1: IMMUNOLOGY OF RENAL DISEASE

IMMUNOLOGICAL PROBLEMS IN KIDNEY DISEASE - AN INTRODUCTION

A. Vercellone

Department of Nephrology
University of Turin
Nephrology and Dialysis Unit
St. John The Baptist and Turin Greater Hospital
Turin (Italy) .

Nephrology is a relatively newly-born science, first recognized as such in 1960, the year of our 1st International Congress. Since then, it has rapidly expanded both in theory and in practice. Progress in nephrology is probably of greater importance than that made in other areas of medical research during the same period of time. Some of the most outstanding contributions are related to renal pathophysiology investigation of kidney function and body electrolytes; new classification of renal diseases; the pathogenesis, management and therapy of hypertension, and stone formation and allied disorders. Finally, the new revolutionary approaches to replace kidney function, both theoretical and technological, deserve special emphasis.

All these advances. have certainly proved to be of major importance in the evolution of nephrology. However, immunology is nowadays felt to have become a major area of research, although in the past its importance - albeit obscure - had already been acknowledged. The pathogenesis of glomerulonephritis, the frustrating chapter of its management as well as the complex area of renal transplantation, are all closely related to immunologic problems. Advances in immunology have greatly improved our knowledge of biology, pathophysiology and clinical aspects of renal diseases.

Over the last few years, we have witnessed an ever increasing body of new revolutionary concepts, deriving from a close contact between basic immunologists and clinical nephrologists.

Seldom has the cooperation between workers of different fields given so much to both.

In fact, immunologists have found in nephrology a most remarkable opportunity to verify a large series of observations developed from the theoretical models of experimental pathology.

On the other hand, clinical nephrologists have become more and more interested in the close relationships between experimental pathology and their own clinical observation. This has led to new interpretations of the pathogenesis of various nephropathies and the approach to treatment and to new trends for future research. The evolution of nephrology has been more rapid than that of many other specialities of clinical medicine mainly due to ready availability of renal biopsy. The unique properties of the renal microvasculature, anatomical functional and hydrodynamic, which all facilitate the trapping of immune complexes in the kidney rather than in other organs, have been more fully clarified in recent years.

Therefore, among the many branches of medical science interdigitating on the common ground of nephrology, immunology in particular has provided a new impulse for development and research in clinical nephrology.

From the clinical nephrologists' standpoint, many stimulating advances have been made. Thanks to increasing knowledge of the immunological basis of renal diseases we are now capable of offering a more precise classification of renal disorders, especially of glomerulonephritis. However, we have become more and more convinced that classification, whilst helpful, must not be accepted too schematically. Today we are inclined to interpret renal diseases in the light of pathogenetic concepts that allow us to understand the many different types of nephritis both on pathogenetic and histo-pathologic grounds. We are also improving our understanding of the often undefined differences between organ and systemic disease and, also in the latter, the recognition of features not yet included in the present classification. The pathogenesis of glomerulonephritis has been further complicated beyond the classical models of anti-kidney

antibodies by Masugi and Stenblay and of circulating immune complexes
by Dixon and Germuth. There has been accumulating evidence both on
experimental and clinical grounds that other immunological mechanisms
may be implicated, such as _in situ_ mounting of complexes, cell-
mediated immunity, and activation of complement pathways by other
factors. Furthermore, the recognition of an increasing number of
mediators found be able to mediate immune complex deposition and
local tissue injury, has opened new ways for further study and new
therapeutic outlooks.

So far, an enormous amount of work has been done, thus allowing
us to establish new concepts on inflammation in general and on the
renal lesions, in particular. From this evolution, besides the amaz-
ing conceptual enrichment, has resulted the renewal of our diagnostic
and therapeutic schemes. More particularly there has been increasing
emphasis on the relevance of immunohistologic techniques in diagnosis:
the recognition of the site and the type of the immune deposits has
allowed a deep insight into pathogenetic mechanisms and has led to
a significant augmentation of our diagnostic capabilities as compared
to classical histology.

Also, perhaps in a more gradual way, new methods have been developed that allow us to follow the humoral changes concomitant with
and responsible for the several immunopathological mechanisms at
work.

On the wave of these recent advances, the detection of circulating immune complexes, the wide range of abnormalities in the antibody production, the serum levels of different complement components
and the detection of several mediators and of their catabolites,
that in the past were sophisticated topics in highly-specialized
centers of basic research, have nowadays become invaluable markers
for the diagnosis and management of immunological renal diseases in
their different clinical presentations.

These techiques, as they represent markers for detection of the
disease and for monitoring of their activity, can be used for the
longitudinal follow-up of patients. This has guided us to a better
interpretation of clinical variations from the natural history of
the disease. It is also very helpful to follow the response to therapy as well as to offer a more refined evaluation also for the
antigen involved. Furthermore, their application to various immu-

nologically-mediated nephropathies has suggested that, in the absence
or eclipse of the already known mechanisms, others may be envisaged,
such as cell-mediated immunity.

The overall effect is the availability of new tests and techniques, the results of which may all in some cases point to a well-
defined pathogenetic entity. In other cases, the results remain
unexplained and contradictory but, in any case, they stimulate
further research.

Facing the new concepts and methodologies, we have witnessed
therapeutic advances although these have been less exciting. However,
looking objectively at the new therapeutic goals so far achieved, we
may say that a positive evolution exists. In fact, we are now capable
of understanding more completely the effects of long-established drugs
which until a few years ago were used empirically. At the same time,
our therapeutic schemes designed to modify the immune response have
been based on more rational grounds in relation to the disease, which
has resulted in the reduction of their side effects and/or in the
enhancement of their pharmacologic activity. Nevertheless, there
have been novelties such as the increasing knowledge and technology
for the artificial clearance of immune complexes out of the circulation by means of plasma exchange and cryopheresis, the induction of
immune unresponsiveness to the transplanted kidney by blood transfusion, the still experimental hypothesis of immunological "traps"
and the proposal for the new strategy of immune stimulation although
this still awaits confirmation in clinical nephrology.

All these new concepts are nowadays an indispensable patrimony
for the clinical nephrologist.

PATHOPHYSIOLOGY OF INFLAMMATION

G. Camussi, C. Tetta, F. Bussolino, C. Masera and
A. Vercellone

Laboratorio di Immunopatologia
Cattedra di Nefrologia Medica
Universita di Torino
Divisione di Nefrologia e Dialisi
Ospedale Maggiore S. G. Battista
Turin (Italy)

ABSTRACT

Inflammation is a "surface" phenomenon triggered by immunological and non-immunological stimuli, capable of activating both
humoral and cellular systems, normally present in the body in an
inactive state and regulated by systemic inhibitors. Among the
cellular systems, we concentrated on the mechanisms responsible
for the activation of polymorphonuclear neutrophils (PMN), basophils, platelets and mononuclear phagocytes in inflammation.

As an early event in immune complex (IC)-induced inflammation,
PMN aggregation and increased adhesiveness to vessel-lining endothelium appears crucial to their recruitment and sequestration at
the inflammatory site. The process of PMN aggregation is triggered
by several membrane-active mediators such as C5a anaphylotoxin,
neutrophil-derived cationic proteins (CP) and particularly their
serum carboxypeptidase-derived, desarginated products as well as a
phospholipid mediator (1-O-alkyl-2-acetyl- glyceryl-3-phosphoryl-
choline), released from stimulated PMN, undistinguishable both
structurally and functionally from the platelet-activating factor

(PAF). The latter mediator may be suggested as the final common .
mediator of IC-induced PMN aggregation. PMN response to chemotactic
stimuli results in their emigration out of the vessels in the
inflammed tissues. PMN membrane activated by specific receptors for
Fc and C3b signals the initiation of metabolic and functional
activities, finally leading to the discharge of their inflammatory
constituents (oxygen-derived radicals, products from arachidonic
acid and a series of enzymes).

    Basophils are responsive in vitro to various stimuli occuring
in vivo in inflammation. Besides the IgE-mediated mechanism, C3a
and C5a anaphylotoxins as well as neutrophil CP have been shown to
have degranulating properties on basophil-rich mixed leukocyte
populations. Besides the preformed mediators such as histamine,
serotonin and heparin, other substances, such as the slow-reacting
substances of anaphylaxis (SRS-A) and arachidonic acid-derived
metabolites, are generated as the secretory process start. As for
the latter substances, PAF is released from mixed leukocyte populations challenged with C5a, CP and specific antigen concomitantly
with basophil degranulation and histamine liberation.

    Platelets are instrumental both in the onset and the amplifi-
cation of vascular permeability during IC diseases. Their involvement
may be looked at as the final end-result of many interrelating
mechanism of inflammatory injury. Platelets interact with IC either
directly through membrane receptors for Fc or indirectly with PAF
released from IC-stimulated leukocytes as well as with injured
endothelium, thus leading to platelet aggregation and "release
reaction".

    Several stimuli transform circulating monocytes into their
activated counterpart, the macrophages. The mechanisms responsible
for the recruitment and sequestration of macrophages appear to be
much like those described earlier for PMN. Macrophages infiltrate
the inflammed tissues and greatly contribute as they discharge
lysosomal enzymes to tissue injury. Furthermore, the recent demonstration by our laboratory that stimulated macrophages release PAF
and, more interestingly, represent the richest leukocyte reservoir
of this mediator, suggests that these cells also may involve platelets in inflammation.

IC   =   immune complexes
PMN   =   polymorphonuclear leukocytes
AA   =   arachidonic acid
CP   =   cationic proteins
cGMP =   3'-,5'-cyclic guanosine monophosphate
cAMP =   3'-,5'-cyclic adenosine monophosphate
HETE =   hydroxy 5'-8'-11'-14'-eicosatetranoic acid
SRS-A =   slow reacting substances of anaphylaxis
ECF-A =   eosinophil chemotactic factor of anaphylaxis
PG   =   prostaglandin
PAF   =   platelet-activating factor
SLE   =   Systemic Lupus Erythematosus
APSGN =   Acute Post-Streptococcal Glomerulonephritis
DNA   =   deoxyribonucleic acid

INTRODUCTION

The complexity of phenomena present in a wide variety of combinations to produce inflammation appears to meet the requirement for self-defense and reparation in the body economy. Although the process may result in irreversible tissue damage and total loss of functional integrity, it operates in immunologically-mediated processes directed to the preservation of self and elimination of foreign or altered antigens from the body. Nevertheless, inflammation may be triggered non-immunologically. Moreover, the non-specific aspects of the response of the host to infection or injury are a critical first line of defense, which precedes the specific recognition of antigens by antibodies or reactive lymphocytes. In fact, the biological processes involved in inflammation are not only triggered by the immune system, but are also readily activated by bacteria, fungi, damaged cells and tissues. Henson and co-workers have recently suggested that initiation of the inflammatory process is primarily a surface phenomenon (1). Humoral and cellular systems involved in inflammation exist in the circulation as precursors or in an inactive state and need to be triggered or activated to be operating in inflammation. In the fluid phase, biological activities generated from activated molecules and reactive cells are modulated or totally abolished by many serum inhibitors. However, the system appears to be fully operating on "surfaces" of connective tissue elements, basement membranes, bacteria, fungi, protozoa and immune complexes (IC) rather than in the fluid phase.

HUMORAL SYSTEMS

The activation of complement cascade, coagulation and fibrinolytic systems (2, 3) is necessary to display many features of the
inflammatory process. These processes occur optimally on "surfaces".
Presumably, the effect of the surface can happen with a negatively
charged particle (- interacting with factors XII, XI, kallicrein
and kininogen - or with IC - interacting with Clq - and with the
plasma membrane - interacting with C5a, C3a, C3b, C3d) and determines
a new stereochemical configuration of molecules. The changes of
spatial orientation of the molecules are necessary to develop the
entire sequence of these humoral systems. The complement, and the
coagulation and the fibrinolytic systems are present in the soluble
phase, i.e. the plasma, and their activation can be disseminated
to the whole body or otherwise limited to isolated districts. This
difference is due to homeostatic mechanisms of the microenviromment
and can be a focus for future study.

CELLULAR SYSTEMS

Mechanisms Of Polymorphonuclear Leucocyte (PMN) Accumulation in Tissues

In inflammation, the initial interaction is the interaction
of neutrophils with the endothelial cells lining the blood vessels.
The first microscopically-detectable change is the reduced flow of
an increased proportion of granulocytes, which are seen to roll
along the vessel walls and finally to stop on the endothelial cell
lining. As the inflammatory abnormality increases, the number of
granulocytes adhering to the vessel wall and among themselves
progressively increases (4). In the end, adherent neutrophils leave
the vessel through junctions between adjacent endothelial cells
and massively infiltrate the inflamed tissue. The sequence of these
events is also implicated in lesions of immunologic origin such as
that exemplified by the Arthus reaction (5). A sequential two-step
process may be advocated that both restricts and localizes the site
of inflammation: 1) the recruitment and sequestration of PMN in
vessels around the inflammation site and 2) the emigration of PMN
out of the vessels into the tissues from which the chemotactic
stimulus originates (6). This sequence is implicated in the local

accumulation of granulocytes in the inflamed tissues. We studied
in vitro and in vivo the mechanisms at work in the aggregation of
PMN, a step crucial to their accumulation in tissues (6, 7, 8, 9).
As for platelets, adenosine diphosphate (ADP)- and arachidonic acid
(AA)/thromboxane A_2-mediated pathways are operating in inducing
aggregation of PMN in augmenting their adhesiveness to endothelial
cells (6). In infarcted areas, tissue necrosis determines an important release of ADP and AA, which may play a role in PMN accumulation. However, the aim of our studies was to elucidate the immunologic triggers of neutrophil accumulation in inflamed tissues.

It is known that PMN possess surface receptors for the C3b
component of complement and the Fc fragment of complexes of aggregated immunoglobulin (10, 11). The interaction between immune
complexes (IC) and PMN surface receptors triggers the release of
lysosomal enzymes (12) and cationic proteins (13, 14). Beside, IC
are able to aggregate PMN in vitro (6, 7). We focused our studies
on the latter phenomenon, which appeared to be related either to
cationic proteins released from IC-stimulated PMN (6, 7) or to
C5a anaphylatoxin generated in serum (15). In fact, both purified
cationic proteins and C5a are cleaved by serum carboxypeptidase B
to their desarginated products, CP des Arg (6, 7) and C5a des Arg
(16), which although inactive in terms of anaphylactic activity,
show a striking potentiation of PMN aggregation in vitro (6, 7).
Although the precise in vivo role of the des Arg fragments is as
yet unknown, the marked PMN aggregating activity and increased
adhesiveness to endothelial cells by CP des Arg and C5a des Arg,
that we studied in vitro (6, 7), may explain the induction of lung
inflammation by C5a des Arg, which is more potent than the native
molecule (17). Our in vitro experiments on C5a, cationic proteins,
C5a des Arg and CP des Arg -induced PMN aggregation show that this
is an active process requiring metabolic energy and Calcium and
Magnesium in the reaction medium (6, 7). Electron micrographs of
PMN aggregates indicate distinctive features consisting in the close
association of parts of cell membranes, that in some tracts are in
closer approximation and regularly parallel (6, 7). Subsequent
studies in our laboratories have shown that in vitro PMN aggregation
induced by IC, C5a, cationic proteins and their des Arg products
is mediated by a low-molecular weight polar phospholipid (1-0-
alkyl-2acetyl-glyceryl-3-phosphorylcholine) (8), previously described
as a mediator of anaphylaxis and named platelet-activating factor
(PAF) after its potent aggregating activity on platelets (18, 19,

20). PAF is released in vitro from PMN during phagocytosis of baker's
yeast particles or IC and upon stimulation with C5a anaphylotoxin,
cationic proteins and their des Arg fragments (8, 21, 22). From
these studies, it emerges that aggregation and secretion are two
strictly interrelated process and that PMN aggregation is dependent
upon PAF secreted by the neutrophils themselves. In fact highly
purified PAF is per se able to induce PMN aggregation which is
metabolically indistinguishable from that induced by IC, C5a,
cationic proteins and their des Arg products (8). Our in vivo studies
show that the acute leucopenia, occurring after intravenous injection of IC, C5a and cationic proteins into rabbits, temporally
correlates with the intravascular PAF release and is secondary to
PMN aggregates in the peripheral capillary network (8). Nevertheless,
rabbits injected with purified PAF, become markedly neutropenic (8).
Substances known to inhibit in vitro PAF-induced PMN aggregation
such as glucocorticoids (8), which interfere with lipids of cell
membranes (23), due to their liposolubility (24), completely prevent
the leucopenia induced either by PAF or IC, C5a and cationic proteins
(8). An additional mechanism of IC-induced PMN aggregation may be
related to thromboxane A_2-like activity generation, which has been
demonstrated elsewhere from PMN after phagocytosis of IC (6, 25).
Our preliminary experimental evidence would suggest that all these
mechanisms responsible for PMN aggregation also operate in the
adhesion of granulocytes to the endothelial cells.

PMN: SECRETORY ORGAN IN INFLAMMATION

As for most inflammatory reactions, phagocytosis of PMN is a
"surface" phenomenon that can be divided into two distinct steps:
attachment and internalization. The attachment of particles to PMN
surface involves the membrane receptors for the Fc fragment of
complexes immunoglobulin and the C3b component of complement (10,
11, 26). Stimulation of the receptors for the Fc fragment leads to
prompt internalization (27), whereas the receptors for the C3b act
primarily by enhancing immunoadherence, thus stimulating the internalization (28). Internalization of phagocytable particles such as
baker's yeast spores or IC is accomplished through a complex process
starting with the phagocytic vacuole, which becomes pinched off,
then gradually drawn into the cell interior and finally merged with
lysosomes flowing from the organelle-rich cytocenter. If the
particle-cell ratio is great, the lysosomal content leaks out of
the cell and at least three categories of inflammatory material

are released (29): 1) collagenase, elastase, acid and neutral
proteases, kininogens, cathepsin G and serine protease, all capable
of directly dearranging tissue collagen structures, 2) toxic products of molecular oxygen such as O_2, H_2O_2, $OH\cdot$ and $^1O_2$ generated
by the burst of oxidative metabolism and finally 3) plasma membrane
phospholipase-derived substances of arachidonate such as hydro-
peroxides, endoperoxides, thromboxanes and stable prostaglandins.
Recently, the list has comprised PAF as this mediator is released
from immune stimulated PMN (8, 21, 22). Electron microscopy studies
allow us to elucidate the morphologic details of the cell internalization (29). As the particle engages the surface of PMN by means
of the appropriate receptor, microfilaments become condensed in the
subplasmalemmal area. Following ingestion, the phagocytic vacuole
merges with lysosomes, which in turn pour out their enzymatic
content. This process is generally started when fusion of the
phagocytic vacuole is still incomplete so that a portion of lysosomal
enzymes such as ß-glucuronidase and neutral proteases happens to
be secreted into the medium. However, this is not the case with
cytoplasmatic enzymes, such as lactate dehydrogenase since the
plasma membrane is intact (29). Nevertheless, a major escape of
lysosomal enzymes from the cell may indeed take place as the consequence either of overstimulation (i.e. non-phagocytable particles)
or impairment of the cell subplasmalemmal microfilaments by Cito-
chalasin B (1). Therefore, PMN plasma membrane plays a pivotal role
in triggering internalization as it is engaged through its receptors
by phagocytable complement-opsonized particles and IC or by soluble
stimuli such as C5a, cationic proteins, Ionophore A 23187. Following
the ligand binding to surface receptors, the secretory response of
stimulated-PMN is accomplished through a sequence of events here
summarized: an early sharp hyperpolarization, followed by depolari-
zation with a gradually return to a hyperpolarized state, concomitantly with the local loss of Calcium from the cell (29). At this
stage, the cell generates superoxide anions and boosts its oxidative
metabolism via the hexose-monophosphate shunt leading to the generation of thromboxane A_2 and prostaglandins from membrane fatty acids.

BASOPHILS

A variety of membrane-active stimuli lead to secretion from
circulating basophils and tissue-sited mastocytes of mediators
which play an important role in inflammation. Basophils and masto-

cytes are known to have specific receptors for the immunoglobulin
of the IgE class (30), which on interacting with the antigen triggers
the degranulation process. Other receptors specific for C3a and C5a
anaphylotoxins as well as for neutrophil cationic proteins are
present and capable of starting the membrane signal for degranulation
to take place (14, 31). Recently, receptors for IgG_{2a} (32) and Fcγ
(33) have been described on rat and human basophils, respectively.
Following the signal from "activated" cell membrane by various
stimuli, intracellular cAMP levels decrease (34) and Calcium ions
flow into the cell (35). These two events are fundamental in the
process of assembly and subsequent release of contractile proteins.
Intracellular cGMP counteracts cAMP as it antagonizes the effect
on Calcium ion flow, thus modulating mediator release (36). Concomitantly with these processes, the stimulus-receptor interaction
activates a serine protease (37) and a phospholipase A_2 (38) with
subsequent activation of an arachidonate-dependent metabolic pathway.
Recently, AA-derived, lipoxygenase-generated metabolites appear to
play an important role in regulating the release reaction (39).
Several mediators are released from the cell during the degranulation
process, such as histamine and serotonin, a polypeptide, which is
chemotactic for eosinophils (ECF-A), products of AA-derived metabolism such as HETE, PGD_2, thromboxane A2, PGF2, PGE2 (40), the slow-
reacting substances of anaphylaxis (SRS-A) and heparin. Although
many of these mediators appear to exist preformed within the cell,
at least SRS-A and AA-derived metabolites require the initiation of
the secretory process for the generation of their biological activity. Similarly, PAF (41) is generated and released only after
challenge. PAF, a low-molecular weight phospholipid (41), recently
identified as a 1-O-alkyl-2-acetyl-sn-glyceryl-3-phosphorylcholine
(18, 19) is released from human mixed leukocyte suspensions, concomitantly with basophil degranulation in response to different
stimuli such as C5a and neutrophil cationic proteins (14, 21, 22),
in addition to the IgE-antigen mechanism (20). More recently, we
have provided further evidence for the release of PAF from human
basophils challenged with the above-mentioned stimuli (21, 22).
Furthermore, we have shown that other leucocytes capable of under-
going the degranulation process such as PMN and monocytes-macrophages
also release PAF following various stimuli (21, 22). PAF induces
platelet aggregation and release of their inflammatory constituents
(42, 43). Our studies show that PAF is not only a membrane-active
mediator of leucocyte-platelet interaction but also of leucocyte-
leucocyte interaction, a crucial step in triggering and amplification

of the inflammatory process.

PLATELETS

In the inflammatory process, platelets may be instrumental
both in the onset and in the amplification of vascular permeability
since stimulated platelets release their lysosomal enzymes injuring
vessel walls similarly to PMN (43). Furthermore, it has been reported
that platelets may be implicated in enhanced vascular permeability,
leading to IC deposition in vessel walls and glomeruli (44). In
immunopathological states, platelets may be involved either directly
through their membrane receptors for IC (45, 46) or indirectly by
the immune-induced release of PAF from (47) or in consequence of
the endothelial injury related to IC deposition in vessel walls.
Direct interaction between IC and platelet surface receptors is
mediated by ADP- (47) and probalby AA/thromboxane A_2 dependent
pathways of platelet aggregation. Furthermore, platelets may be
involved indirectly by PAF released from leucocytes by an IgE- or
IC complement- dependent mechanism (14). The biological activity
of PAF on platelets is quite distinct (47, 21, 23) from that of
ADP and AA, as it is unaffected by ADP scavengers and cycloxygenase
inhibitors, respectively, thus representing the mediator of the
putative third pathway of platelet aggregation (48).

MONONUCLEAR PHAGOCYTES

Monocytes, representing the relatively quiescent precursors
of tissue-localized macrophages in the circulation, are accumulated
in inflamed tissues through "surface" mechanisms similar to those
described earlier for PMN (1). The inflammatory macrophages are a
feature of chronic inflammation and of the latter stages of acute
inflammatory events (49). Monocytes adhere to endothelial cells
then enter the inflamed tissues by ameboid migration through the
walls of the venules. The first step in the recruitment of monocytes
appears to be the surface interaction with platelets adhering to
injured endothelium and to the underlying connective tissue elements.
Monocytes recognize platelets or their products and become adherent
to platelet aggregates through an as yet unknown mechanism (1). The
emigration out of vessels under chemotactic stimuli results in the
differentiation into macrophages. This transformation is conditioned

either by the offending stimulus, by immunological mechanisms, or by the mere act of leaving the vascular compartment. Activation comprises cellular enlargement and an attendant increase in the number of cytoplasmatic organelles, lysosomal content, and density of surface receptors for Fc C3b and C3d (49). Similarly to PMN, macrophages release lysosomal enzymes and produce oxygen-derived radicals as well as AA-derived metabolites during phagocytosis or upon stimulation by soluble stimuli such as C5a (1).

IMMUNOPATHOLOGY

The role of neutrophils in IC-mediated tissue injury has been investigated in experimental pathology (50, 51). The determinant role of PMN accumulation if the arteritis of acute IC disease has been inferred from the fact that tissue damage can be prevented by depletion of circulating PMN or complement (C3 and late components) with the anti-complementary factor from cobra venom (44, 52). This arteritis is a necrotizing and is characterized by massive neutrophil accumulation consequent upon complement-mediated recruitment of PMN and their emigration out of vessel lumina under chemotactic stimuli, phagocytosis of IC deposits and release of their injurious constituents capable of digesting structures within the vessel walls (50). In contrast, the deposition of IC and the development of the acute disease model appear to be independent of PMN, because they are not prevented by PMN or complement depletion (44, 52). However, the latter prevents the PMN accumulation in glomeruli and reduces the extent of tissue injury. In glomerulonephritis produced by anti-glomerular basement antibody, such as in experimental nephro-toxic nephritis and in Goodpasture's disease in man, antibody binds along the membrane and, in the experimental system, produces injury by both neutrophil-dependent and indipendent processes (50, 51). PMN fill the glomerular vessel lumen, having pushed aside the endothelial lining and having become closely adherent to antibody and complement bound to the basement membrane. They then release their inflammatory constituents. The depletion of PMN or serum complement prevents the neutrophil accumulation and the subsequent tissue injury. In the rabbit, the injection of preformed IC results in leucopenia and in the sequestration of PMN in the lung capillary network and in glomeruli in consequence of the interaction between IC and PMN membrane receptors for Fc and C3b (8). Preliminary evidence in man indicates an _in vivo_ interaction between IC and surface

receptors of PMN in the acute phases of Systemic Lupus Erythematosus
(SLE) (53, 54) in at least some cases of idiopathic neutropenia
(55), and, to a minor extent, in Acute Post-Streptococcal Glomerulonephritis (A.PS.GN) (54).

Several types of human and experimental "proliferative" glomerulonephritis are characterized by glomerular hypercellularity,
the nature of which has been recently investigated. Apart from
neutrophils, which are limited to acute and necrotising glomerulonephritis, the increased numbers of cells could be endothelial,
mesangial, epithelial or blood-derived infiltrating mononuclear
cells. Recent studies have lent support to the concept of mononuclear
cells infiltrating the glomerular tuft and, in various types of
rapidly progressive glomerulonephritis, forming glomerular crescents
(56). Experimental studies carried out in rabbit nephrotoxic nephritis (57) and in rabbit acute serum sickness (58) have established
that macrophages play a major role in glomerular hypercellularity.
Macrophage infiltration in glomeruli has been also shown in human
pathology (11). Finally, it has been postulated that macrophages
play an important role in glomerular inflammation as they (60, 61)
and in particular platelets (62) may secrete factors which induce
proliferation of fibroblasts and smooth muscle cells and, in the
case of macrophages, vascular endothelium.

IgE-sensitized basophil degranulation and PAF release have
been implicated in vasopermeabilization leading to IC deposition in
glomeruli and vessel walls in rabbit acute serum sickness (18, 63).
In man, several immunological mechanisms leading to enhanced vascular
permeability have been recently identified (14, 21, 7). The deposition of circulating IC may be envisaged as the consequence of a
cascade reaction involving subsequently IC/complement, PMN, basophil/
mastocytes and platelets (14, 21, 7). Besides IgE-dependent PAF
release and basophil degranulation, IC/complement can release PAF
from PMN and monocytes during phagocytosis and indirectly from
basophils (9, 22). The antigen-antibody ratio conditions the predominant involvement of one or more pathways of vasoactiva amine
release. In a.PS.GN., a human model of IC pathology in antigen
excess, the IgE-basophil/mastocyte plays an important role in
enhancing the permeability of vessel walls and glomeruli (64). We
have shown a drop in circulating, metachromatically-staining basophil,
platelet and mastocyte counts and low amounts of mediators that
can be released from circulating basophils during the acute phases

of the disease (64). After recovery, in vitro basophil degranulation
caused by exogenous streptococcal antigen can be detected and PAF
is recovered from mixed leucocyte suspensions (64). Furthermore, we
have recently shown the release of PAF from renal tissue itself
(14). In SLE, a model of IC in antibody excess, platelets and PMN
are mainly involved by IC through their membrane receptors (53, 54).
The IC-PMN interaction leads to the release of lysosomal enzymes
and cationic proteins from the cells (4, 13). In addition, IC
generate C5a anaphylotoxin in serum. Both cationic proteins and
C5a are able to degranulate basophils and release PAF from mixed
leucocyte suspensions (14). Basophils are degranulated in vivo and
the PAF reservoir of leucocytes is depleted in the blood during the
acute phases of the disease (65, 66). During the latent phases of
the disease, metachromatically staining basophils return to normal
values and may be shown to degranulate – and PAF is released from
mixed leucocyte suspensions – in the presence of native DNA and Sm
antigen (21).

Platelets are also involved directly through their membrane
receptors for Fc in immunopathologic states and several studies
have clearly established the implication of platelets in glomerular
injury (67).

However, much work is needed to define the precise role of
direct involvement of platelets in human pathology.

REFERENCES

1. Henson, P.M., Hollister, J.R., Musson, R.A., Spears, P.,
    Henson, J.E. and Mc Carthy, K.M.: Inflammation as a surface
    phenomenon: Initiation of inflammatory processes by surface
    bound immunologic components. Adv. in Inflammation Res. (Ed.
    Weissmann et al.), 341. Raven Press, New York, (1979).
2. Müller-Eberhard, H.Y.: Complement. A. Rev. Biochem. 38: 389–
    413, (1969).
3. Cochrane, C.G., Wiggins, R.C., Rewak, S.D. and Griffin, Y.H.:
    The Hageman factor system in inflammation. Adv. in Inflammation
    Res. (Ed. Weissmann et al.) 341. Ravern Press, New York, (1979).
4. Born, G.V.R. and Planker, M.: Toward the mechanism of the
    intravascular adhesion of granulocytes in inflamed vessels.
    Adv. in Inflammation Res. (ED. Weissmann et al.) 11. Raven

Press, New York, (1979).

5. Brown, D.C.: Biological activities of Complement. In "The
    immune system. A course on the molecular and cellular basis
    of Immunity". (Ed. Hobart et al.) 274. Blackwell, Oxford,
    (1975).

6. Camussi, G., Tetta, C., Bussolino, F., Caligaris Cappio, F.,
    Coda, R., Masera, C. and Segoloni, G.: Mediators of immune
    complex induced aggregation of polymorphonuclear neutrophils.
    C5a anaphylotoxin, neutrophil cationic proteins and their
    cleavage fragments. Int. Arch. Allergy appl. Immun. 62: 1-15,
    (1980).

7. Camussi, G., Coda, R., Tetta, C., Bussolino, F., Stratta, P.,
    Segoloni, G., Coppo, R. and Vercellone, A.: Immune mechanisms
    of PMN involvement in inflammation. In "Renal pathophysiology"
    (Ed. Leaf et al.) Raven Press, New York, (1980).

8. Camussi, G., Tetta, C., Bussolino, F., Caligaris Cappio, F.,
    Coda, R., Masera, C. and Segoloni, G.: Mediators of immune-
    complex induced aggregation of PMN. II Platelet-activating
    factor (PAF) as the effector substance of immune-induced aggre-
    gation. Int. Arch. Allergy appl. Immun. (submitted for publ.),
    (1980).

9. Camussi, G., Bussolino, F., Tetta, C., Caligaris, F., Benve-
    niste, J., Emanuelli, G. and Vercellone, A.: Platelet-activating
    factor (PAF): A mediator of immune dependent vasopermeabi-
    lization. Min. Nefr. 1: 15-18, (1980).

10. Henson, P.M.: The adherence of leucocytes and platelets induced
    by fixed IgG antibody or complement. Immunology 16: 107-121,
    (1969).

11. Lay, W.H. and Nussenrweig, V.: Receptors for complement on
    leucocytes. J. Exp. Med. 128: 991-1007, (1968).

12. Weissmann, G. and Dukor, P.: The role of lysosomes in immune
    response. Adv. Immunol. 12: 283-295, (1970).

13. Camussi, G., Segoloni, G., Stratta, P., Mencia-Huerta, J.M.,
    Ragni, R., Piccoli, G. and Vercellone, A.: In vivo fixation
    of immune complexes on polymorphonuclear neutrophils and
    release of neutrophil cationic proteins in systemic lupus
    erythematosus. EDTA Proc., Pitman Medical (Ed. Robinson B.H.),
    14: 478-482, (1977).

14. Camussi, G., Mencia-Huerta, J.M. and Benveniste, J.: Release
    of Platelet-activating factor and histamine. I. Effect of
    immune complex, complement and neutrophils on human and rabbit
    mastocytes and basophils. Immunology 33: 523-533, (1977).

15. Craddock, P.R., Hammerschmidt, D., Whyte, J.G., Dalmasso, A.P.
     and Jacob, H.: Complement (C5a)-induced granulocytes aggregation
     in vitro. J. Clin. Invest. 60: 260-264, (1977).
16. Bokisch, V.A. and Müller-Eberhard, H.J.: Anaphylotoxin inac-
     tivator of human plasma: its isolation and characterization
     as a carboxypeptidase. J. Clin. Invest. 49: 2427-2436, (1970).
17. Mc Carthy, K. and Henson, P.M.: Induction of lysosomal enzyme
     secretion by alveolar macrophages in response to purified
     complement fragments C5a and C5a desArg. J. Immunol. 123: 2511-
     2517, (1979).
18. Benveniste, J., Henson, P.M. and Cochrane, C.G.: Leucocyte-
     dependent histamine release from rabbit platelets. The role
     of IgE, Basophils and Platelet-activating factor. J. Exp. Med.
     136: 1356-1377, (1972).
19. Pinckard, R.N., Farr, R.S. and Hanahan, D.: Physiochemical
     and fuctional identity of rabbit platelet-activating factor
     (PAF) released in vivo during IgE anaphylaxis with PAF released
     in vitro from IgE-sensitized basophils. J. Immunol. 123: 1847-
     1857, (1979).
20. Benveniste, J., Tence, M., Bidault, J., Boullet, C. and Polon-
     sky, J.: Semi-synthese et structure proposée du facteur activant
     les plaquettes (PAF): PAF-acether, un alkyl ether analogue de
     la lyso-phosphatidylcholine. C.R. Acad. Sci. Paris D 289: 1037-
     1040, (1979).
21. Camussi, G., Bussolino, F., Tetta, C., Benveniste, J. and
     Vercellone, A.: Platelet-activating factor and glomerulone-
     phritis. In "Endothelium, platelets and prostaglandins: fresh
     insight into renal diseases." Raven Press, New York, (1980),
     (in press).
22. Camussi, G., Aglietta, M., Coda, R., Bussolino, F., Piacibello,
     W. and Tetta, C.: The cellular origins of human Platelet-
     activating factor. Immunology. (submitted for publ.) (1980).

23. Cazenave, J.P., Benveniste, J. and Mustard, J.F.: Aggregation
     of rabbit platelets by Platelet-activating factor is independ-
     ent of the release reaction and the arachidonate pathway and
     inhibited by membrane-active drugs. Lab. Invest. 41: 275-285,
     (1979).
24. Seeman, P.: The membrane action of anesthetics and tranquil-
     lizers.Pharmacol. Rev. 24: 583-594, (1972).
25. Higgs, G.A., Bunting, S., Moncada, S. and Vane, J.R.: Poly-
     morphonuclear neutrophils produce TXA_2-like activity during

phagocytosis. Prostaglandins. 12: 749–757, (1975).

26. Stossel, T.P.: Phagocytosis, recognition and ingestion. Sem. Haematol. 12: 1–28, (1975).

27. Scribner, D. and Fahrney, D.: Neutrophil receptors for IgG and complement. J. Immunol. 116: 892–917, (1976).

28. Ehlessberg, A.G. and Nussenzweig, V.: The role of membrane receptors for C3b and C3d in phagocytosis. J. Exp. Med. 145: 357–371, (1977).

29. Weissmann, G., Korchak, H.M., Perez, D., Smolen, J.E. and Hoffstein, S.T.: Leucocytes as secretory organs in inflammation. Adv. in Inflammation Res. (Ed. Weissmann G. et al) 95. Raven Press, New York, (1979).

30. Ishizaka, K., Tomioka, H. and Ishizaka, T.: Mechanisms of passive sensitization. Presence of IgE and IgG molecules on human leucocytes. J. Immunol. 105: 1458–1466, (1970).

31. Hook, W.A., Siraganian, R.P. and Wahl, S.M.: Complement-induced histamine release from human basophils. J. Immunol. 114: 1185–1191, (1975).

32. Halper, J. and Metzger, H.: The interaction of IgE with rat basophilic leukemia cells. Immunochemistry 13: 907–910, (1976).

33. Ishizaka, T., Sterk, A.R. and Ishizaka, K.: Demonstration of Fc receptors on human basophil granulocytes. J. Immunol. 123: 578–593, (1979).

34. Bussolino, F. and Benveniste, J.: Pharmacologic modulation of platelet-activating factor release from rabbit leucocytes. Role of cAMP. Immunology (in press), (1980).

35. Lichtenstein, L.M.: The mechanism of basophil histamine release induced by antigen and by calcium ionophore A23187. J. Immunol. 114: 1692–1699, (1975).

36. Sullivan, T.J. and Parker, C.W.: Pharmacologic modulation of inflammatory mediator release by mast cells. Am. J. Pathol. 85: 437–468, (1976).

37. Becker, E.L. and Henson, P.M.: In vitro studies of immunologically induced secretion of mediators from cells and related phenomena. Adv. Immunol. 17: 93–106, (1973).

38. Ho, P.G. and Orange, R.P.: Indirect evidence of phospholipase A activation in purified mast cells during reserve anaphylactic challenge. Fed. Proc. 37: 1667, (1979).

39. Sullivan, T.J. and Parker, C.W.: Possible role of arachidonic acid and its metabolites in mediator release from rat mast cells. J. Immunol. 122: 431–440, (1979).

40. Roberts, L.J., Lewis, R.A., Austen, K.F. and Oades, J.A.:

Arachidonic acid metabolism by rat mast cells. Prostaglandins
15: 717-720, (1978).

41. Benveniste, J., Camussi, G. and Polonsky, J.: Platelet-activa-
     ting factor. Monogr. Allergy 12: 138-144, (1977).

42. Bussolino, F., Camussi, G. and Tetta, C.: Aggregation of human
     platelets by platelet-activating factor (Submitted to Agents
     and Actions), (1980).

43. Nachman, R.L. and Polley, M.: The platelet as an inflammatory
     cell. Adv. in inflammation Res. (Ed. Weissman G. et al), 169.
     Raven Press, New York, (1979).

44. Henson, P.M. and Cochrane, C.G.: Immune complex disease in
     rabbit. The role of complement and of a leucocyte-dependent
     release of vasoactive amines from platelets. J. Exp. Med. 133:
     554-571, (1971).

45. Müller-Eckardt, L.H. and Luscher, E.F.: Immune reactions of
     human blood platelets. A comparative study on the effects on
     platelets of heterologous anti-platelet antiserum, antigen-
     antibody complexes, aggregated gamma globulins and thrombin.
     Thrombos. Diathes. Haemorrh. 20: 821-826, (1968).

46. Müller-Eckardt, L.H. and Luscher, E.F.: Immune reactions of
     human blood platelets. The effect of latex particles with gamma
     globulin in relation to complement activation. Thrombos. Diathes
     Haemorrh. 20: 168-176, (1968).

47. Camussi, G., Tetta, C., Segoloni, G. and Vercellone, A.:
     Immunological mechanisms of human platelet involvement. La
     Ricerca Clin. Lab. 8: 262-272, (1978).

48. Chignard, M., Le couedic, J.P., Tence, M., Vargaftig, B.B. and
     Benveniste, J.: The role of Platelet-activating factor in
     platelet aggregation. Nature 279: 799-800, (1979).

49. Spector, W.G.: The macrophage and its derivatives. Adv. in
     Inflammation Res. (Ed. Weissmann G. et al.) 113. Raven Press,
     New York, (1979).

50. Cochrane, C.G.: Immunologic tissue injury mediated by neutro-
     philic leucocytes. Adv. Immunol. 9: 97-161, (1968).

51. Henson, P.M.: Interaction of cells with immune complex: aherence
     release of constituents and tissue injury. J. Exp. Med. 134:
     1145-1335, (1971).

52. Kniker, W.T. and Cochrane, C.G.: Pathogenetic factors in
     vascular lesion of experimental serum sickness. J. Exp. Med.
     127: 119-129, (1968).

53. Camussi, G., Tetta, C. and Caligaris Cappio, F.: Detection of
     immune complexes on the surface of polymorphonuclear neutro-

phils. Int. Arch. Allergy appl. Immun. 58: 135-139, (1979).

54. Camussi, G., Caligaris Cappio, F., Messina, M., Coppo, R., Stratta, P. and Vercellone, A.: The polymorphonuclear neutrophil (PMN) immunohistological technique: detection and characterization of the immune complexes bound to the PMN membrane in acute poststreptococcal and lupus nephritis. Clinical Nephrology (in press).

55. Caligaris Cappio, F., Camussi, G. and Gavosto, F.: Idiopathic neutropenia with normocellular bone marrow: an immune complex disease. British J. of Haematology 43: 595-605, (1979).

56. Cotram, R.S.: Monocyte proliferation and glomerulonephritis. J. Lab. Clin. Med. 92: 837-840, (1978).

57. Screiver, G.F., Cotram, R.S., Pardo, V. and Unanue, E.R.: A mononuclear cell component in experimental immunological glomerulonephritis. J. Exp. Med. 147: 369-384, (1978).

58. Hunsicker, L.G., Plattner, S.B. and Weisenburger, D.: The role of monocytes in serum sickness nephritis. J. Exp. Med. 150: 413-425, (1979).

59. Monga, G., Mazzucco, G., Barbiano di Belgiojoso, G. and Busnack, C.: The presence and possible role of monocyte infiltration in human chronic proliferative glomerulonephritis. Light microscopy immunofluorescence and histochemical correlation. Am. J. Pathol. 94: 271-276, (1979).

60. Leibovich, S.J. and Ross, R.: A macrophage-dependent factor that stimulates the proliferation of fibroblasts in vitro. Am. J. Pathol. 84: 501-506, (1976).

61. Polverini, P.G., Cotran, R.S., Gimbrone, M.A. and Unanue, E.R.: Activated macrophages induce vascular proliferation. Nature 269: 804-805, (1977).

62. Ross, R. and Vogel, A.: The platelet-derived growth factor. Cell 14: 203-207, (1978).

63. Camussi, G., Mencia-Huerta, J.M., Segoloni, G. and Benveniste, J.: Platelet-Activating-Factor release during serum sickness. Abs. VII Congress of Nephrology, G2, (1978).

64. Camussi, G., Bosio, D., Segoloni, G., Tetta, C. and Vercellone, A.: Evidence for the involvement of the IgE-basophil-mastocyte system in human acute post-streptococcal glomerulonephritis. La Ricerca Clin. Lab. 8: 56-54, (1978).

65. Camussi, G., Benveniste, J., Tetta, C. and Vercellone, A.: Immediate hypersensitivity in Systemic Lupus Erythematosus. Abs. VIIth Int. Congress of Nephrology. G3, (1978).

66. Camussi, G., Tetta, C., Coda, R. and Benveniste, J.: Release

of Platelet-Activating-Factor in human pathology. I Evidence
for the occurrence of basophil degranulation and PAF release
in Systemic Lupus Erythematosus (SLE). Lab. Invest. (Submitted
for publication).

67. Cameron, J.S.: Platelets and glomerulonephritis. Nephron 18:
253-265, (1977).

THE ABNORMALITIES OF THE COMPLEMENT SYSTEM IN HUMAN GLOMERULO-NEPHRITIS

F.C. Berthoux, B. Laurent, C. Genin and J.C. Sabatier

Groupe de Recherches sur le Complément et les Glomérulonéphrites Humaines
Service de Nephrologie
UER de Médecine et Hôpital de Bellevue
42023 Saint-Etienne (France)

supported by grants: INSERM CRL n° 76.5.165.5,
                     INSERM CRL n° 77.5.022.5,
                     and INSERM ATP 78/79.110.

INTRODUCTION

The implication of the complement system in the pathogenesis
of human glomerulonephritis (GN) has been suspected since the work
of Zinsser and Johnson in 1911 (31) and of Gunn in 1914 (14).

The complement system plays a physiological role in the humoral
host defense against various infectious agents (lysis in association
with antibody), and a pathological role as an important humoral
mediator of inflammation. Knowledge of it is now at the molecular
level (23).

LIST OF ABBREVIATIONS

AGN =   acute glomerulonephritis
CH50=   total complement hemolytic activity

CIC = circulating immune complexes
EGTA= ethyleneglycol-tetraacetate
I-C = immune complexes
GN = glomerulonephritis
MPGN= membrano-proliferative glomerulonephritis
NF or NeF= nephritic factor
PEG = polyethyleneglycol
SLE = systemic lypus erithematosus.

ACTIVATION OF THE COMPLEMENT SYSTEM

     The complement system is composed of three different pathways:
the classical, the alternate, and the final common pathways (30,
23).

     The classical pathway is composed of the sequential proteins
C1q, C1r, C1s, C4, C2 and of the control proteins C1sINA, C4bp -
and C3b-C4bINA. It can be activated by immune complexes (I-C) with
IgM or IgG (subclasses 3, 1, and possibly 2) antibodies. Such activation produces a C3 cleaving enzyme, C4b2a ($\overline{C42}$), or classical
pathway C3 convertase.

     The alternate pathway (16) is composed of the proteins C3b, B,
D, P and the regulatory proteins C3bINA and ß1H. The initial inter-
action of $\overline{D}$, B and C3 produces a C3 cleaving enzyme, C3Bb (C3$\overline{B}$) or
initial alternate pathway C3 convertase. The C3b molecules liberated
by this enzyme, allow the formation of the amplifying alternate
pathway C3 convertase, C3bBb ($\overline{C3B}$). The activity of this latter
enzyme is controlled by C3bINA and ß1H proteins. This alternate
pathway is mainly activated by negative feedback (deregulation)
such as by polysaccharides, properdin itself, and nephritic factor
(NF). The positive activation remains controversial and may be
achieved by IgA or peculiar I-C.

     The final common pathway is composed of the sequential proteins
C3, C5, C6, C7, C8, C9 and the regulatory proteins C3bINA, C6INA
and AI. The key protein, C3 or ß1C, can be cleaved by three different
convertases ($\overline{C42}$, C3$\overline{B}$, $\overline{C3B}$) into a major fragment C3B and a minor
fragment C3a (anaphylatoxin I). The activation of this pathway
liberates fragments with phlogistic properties, C3a, C5a, $\overline{C567}$,
which are responsible for inflammatory lesions.

PRACTICAL METHODS FOR COMPLEMENT STUDY

Sophisticated techniques have been used to study the complement
system in order to establish its role in the pathogenesis of GN:
measurement of the fractional catabolic rate of labelled proteins
(C3, C4, B ...); synthesis of complement components by human cultured
cells (liver, fibroblasts, ...) and detection of different complement
components deposited in tissues, by immunofluorescent microscopy
(30).

However only complement component measurements in biological
fluids (mainly plasma and serum) have achieved usefulness for routine
clinical investigation. The techniques are either functional (hemolytic assay) or immunochemical (antigenic assay). The former is
more sensitive but more difficult (22).

Only three techniques are necessary for GN investigation (7):
total serum complement hemolytic activity; serum antigenic levels
of C3, C4 and B, and measurement of alternate pathway C3 cleaving
activity.

1. Total Complement Hemolytic Activity (CH50)

This technique was been initially developed by Kabat-Meyer.
Sensitized sheep red blood cells (EA) are lysed according to the
amount of complement in the reaction. The relation is a sigmoidal
curve and the 50% lysis point (CH50) expresses the amount of hemolytic complement in serum. Components from C1 to C9 should be present
to get lysis. The results are expressed in CH50 units per ml of
serum, or better in percent of a large pool of normal human sera
(100%).

2. Serum Antigenic Levels of C3, C4 and B

The proteins C4, B and C3 are respectively representative of
the normal, the alternate, and the final common pathways. This can
be measured by any immunochemical techniques, usually by Mancini's
method.

3. Serum Alternate Pathway C3 Cleaving Activity (8)

The nephritic serum is incubated for 30 min at 37° C with a
fresh normal human serum containing Mg EGTA (source of C3 with
classical pathway blockade). The C3 cleavage is then measured by
hemolytic or antigenic techniques. We are measuring the disappearance
of the native antigen of C3 by the Mancini method expressed as a
percentage of C3 conversion (5, 8).

ACQUIRED ABNORMALITIES OF THE COMPLEMENT SYSTEM IN HUMAN GN

The abnormalities observed are hypocomplementemia or hypercomplementemia. In these acquired abnormalities, there is a good
correlation between serum CH50 and serum C3 levels.

1 - Hypocomplementemia is a decrease in CH50 and/or C3. It is
the result of hypercatabolism, low synthesis or both factors. Hypercatabolism may be associated with a C3 level in the normal range.
Hypocomplementemia is a relatively insensitive indicator of an
ongoing immunopathological process.

2 - Hypercomplementemia is an increase in CH50 or/and C3. Very
little is known about it. It probably reflects an inflammatory
process with over synthesis of acute phase reactant proteins, including complement components.

Most of the hypocomplementaemic GN are immune-complex diseases:
acute GN (AGN), lupus GN (SLE), membrano-proliferative GN (MPGN),
shunt nephritis, subacute bacterial endocarditis, serum sickness
nephritis, mixed-cryoglobulin nephritis, rheumatoid nephritis,...
We will focus only on AGN, SLE and MPGN.

1. The Complement Profile in AGN (7, 13, 14, 15, 17, 21, 26)

When all the signs of acute GN are present (proteinuria,
hematuria, hypertension, edema, renal failure), the disease is
always hypocomplementemic with low levels of serum CH50 and serum
C3 (17). Thus the diagnostic value of complement measurement is
very strong, and in the absence of hypocomplementemia one should
suspect another disease starting acutely. In our experience (7, 8)

C3 is mainly activated by the alternate pathway (normal serum C4 in 23 out of 29 cases) with the presence of specific serum activators in 45% of cases (13 of 29). Classical pathway activation with low C4 was seen in only 6 cases. There is no correlation between the severity of the disease and the intensity of hypocomplementemia. However, a persistent hypocomplementemia after 3 months in children or after 6 months in adults, indicates a still active disease and therefore renal biopsy is indicated. It often shows a membrano-proliferative GN either as the primary process or secondary to a typical endocapillary GN.

2. The Complement Profile in SLE (6, 7, 13, 15, 21, 24, 29)

SLE frequently presents with nephritis (60% or more). In our experience (6) any active non-treated lupus GN (22 cases) is hypo-complementemic, and therefore complement measurement has diagnostic value.

Serum C4, C1q, and B were decreased in 9/17 (53%), 5/9 (56%), and 7/13 (54%) patients respectively. C3 activation proceeds mainly through the classical pathway with recruitment of the amplifying C3b loop. Serum activators of the alternate pathway (8) are usually absent (0/30). There is a clear correlation between complement level and clinical outcome under Prednisone and Azathioprine/Cyclophosphamide treatment: a favorable outcome is preceded by normocomplementemia and renal failure by persistent hypocomplementemia.

3. The Complement Profile in MPGN (2, 5, 7, 8, 13, 15, 21, 28)

Persistent hypocomplementemic MPGN has been recognized since 1965. The frequency of hypocomplementemia is about 70% in our experience (47/65 cases) (8). All cases of MPGN type II (dense deposits) were hypocomplementemic (8/8) versus 68% (39/57) in MPGN type I (subendothelial deposits).

In our hypocomplementemic cases, serum C4 was normal in 85% (40/47). Hypocomplementemia is the consequence of hypercatabolism of C3 through an alternate pathway associated with a low C3 synthesis. Such alternate pathway activation is confirmed by the presence of specific serum activators (8) in 53% (25/47) of the cases (7/8 in

MPGN type II and 18/39 in MPGN type I). These alternate pathway
activators are similar to the C3 NeF activity, initially described
in 1969 (27). Some of these activators (5, 8) are probably nephritic
factor (NF) (19) which is known to be an IgG antibody (auto) against
a conformational antigen on the B part of the C3bBb ($\overline{C3B}$) complex (11).

Hypocomplementemic MPGN (3, 4) and type II MPGN (18) are more
likely to recur after renal transplantation (10).

Hypocomplementemia may also be associated with partial lipo-
dystrophy and secondary nephritis of the MPGN type II (25).

Correlation between circulating immune complexes (CIC) and
hypocomplementemia. In a prospective study (9) of 128 cases of GN
(17 AGN, 15 SLE nephritis, 24 MPGN, 16 membranous GN, 16 segmental/
focal hyalinosis, 10 minimal lesions and 30 mesangial IgA GN), we
found 27% of hypocomplementemia and 24% of CIC by the PEG-EDTA
technique. Regardless of the GN type, the presence of CIC correlated
with low C3 ($P < 0.001$), with low C4 ($P < 0.001$), with low C1q
($P < 0.001$) and with classical pathway activation ($P < 0.001$). But
there is no correlation between CIC and serum alternate pathway
activators (12).

HEREDITARY COMPLEMENT DEFICIENCIES AND HUMAN GN

Hereditary deficiencies (1) have been described for almost all
human complement components: C1q, C1r, C1s, C1s INA, C4, C2, C3,
C3bINA, C5, C6, C7, C8, C9 and B. The complement deficiencies: C1r,
C1s, C1sINA, C4, C2, C5, C7 and C8 have been associated with GN,
either primary or lupus nephritis.

We will focus mainly on C2 deficiency, the most frequent.
The gene controlling C2 synthesis is located on chromosome 6, inside
the HLA complex. The C2 deficient gene is usually associated with
the A 10 - B 18 - DW2 - BfS haplotype and is located outside but
close to D/DR. The homozygous deficient presents with an absent
CH50, a normal C3, and an absent C2, measured by specific hemolytic
titration. The heterozygous has a borderline low CH50, a normal
C3, and a 50% hemolytic C2. The frequency of the heterozygous state
(24) is 1.2% in the general population, 5.9% in the SLE group (X 5),
and 3.7% in the rheumatoid arthritis group (X 3). Such GN resembles

closely lupus nephritis but antinuclear antibodies, DNA antibodies
and LE preparation are often negative, despite a full SLE-like
clinical syndrome. Undoubtedly, such complement deficiency predisposes to the development of various immune complex diseases. The
complement profile is modified by additional acquired abnormalities.

CONCLUSION

The involvement of the complement system in the pathogenesis
of human GN is a complex process in which acquired and inherited
deficiencies can be superimposed. The practical value of complement
component measurements in the follow-up of GN is obvious.

REFERENCES

1. Alper, C.A. and Rosen, F.S.: Genetic aspects of the complement
    system. In "Advances in Immunology" 14: 251, (1971).
2. Berthoux, F.C., Carpenter, C.B., Traeger, J. and Merril, J.P.:
    The C3 Nephritic Factor (C3NeF) and the heat labile complement
    inactivator (HLCI) in chronic hypocomplementemic mesangio-
    proliferative glomerulonephritis. In "Advances in Nephrology".
    Edited by J. Hamburger, J. Crosnier and M.J. Maxwell. 4: 91,
    (1974).
3. Berthoux, F.C., Ducret, F., Touraine, J.L. and Traeger, J.:
    Increased recurrence of mesangioproliferative glomerulonephritis
    (MPGN) on kidney grafts. Transpl. Proc. 7: 699, (1975).
4. Berthoux, F.C., Ducret, F., Colon, S., Blanc-Brunat, N., Zech,
    P.Y. and Traeger, J.: Renal transplantation in mesangioproli-
    ferative glomerulonephritis (MPGN): Relationship between the
    high frequency of recurrent glomerulonephritis and hypocom-
    plementemia. Kidney Int. 7: 323, (1975).
5. Berthoux, F.C., Carpenter, C.B., Freyria, A.M., Traeger, J. and
    Merrill, J.P.: Human glomerulonephritis and the C3 Nephritic
    Factor (C3NeF). Clinical Nephrology 5: 93, (1976).
6. Berthoux, F.C., Olivier, J.A., Freyria, A.M., Colon, S., Zech,
    P.Y. and Traeger, J.: Étude des systèmes complément et proper-
    dine dans le lupus érythémateux disséminé (LED) avec atteinte
    rénale: valeur diagnostique et pronostique. Lyon Médical 235:
    31, (1976).
7. Berthoux, F.C.: Intérêt pratique de l'étude des systèmes com-

plément et properdine en pathologie rénale. Lyon Médical 236: 497, (1976).

8. Berthoux, F.C. and Freyria, A.M.: Dosage spécifique des activateurs de la voie alternative du système complément dans les glomérulonéphrites humaines: 125 cas. Néphrologie, 1980 (In press).

9. Berthoux, F.C., Laurent, B., Genin, C. and Sabatier, J.C.: Detection of circulating immune complexes (CIC) in human glomerulonephritis (GN, 128 cases). 4th International Congress of Immunology. July 1980, Paris (In press).

10. Curtis, J.J., Wyatt, R.J., Bhatena, D., Lucas, B.A., Holland, N.H., Luke, R.G. and Forristal, J.: Renal transplantation for patients with type I and type II membranoproliferative glomerulonephritis. Am. J. Med. 66: 216, (1979).

11. Daha, M.R., Austen, K.F. and Fearon, D.T.: Heterogeneity, polypeptide chain composition and antigenic reactivity of C3 Nephritic Factor. J. Immunol. 120: 1389, (1978).

12. Freyria, A.M., Vincent, C., Berthoux, F.C., Revillard, J.P. and Traeger, J.: C1q binding immune complexes and C3 nephritic factor (C3NeF) in hypocomplementemic glomerulonephritis. In "Protides of the Biological Fluids" (H. Peeters ed.) 26: 301, (1979).

13. Gewurz, H., Pickering, R.J., Mergenhagen, S.G. and Good, R.A.: The complement profile in acute glomerulonephritis, systemic lupus erythematosus and hypocomplementemic chronic glomerulonephritis. Int. Arch. Allergy 34: 556, (1968).

14. Gunn, W.C.: The variation in the amount of complement in the blood in some acute infectious diseases and its relation to the clinical features. J. Path. Bact. 19: 155, (1914).

15. Hunsicker, L.G., Ruddy, S., Carpenter, C.B., Schur, P.H., Merrill, J.P., Müller-Eberhard, H.J. and Austen, K.F.: Metabolism of the classical and alternate (properdin) pathways for complement activation. New Engl. J. Med. 287: 837, (1972).

16. Kazatchkine, M., Nydegger, U. and Fearon, D.T.: La voie alterne du complément. Nouv. Presse Med. 8: 2187, (1979).

17. Lange, K., Wasserman, E. and Slobody: The significance of serum complement levels for the diagnosis and prognosis of acute and subacute glomerulonephritis and lupus erythematosus disseminatus. Ann. Int. Med. 53: 636, (1960).

18. Leibowitch, J., Halbwachs, L., Wattel, S., Gaillard, M.H. and Droz, D.: Recurrence of dense deposits in transplanted kidney: II Serum complement and nephritic factor profiles. Kidney Int.

15: 396, (1979).

19. Leibowitch J. and Lesavre, P.: Progrès dans la connaissance du complément II. Les facteurs néphritiques. Nouv. Presse Méd. 8: 2447, (1979).

20. Lesavre, P. and Leibowitch, J.: Progrès dans la connaissance du complément: I – Structure, activation et contrôle du systéme complémentaire. Nouv. Presse Méd. 8: 2385, (1979).

21. Lewis, E.J., Carpenter, C.B. and Schur, P.H.: Serum complement component levels in human glomerulonephritis. Ann. Int. Med. 1971: 75: 555, (1971).

22. Roncato, M., Vial, M.C., Maillet, F. and Peltier, A.: Hypo-complementémies de consommation: intérêt comparé des dosages hémolytiques et protéiques des composants de la voie classique du complément. Nouv. Presse Méd. 8: 1657, (1979).

23. Ruddy, S., Gigli, I. and Austen, K.F.: The complement system of man. New Engl. J. Med. 287: 489, (1972).

24. Schur, P.H.: Lupus érithémateux. In "Actualités Néphrologiques de l'Hôpital Necker", pp. 58–70, E.M. Flammarion Ed., (1976).

25. Sissons, J.G.P., West, R.J., Fallows, Williams, D.G., Boucher, B.J., Amos, N., Peters, D.K.: The complement abnormalities of lypodystrophy. New Engl. J. Med. 294: 461, (1976).

26. Sjöholm, A.G.: Complement components and complement activation in Acute Post–Streptococcal Glomerulonephritis. Intern. Arch. Allergy Appl. Immunol. 58: 274, (1979).

27. Spitzer, R.E., Vallota, E.M., Forristal, J., Sudora, E., S itzel, A., Davis, N.C. and West, C.D.: Serum C3 lytic system in patients with glomerulonephritis. Science 164: 436, (1969).

28. Whaley, K., Ward, D. and Ruddy, S.: Modulation of the properdin amplification loop in membranoproliferative and other forms of glomerulonephritis. Clin. Exp. Immunol. 35: 101, (1979).

29. Whaley, K., Schur, P.H. and Ruddy, S.: Relative importance of C3b inactivator and 1H globulin in the modulation of the pro-perdin amplification loop in systemic lupus erythematosus. Clin. Exp. Immunol. 36: 408, (1979).

30 Wyatt, R.J., Mc Adams, A.J., Forristal, J., Snyder, J. and West, C.D.: Glomerular deposition of complement, control proteins in acute and chronic glomerulonephritis. Kidney Int. 16: 505, (1979).

31. Zinsser, H. and Johnson, W.C.: On heat sensitive anticomplementary bodies in human blood serum. J. Exp. Med. 13: 31, (1911).

ACKNOWLEDGEMENTS

We thank Chantale Bory and Nicole François for technical assistance, Jacqueline Janutolo and Eliane Grari for typing the manuscript, and the Association Stéphanoise pour l'Étude et le Traitement des Affections Rénales (ASSETAR) for financial support.

THE SIGNIFICANCE OF THE DETECTION OF CIRCULATING IMMUNE COMPLEXES IN HUMAN GLOMERULONEPHRITIS

R. Coppo, D. Roccatello, C. Rollino, G. Martina,
A. Aprato, G. Camussi, M. Messina, and G. Piccoli

Cattedra di Nefrologia Medica dell'Università
Turin (Italy)
and Divisioni di Nefrologia e Dialisi
Ospedale Maggiore di San Giovanni Battista
Turin (Italy)

ABSTRACT

Much attention has been devoted to the development of clinically applicable ways of detecting immune complexes (IC) in the
circulation. Through a reliable IC assay it is hoped to obtain
information about the pathogenesis, diagnosis and prognosis of
human glomerulonephritis. The authors present a critical re-evalu-
ation of their own data, with a follow-up of 1-42 months. The
authors' experience refers to 428 sera from 207 patients affected
by idiopathic or secondary glomerulonephritis. In the last 4 years
3 tests for the detection of circulating IC were employed: solid
phase C1q test, PEG precipitation test and PMN immunohistological
technique. The limits and specificity of each test and the diagnostic
significance of the data obtained in nephropathic patients are
discussed.

LIST OF ABBREVIATIONS

IC = immune complexes.
AHG = aggregated human γ -globulins.

PEG = polyethylene glycol.
EM = electron microscopy.
IF = immunofluorescence.
GN = glomerulonephritis.
N.S.= non significant.
PMN = polymorphonuclear neutrophil.
SLE = systemic lupus erythematosus.
SP - C1q = solid phase - C1q.
O.D. 280 nm = optical density (asorbance) at 280 nm.

INTRODUCTION

The pathogenetic role of circulating immune complexes (IC) in human glomerulonephritis (GN) was not settled until 1950 when IC were identified by Germuth and Dixon (18, 12), as being the toxic products indicated as responsible for serum sickness first described by Von Pirquet in 1911 (31).

The fate of IC formed in the circulation depends on a number of variables, including the rate of production, the rate of removal, the condition of the phagocytic system, the degree of lattice formation and the nature of antigens and antibodies. Physico-chemical and biological properties of serum IC seem to determine their pathogenetic role in glomerulonephritis, resulting in complement-mediated and polymorphonuclear cell-mediated tissue injury.

In addition, IC may interact with cell receptors resulting in release of biologically active mediators or in interference with the cell recognition mechanism.

The renal localization of circulating IC is moreover enhanced by local factors, such as filtration pressure, increased vascular permeability and C3b receptors in epithelial and mesangial cells (17).

Considering the immunoglobulin and complement granular pattern detected by immunofluorescence microscopy (IF) and by electron microscopy (EM) in human glomerulonephritis, which are similar to those observed in the experimental models, there is good reason to believe that IC deposition causes approximately 80% of human GN (13).

The presence of IC in renal deposits can be confirmed by the

Table 1.   Antigens involved in human IC-induced GN.

EXOGENOUS

- pharmaceuticals   (foreign serum, toxoids, penicillamine, sulfa
                       compounds, trimethodione, mercury, gold,
                       heroin, etc.)

- infectious agents   (bacterial, viral, parasitic, fungal)

ENDOGENOUS

- nuclear antigens
- renal tubular brush border antigens
- thyroglobulin
- carcinoembryonic antigen
- tumour antigens
- immunoglobulins

---

analysis of renal eluates. The antigens detected up to date are
exogenous or endogenous (table 1): it must be stressed that the
presence of antigens bound to immunoglobulins and to complement in
the renal deposits can be recognized only in a few cases. The antigen
or antigens responsible for the great majority of human GN are still
unknown. The occurrence of IC in renal deposits is therefore difficult to identify.

Moreover, the circulating IC trapping mechanism does not fully
account for the marked variability in histologic and clinical
features of different forms of human GN not always reproducible by
experimental models.

Recent studies (14) provide evidence of the mechanisms of renal
in situ complex formation, as a result of antibodies binding to
antigens normally present or previously localized in the glomeruli.
This may be due either to mesangial uptake or physico-chemical
affinity for normal renal structures. The binding of DNA to the
basement membrane collagen (20) or that of Be hepatitis antigen or
of tumour antigens to the slit membranes (29) may be due to their
electrical charges.

Table 2. Tests based on IC physico-chemical properties.

- Size and solubility changes of antibody combined with antigen.

- Cryoprecipitation.

- Analytical ultracentrifugation (Kunkel).

- Sucrose gradient ultracentrifugation (Kunkel).

- Gel filtration.

- Percipitation in polyethylene glycol (Digeon)

The IC in renal deposits are probably in dynamic balance with the serum components. Their amount can increase or they can solubilize after reaction with free antigen or antibody. The renal biopsy provides only a static picture of this balance at one particular moment and the indications given may reflect only a transient situation. Therefore much attention has been devoted to the development of clinically useful ways of detecting immune complexes in the circulation in order to obtain reliable pathogenetic, diagnostic or prognostic information.

More than 30 methods have been described for the detection of circulating IC.

The direct demonstration of serum IC is possible only in experimental disease, injecting labelled antigens, or when the antigen, such as a virus, is visible on electron microscopy.

In most cases one must employ indirect methods, allowing the detection of molecules having physico-chemical or biological properties similar to IC.

The tests based on the physico-chemical characteristics of solubility or on size of IC (table 2), are scarcely sensitive and hardly employable in routine assays, but they may be successfully applied to further analytical studies of IC.

The polyethylene glycol (PEG) precipitation test (11) is based

Table 3.   Tests based on IC biological properties: binding to
free molecules.

---

1) C1q fixation or complement activation

   a) complement consumption (Mowbray)

   b) heavy C3 (Williams)

   c) C1q gel diffusion (Agnello)

   d) C1q binding (Zubler)

   e) C1q deviation (Sobel)

   f) C1q solid phase (Hay)

   g) C1q latex agglutination inhibition assay (Cambiaso)

   h) C1q binding inhibition radioimmunoassay (Gabriel)

---

on the selective precipitation of IC by low concentrations of this
uncharged linear polymer. In theory, this assay may be influenced
by molecules of heavy molecular weight in the tested serum.

The tests based on the biological properties of IC are more
sensitive. The main limitation of the methods based on the interaction of IC with the complement system (26) consists of the reactivity of the various substances other than IC: C reactive
protein, DNA, endotoxins, C3 nephritic factor, plasmin. The tests
based on complement activation (1, 32, 28, 19, 22, 16) (table 3)
are able to detect only C1q-fixing immunocomplexes IgG1, IgG3, or IgM.
In some cases, non specific reactions with DNA and endotoxins can
occur. Another test, employing the binding to conglutinins, (3) is
able to detect only C3b-fixing IC (table 4).

Some methods for detecting circulating IC employ the reactivity
of IC with cellular receptors (Fc IgG, C3D, C3b) (table 5) of platelets
(25), K cells (24), macrophages (27), lymphoblasts (30) or PNM (2).
Such tests are subject to possible interference by various factors
related to cell metabolism, as well as by antibodies reacting with
cell surface antigens, by Ig levels, and by bacterial lipopoly-
saccharides.

Table 4. Tests based on IC biological properties: binding to
          free molecules.

---

2) Conglutinin

    conglutinin solid phase (Casali)

3) Rheumatoid factor (RF) interaction

    a) monoclonal RF precipitation (Agnello)

    b) monoclonal RF binding inhibition RIA (Gabriel)

    c) monoclonal RF binding inhibition (Luthra)

    d) polyclonal RF latex agglutination inhibition (Lurhuma)

    e) polyclonal RF inhibition radioimmunoassay (Cowdery)

It is clear that it is impossible to select one best single
test which is able to satify all the needs for specificity, repro-
ducibility, sensitivity and easy performance. It must be stressed
that in different diseases, and also in different stages of a single
disease, various kinds of IC may play a role. Furthermore, each

Table 5. Tests based on IC biological properties: binding to
          cell surface receptors.

- Raji cell RIA (Theofilopoulos)

- Inhibition of complement-dependent lymphocyte rosette formation
  (Ezen)

- Platelet aggregation (Myllyla)

- Macrophage uptake inhibition (Onyewotu)

- K lymphocyte cytotoxicity inhibition (Mc Lennan)

- IF on polymorphonucleates (Camussi)

---

test detects only a few IC, with particular physico-chemical or
biological properties.

A WHO paper has tried to compare the results of 18 differents
tests for serum IC (21): the data were often conflicting. This is
not surprising since the tests are based on different properties
of IC.

These latter data have engendered a great skepticism following
the enthusiasm of the years when new tests were discovered. Some
clinicians even denied the specificity of these tests and the clinical
significance of reported data.

PERSONAL EXPERIENCE

We would like to present a critical re-examination of this
problem based on our data resulting from a 4-year experience, with
long-term follow-up in most cases, with special reference to the
limitation and specificity of the various tests, and to the diagnostic
significance of these data in nephropathic patients.

We routinely employed 3 tests for the detection of serum IC
in GN: the PEG precipitation test (6, 5, 7), the SP-C1q test (8, 4)
and the PNM immunohistological technique (2).

The PEG test is very easy to perform and is not influenced by
the Ig class of IC, but it shows a very low sensitivity (200 μg
AHG eq/ml) and it is probably influenced by other serum molecules.
The serum lipid content or the serum alpha-2-macroglobulin levels
do not seem, however, to affect the results of the PEG test(Student's
test: p not significant). We found IgG, IgM, C3, C4 in the precipitates (table 6). The amount of C4 significantly increased in patients with high PEG test data, in accordance with the data from
other authors (11). There is however no correlation between the IgG
(r= 0.19; p= N.S.), IgA (r= 0.12; p= N.S.) and IgM (r= 0.07; p= N.S.)
levels and the PEG test data.

The second test we employed was the SP-C1q. The IC bound to
C1q, adsorbed to the plastic surfaces of the test tubes, are
detected by staphylococcal A protein (able to bind to IgG) labelled
with alkaline phosphatase (enzyme linked immune sorbent assay:

Table 6.   Qualitative analysis of 42 PEG precipitates.

| | Increased in comparison with healthy people | | | | |
	IgG	IgM	IgA	C3	C4
PEG test + sera	32%	60%	13%	40%	64%
PEG test − sera	32%	50%	26%	11%	10%

ELISA). This test can detect as much as 8 μg AHG eq/ml, but its spectrum is only limited to those IgG1 which are capable of fixing C1q and reacting with staphylococcal A protein.

The third test employed is the PNM immunohistological technique (IF − PMN): the IC bound to Fc or C3b receptors located on the PMN are detected by immunofluorescence.

We studied 428 sera from 207 patients affected by various types of idiopathic glomerulonephritis or related to systemic diseases, of whom 188 were biopsied and 107 were followed with a clinical follow-up up to 42 months (table 7). We observed that the levels of serum IC, using the C1q-SP and the IF-PMN tests, are significantly lower in idiopathic GN than in GN related to systemic diseases, such as lupus nephritis, where we found the highest values. Among some groups of idiopathic GN (minimal change GN, membranous GN and focal sclerosing GN) the C1q-SP rarely provided us with positive data. On the contrary, relatively low levels of IC could success-fully be detected by the IF-PMN and PEG tests.

Significant levels of serum IC (p $<$ 0.01) were observed in GN with prevalent mesangial IgA deposits (Berger's GN). We recently applied to these sera a new test, developed by our group (9), i.e. the conglutinin solid phase test with anti IgA, which is able to detect IgA IC or aggregated IgA. These data suggest the presence of two kinds of IC, some able to fix C1q and having IgG as antibody, and some conglutinin-fixing and IgA-containing IC.

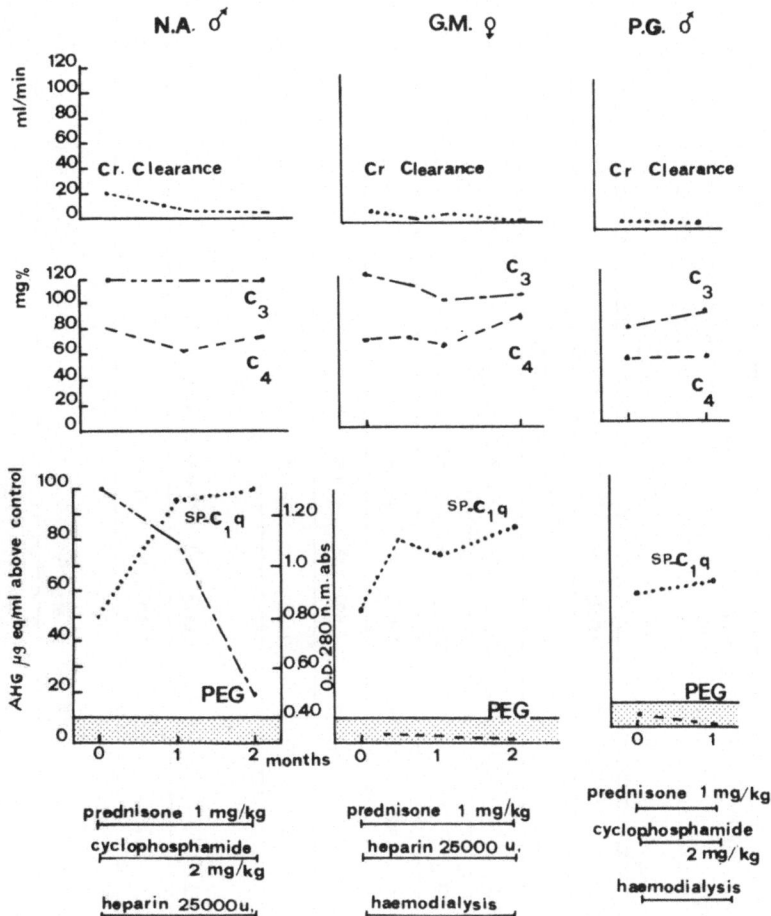

Fig. 1. Renal function and levels of serum immune complexes
         in three patients with rapidly progressive glomerulo-
         nephritis.

In mesangiocapillary GN and, moreover, in extracapillary.
proliferative GN with granular IF we found IC levels significantly
higher than those detected in healthy people. On the contrary, 2
cases with linear IF pattern displayed negative data for every test
employed. It is noteworthy to observe that, in rapidly progressive
GN with granular IF pattern, constantly high levels of serum IC
always represented a sign indicating poor prognosis: 3/3 patients
with repeatedly positive SP - C1q data, progressed to chronic uraemia,
in spite of corticosteroid and immunosuppressive therapy (Fig. 1).

Table 7. Levels of serum immune complexes, detected by three different methods, in 207 patients affected by primary and secondary glomerulonephritis.

	n° cases	n° samples	SP - C1q Test		PEG Test		IF on PMN		
			mean AHG* µg eq/ml	positive data	mean O.D 280 nm	positive data**	mean of PMN showing IgG fluorescence	mean of PMN showing C3 fluorescence	positive data
Healthy people	65	104	1.3 ± 3.8	6.8%	0.26+0.10	8.8%	3.2 ± 3.5	1.5 ± 2.2	0
Acute post-infectious GN	14	53	10.2 ± 13.6	42.8%	0.50+0.36	78.5%	13.9 ± 16	5 ± 13	46.6%
Membranous GN	19	27	3.8 ± 7.4	31.5%	0.44+0.25	57.8%	14.3 ± 14.4	3.3 ± 8	72.7%
Focal sclerosing GN	16	18	2.6 ± 5.2	18.7%	0.54+0.41	50.0%	17 ± 23.8	13 ± 23	33.3%
Minimal change GN	23	75	0.8 ± 3.9	8.9%	0.64+0.64	55.0%	12.8 ± 18	4.8 ± 8.5	62.5%
Berger's disease	27	40	9.6 ± 15.7	42.9%	0.64+0.38	35.0%	13.7 ± 18	5.8 ± 8.3	54.5%
Schönlein-Henoch disease	7	21	2.4 ± 5.14	40.0%	0.59+0.45	75.0%	10 ± 13.8	2.6 ± 3	50%
Mesangiocapillary GN	33	39	11.9 ± 21.6	33.3%	0.48+0.50	39.3%	21.9 ± 27	8.4 ± 14.7	57.8%
Extracapillary GN									
with IF granular pattern	4	9	54.1 ± 35.2	75.0%	0.49+0.40	25.0%	20.7 ± 20	4.5 ± 4.9	71.4%
with IF linear pattern	2	2	4.0 ± 4.0	0 %	0.19+0.05	0 %	-	-	-
Proliferative mesangial and stalk mesangial GN	6	6	9.3 ± 10.7	16.6%	0.53+0.42	66.6%	8.5 ± 10.3	4.2 ± 5.4	52.9%
Schönlein-Henoch	7	21	2.4 ± 5.14	40.0%	0.59+0.45	75.0%	10 ± 13.8	2.6 ± 3	50%
Mesangiocapillary GN	33	39	11.9 ± 21.6	33.3%	0.48+0.50	39.3%	21.9 ± 27	8.4 ± 14.7	57.8%

	n	n							
Extracapillary GN with IF granular pattern	4	9	54.1 ± 35.2	75.0%	0.49+0.40	25.0%	20.7 ± 20	4.5 ± 4.9	71.4%
with IF linear pattern	2	2	4.0 ± 4.0	0%	0.19+0.05	0%	–	–	–
Proliferative mesangial and stalk mesangial GN	6	6	9.3 ± 10.7	16.6%	0.53+0.42	66.6%	8.5 ± 10.3	4.2 ± 5.4	52.9%
Lupus nephritis (LN) biopsied and not biopsied	38	109	45.7 ± 91.8	68.4%	0.55+0.46	68.4%	–	–	–
LN CLASSES: II	3	7	22.3 ± 61.4	100%	0.31+0.46	33.3%	14 ± 8	0	50%
III	4	30	44.2 ± 75.3	100%	0.46+0.33	75.0%	25 ± 7.5	3 ± 2.7	60%
IV	12	44	46.1 ± 89.8	75.0%	0.51+0.50	75.0%	22 ± 21	7 ± 8	70%
V	2	9	22.7 ± 54.7	100%	0.27+0.06	0%	38 ± 31	4 ± 3.1	50%
Polyarteritis nodosa	3	12	18.5 ± 29.6	66.6%	0.42+0.85	66.6%	24 ± 29.7	10 ± 8.7	100%
Renal amyloidosis	4	4	4.0 ± 6.9	25.0%	0.58+0.38	50.0%	3.3 ± 4.1	8 ± 8.9	33.3%
Cryoglobulinemic GN	11	13	3.3 ± 6.9	36.3%	0.77+0.64	90.9%	31.5 ± 24.8	10 ± 14.6	70%

* AHG aggregated human gammaglobulins.

** Positive data = n% of patients having at least once positive result (mean + 2 standard deviations) during follow-up.

___ Significant difference (p < 0.01) from healthy people data.

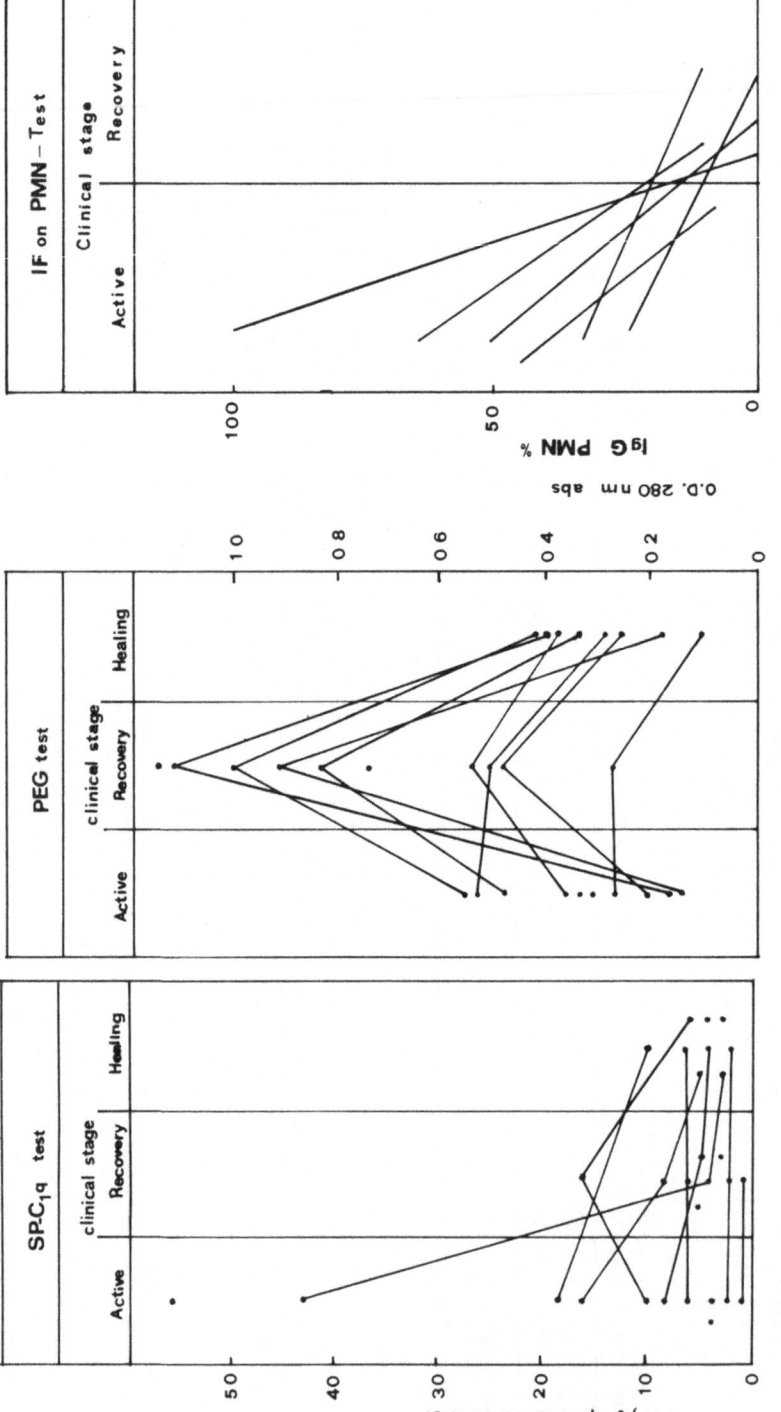

Fig. 2. Correlation between clinical disease and levels of serum immune complexes in
acute postinfectious glomerulonephritis.

   Five out of eleven patients with acute postinfectious GN
followed from the first manifestation of disease, showed positive
SP - C1q data: every patient displayed negative data later in the
healing stage. Similar results were provided by IF - PMN test
(Fig. 2). In contrast to the SP - C1q data, the PEG test values
were usually negative or slightly positive in the acute phase and
markedly increased during improvement. They finally became negative
in the recovery phase. The rise of PEG-positivity during recovery
may be due to an increase of some still undetermined unrelated
protein, or to less nephropathogenic immune complexes, less soluble
in PEG. Its interpretation is puzzling, but as this effect was
observed in 6 out of 8 cases, it does not appear to be only due to
chance.

   It is our belief that this phenomenon, which appears early in
the clinical course of the disease, may provide us with data of
clinical and prognostic significance.

   Very high levels of serum IC were found in SLE (table 8),
especially in cases displaying cellular proliferation and subendothelial deposits at renal biopsy; lower values were observed
in mesangial and membranous GN. The results of SP - C1q tests were
positive in 17 out of 18 sera from lupus nephritis with a clinical
presentation of acute nephritic syndrome (Fig. 3). The dissociation
between the SP - C1q and the PEG data in the acute phase and during

Table 8.   Conglutinin solid phase test with anti IgA.

	POSITIVE DATA	
	SERA	%
IgA IC or aggregated IgA	11/54	20.4%
IgA and IgG IC	6/54	11%
IgG IC	4/54	7.4%
IgG and or IgM IC	4/54	7.4%
No detectable IC with all the three tests	29/54	53%

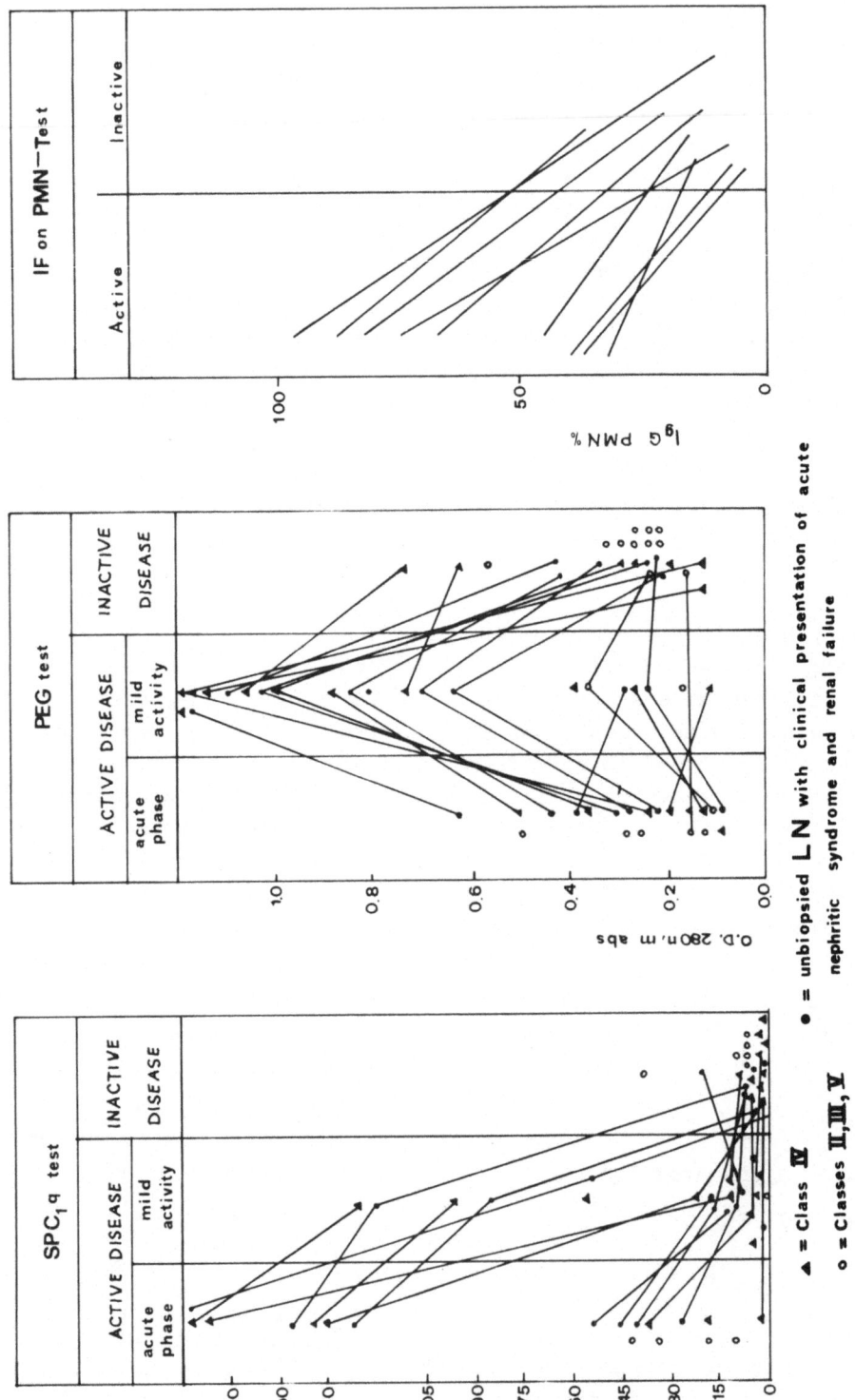

Fig. 3. Correlation between clinical disease and levels of serum immune complexes in systemic lupus erythematosus.

recovery, already mentioned above with reagrd to acute postinfectious GN, was observed also in 9 out of 10 cases of SLE. It always represented a sign of good prognostic value. The results obtained using the IF - PMN test were similar to those obtained by the SP - C1q tests.

We found a good agreement between SP - C1q values and anti DNA levels: (2 = + 0.45, p <0.05). In only 4 cases IC data indicated a clinical activity phase, in contrast to negative anti-DNA antibodies.

CONCLUSIONS

At the beginning of the 1970's the work of Dixon (13) seemed to clarify the pathogenesis of most GN: in 5% of the cases anti-glomerular basement membrane (BM) antibodies were operating. In 80% of cases of human GN pathogenesis was ascribed to circulating IC. However, the great histological and clinical variability of human GN in comparison to the few experimental models, the IF and EM patterns often not typical for a IC disease, as well as the exceptional finding of specific antigens bound to immunoglobulins and complement in renal deposits, have suggested that it may be opportune to reconsider the hypothesis of the pathogenetic role of circulating IC in some cases of human GN.

The application of the new tests for serum IC detection have led to the hope of gaining new data in investigating the pathogenesis of and in clinically monitoring human GN. After 4 years of experience of these tests, we can draw some conclusions. Each of these methods reveals serum material reacting as IC. However, they display limitations of specificity that in some cases, as considered above, greatly restrict their value.

Every test can detect only a few of the IC present in a serum sample, i.e. those which are able to react with a particular system. The sensitivity of each test is probably sufficient only in sera containing high amounts of IC.

In single cases the isolated findings of circulating IC, without consideration of their physico-chemical and biological properties, is a very coarse approach to the study of GN.

In spite of this, it must be admitted that in some GN cases we frequently found high values of serum IC. They were quantitatively and significantly different from normal values, and they were reproducible. Furthermore they present a sequential progress, according to theoretical grounds, so that such a phenomenon is likely to be strictly related to the clinical evolution of the disease.

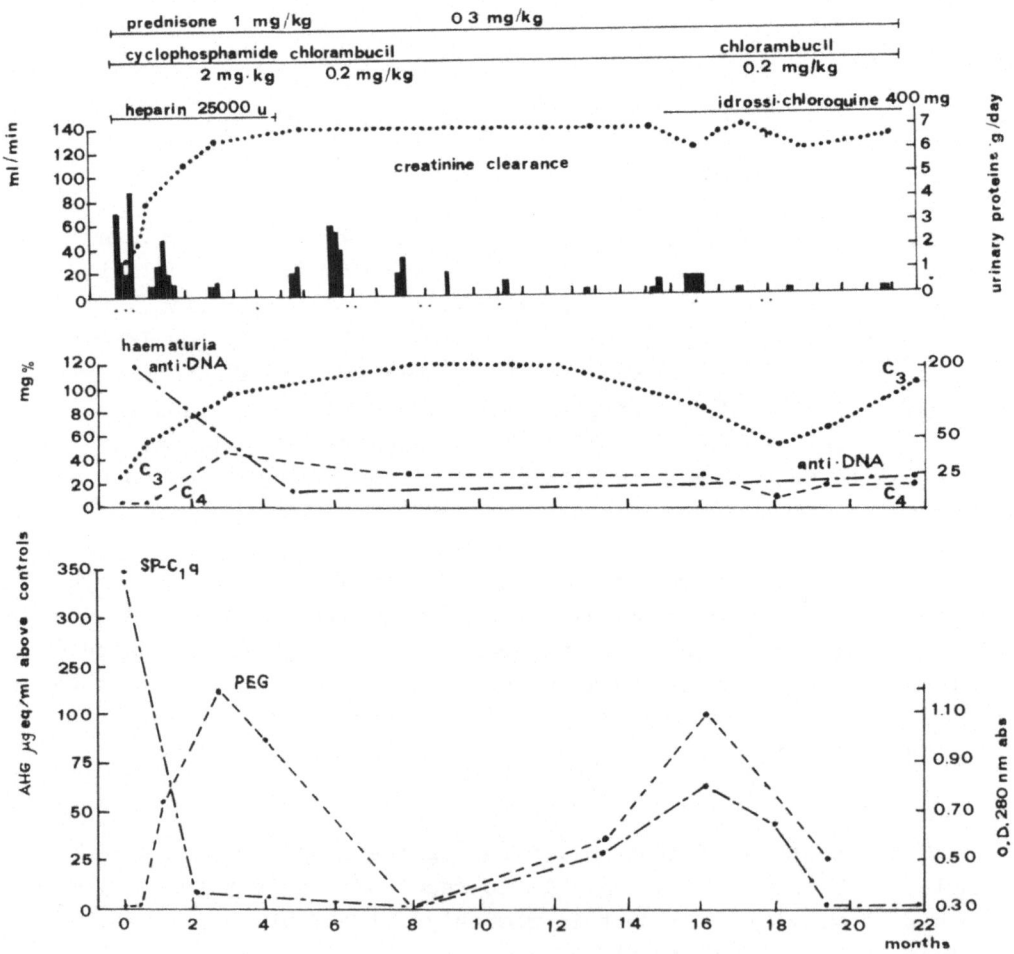

Fig. 4. Levels of serum immune complexes, clinical follow-up
          and treatment in a case of systemic lupus erythema-
          tosus.

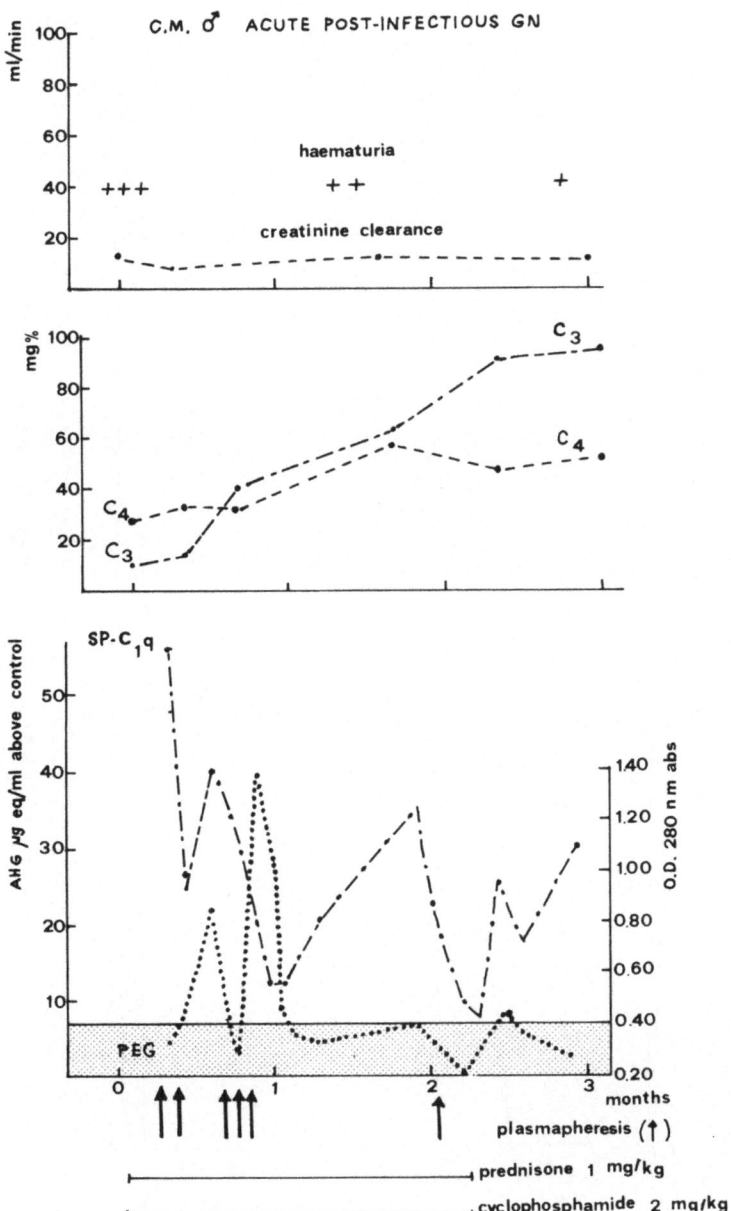

Fig. 5. Levels of serum immune complexes, clinical follow-up and treatment in a case of acute postinfectious glomerulonephritis.

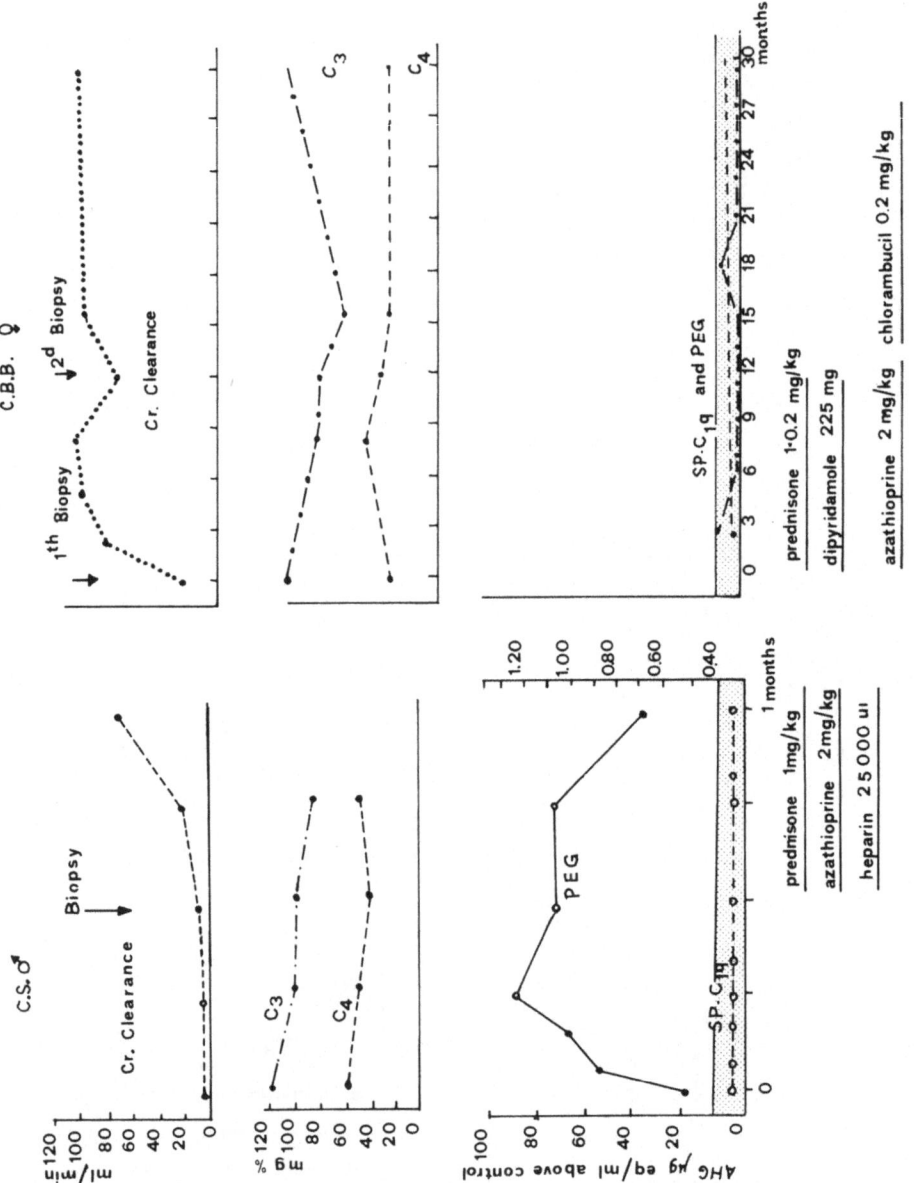

Fig. 6. Low levels of serum immune complexes indicate a good prognosis in two cases of acute post-infectious glomerulonephritis.

When we found abnormal data, multiple controls during the course of the disease generally provided us with precise correlation with the clinical follow-up. Moreover, among some nephropathies, such as acute GN, lupus nephritis, polyarteritis and rapidly progressive GN, it is possible to identify a relatively common pattern of behaviour of serum IC levels. It is so costant that it can assume diagnostic significance, useful for clinical practice. Such data can also provide us with dynamic information, more precise than that obtained with other routinely employed immunological tests. As an example, we may observe the data reported in fig. 4.

In a patient suffering from proliferative diffuse lupus nephritis, with a follow-up of 22 months, the serial evaluations of circulating IC during the acute phase, and during the following remission and reactivation stages, represented very sensitive diagnostic signs. It is worth mentioning also the prognostic significance that can be inferred from the IC data, as in the case reported in Fig. 5. This is a case of acute post infectious GN, where the unusually high levels of serum IC, in spite of corticosteroids' and immunosuppressive treatment and repeated plasmaphereses, represented the only diagnostic sign of active disease, indicating a poor prognosis. Contrariwise, the constantly negative data in the two cases showing a favourable course, reported in Fig. 6, represented a good prognostic sign. Both patients were suffering from acute postinfectious GN with marked extracapillary proliferation (involving more than 50% of the glomeruli) with clinical and humoral features similar to those of the case reported in fig. 5. Finally the observation, initially disappointing, of negative IC data provided in some GN, as membranous GN, where a role of circulating IC have been suspected, were very useful because they stimulated the search for other pathogenetic mechanisms, namely in situ IC formation. In conclusion the evaluation of serum IC, together with other immunological tests, offers a multifactorial study, that can provide important information on the dynamics of the incompletely known pathogenetic mechanisms of human GN.

REFERENCES

1. Agnello, V., Winchester, R.J. and Kunkel, H.G.: Precipitin reactions of the C1q component of complement with aggregated ɣ-globulin and immune complexes in gel diffusion. Immunology 19: 909, (1970).

2. Camussi, G., Tetta, C. and Caligaris Cappio, F.: The detection of
    immune-complexes on the surface of polymorphonuclear neutro-

3. Casali, P., Bossus, A., Carpentier, N. and Lambert, P.H.: Solid
    phase enzyme immunoassay or radioimmunoassay for the detection
    of immune complexes based on their recognition by conglutinin.
    Clin. exp. Immunol. 29: 342, (1977).

4. Coppo, R., De Marchi, M., Quarello, F., Stratta, P., Triolo,
    G., Carbonara, A. and Piccoli, G.: Utilizzazione di alcuni
    tests per la ricerca di immunocomplessi sierici nello studio
    di una casistica nefrologica. Atti del IV Simposio Internazio-
    nale "Applicazioni delle tecniche radioisotopiche in vitro".
    p. 123, Napoli, December 1977.

5. Coppo, R., De Marchi, M., Segoloni, G., Quarello, F. and
    Piccoli, G.: Circulating immune complexes in nephritis. La
    Ricerca, 7: 354, (1977).

6. R. Coppo, Segoloni, G., Alloatti, S., De Marchi, M., Vercello-
    ne, A. and Piccoli, G.: Ricerca di immunocomplessi circolanti
    in nefropatie umane. Min. Nefrol. 24: 155, (1977).

7. Coppo, R., De Marchi, M., Segoloni, G., Alloatti, S., Camussi,
    G., Quarello, F., Rosso, E., Bussolino, F. and Piccoli, G.:
    Analisi qualitative di immunocomplessi sierici (IC s) in glo-
    merulonefriti umane. Min. Nefr. 25: 7, (1978).

8. Coppo, R., De Marchi, M., Mazzucco, G., Carbonara, A.O., Sego-
    loni, G., Monga, G. and Piccoli, G.: Application of solid
    phase Clq assay in sera from nephropathic patients.In: Protides
    of the biological fluids. Pergamon Press, Oxford, p. 313, 1979.

9. Coppo R., De Marchi, M., Carbonara, A.O., Boccazzi, C., Bul-
    zomì, M.R., Messina, M., Martina, G., Rollino, C., Segoloni,
    G. and Piccoli, G.: Presentazione di un nuovo test (conglutinin
    solid phase con anti-IgA) per la evidenziazione di immunocom-
    plessi sierici a componente anticorpale IgA. Min. Nefrol. 27:
    447, (1975).

10. Cowdery, J.S., Treadwell, P.E. and Fritz, R.B.: A radioimmuno-
    assay for human antigen-antibody complexes in clinical material.
    J. Immunol. 114: 5, (1975).

11. Digeon, M., Laver, M., Rizza, J. and Bach, J.F.: Detection of
    circulating immune complexes in human sera by simplified assay
    with polythylene glycol. J. Immunol. Met. 16: 165, (1977).

12. Dixon, F.J.: The role of antigen-antibody complexes in disease.
    Harvey Lecture, 58: 21, (1963).

13. Dixon, F.J.: The pathogenesis of glomerulonephritis. Am. J.
    Med. 44: 493, (1968).

14. Editorial Review. In situ immune complexes formation and
     glomerular injury. Kidney Int. 17: 1, (1980).
15. Franklin, E.C., Holman, H.R., Muller-Eberhard, H. and Kunkel,
     H.G.: An unusual protein serum of certain patients with rheu-
     matoid arthritis. J. Exp. Med. 105: 425, (1957).
16. Gabriel, A. Jr. and Agnello, V.: Detection of immune complexes.
     The use of radioimmunoassay with Clq and monoclonal rheumathoid
     factor. J. Clin. Invest. 59: 990, (1977).
17. Gelpaud, M.C., Frank, M.M. and Gree, I.: A receptor for the
     third component of complement in the human renal glomerulus.
     J. Exp. Med. 142: 1029, (1975).
18. Germuth, F.G.: A comparative histologic and immunologic study
     in rabbits of induced hypersensitivity of the serum sickness
     type. J. Exp. Med. 97: 257, (1953).
19. Hay, F.C., Nineham, L.J. and Roitt, I.M.: Routine assay for
     the detection of immune complexes of known immunoglobulin class
     using solid phase Clq. Clin. exp. Immunol. 24: 396, (1976).
20. Izui, S., Lambert, P.H. and Miescher, P.A.: In vitro demon-
     stration of a particular affinity of glomerular basement
     membrane and collagen for DNA: A possible basis for local
     formation of DNA - Anti - DNA complexes in systemic lupus
     erythematosus. J. Exp. Med. 144: 428, (1976).
21. Lambert, P.H. et alii: A WHO collaborative study for the evalu-
     ation of 18 methods for detecting immune complexes in serum.
     J. Clin. Lab. Immunol. 1: 1, (1978).
22. Lurhuma, A.Z., Cambiaso, C.L., Masson, P.L. and Heremans, J.F.:
     Detection of circulating antigen-antibody complexes by their
     inhibitory effect on the agglutination of IgG-coated particles
     by rheumatoid factor or Clq. Clin. exp. Immunol. 25: 21,
     (1976).
23. Luthra, H.S., McDuffie, F.C., Hunder, G.C. and Samayoa, E.A.:
     Immune complexes in sera and synovial fluids of patients with
     rheumatoid arthritis. Radioimmunoassay with monoclonal rheu-
     matoid factor. J. Clin. Invest. 56: 458, (1975).
24. McLemman, I.C.M.: Competition for receptor for immunoglobulins
     on cytotoxic lymphocytes. Clin. Exp. Immunol. 10: 275, (1972).
25. Myllylä, G.: Aggregation of human blood platelets by immune
     complexes in the sedimentation pattern test. Scand. J. Haematol.
     Suppl. 19, (1973).
26. Mowbray, J.F., Hoffbraud, A.V., Holborow, E., Seah, P.P. and
     Fry, L.: Circulating immune complexes in dermatitis herpeti-
     formis. Lancet, 1: 400, (1973).

27.   Onyerwotu, I.I., Holborow, E.J. and Johnson, G.D.: Detection
      and radioassay of soluble circulating immune complexes using
      guinea pigs peritoneal exudate cell. Nature, 248: 156, (1974).
28.   Sobel, A.T., Bokisch, V.A. and Muller-Eberhard, H.J.: C1q
      deviation test for the detection of immune complexes aggregates
      of IgG and bacterial products in human sera. J. Exp. Med. 142:
      139, (1975).
29.   Takesoshi, Y., Tanaka, M., Miyakawa, Y., Yoshizawa, H.,
      Takahashi, K. and Mayumi, M.: Membranous glomerulonephritis
      induced by Hepatitis Be antigen-antibody complex. New Engl.
      J. Med. 300: 814, (1979).
30.   Theofilopoulos, A.N., Eisenberg, R. and Dixon, F.J.: Use of
      B-type lymphoblastoid cells (Raji) for the isolation of pre-
      formed immune complexes from human sera (abstr.). Fed. Proc.
      Fed. Am. Soc. Exp. Biol. 36: 1213, (1977).
31.   Von Pirquet, C.E. Allergy. Arch. Int. Med. 7: 259, (1911).
32.   Zubler, R.H. and Lambert, P.H.: The ^{125}I-C1q binding test for
      the detection of soluble immune complexes. In: B.R. Bloom and
      J.R. David: In vitro methods in cell-mediated and tumor immunity
      Academic Press, Nex York, p. 565, 1976.

ACKNOWLEDGEMENTS

      The authors acknowledge Prof. A. Carbonara and Dr. M. De Marchi
of the Medical Genetics Institute of the University of Torino, for
their foundamental advice and assistence in setting up the C1qSP,
the PEG and the conglutinin-anti IgA tests.

IMMUNOLOGICAL ASPECTS OF RENAL VASCULAR INVOLVEMENT

J. Egïdo, M. Sanchez Crespo and L. Hernando

Servicio de Nefrologia
Fundacion Jimenéz Diaz
Madrid (Spain)

ABSTRACT

The majority of cases of human vasculitis are thought to be immunologically mediated. The contribution of experimental models, such as acute serum sickness of rabbits, Arthus reaction (including Arthus nephritis) and virus-induced vasculitis, to the knowledge of the problem is reviewed, emphasizing the mechanisms involved in the deposition of immune complexes and mediator systems of vascular injury.

The following immunological aspects of human vasculitis are especially analyzed:

- the data supporting participation of immune complexes and of known exogenous and endogenous antigens;

- the need of an increased vascular permeability to allow the deposition of immune complexes;

- the mechanisms involved in other vasculitis, such as Henoch Schonlein syndrome and Wegener are granulomatosis.

INTRODUCTION

The complex and highly specialized vascular system of the kidney receives more than 20% of the cardiac output. Taking into account the large surface of glomerular capillaries, there is practically no immunological injury to the kidney in which the blood vessels do not participate. For these reasons vasculitis is not infrequently accompanied by some degree of glomerular involvement.

The majority of human vasculitis, pathologically defined by inflammation and necrosis of the vessel walls, are thought to be mediated through immune complex (IC), though the etiologic factors are seldom identified. Hyperacute renal allograft rejection caused by preexisting immunization is the most characteristic example of antibody-mediated vascular damage. The role of cell-mediated immunological reaction in blood-vessel injury is, as we shall mention, less clearly defined.

1. EXPERIMENTAL MODELS OF VASCULITIS

Most of our present knowledge on the immunological aspects of human vasculitis has been obtained from studies in animals. The best known models of experimental vasculitis are acute serum sickness of rabbits, the classical Arthus reaction and the virus-induced vasculitis especially in mice (Table 1).

One-shot experimental serum sickness is a model for acute glomerulonephritis and vasculitis, whose sequence of events has already been fully reviewed (9). Following intravenous injection of bovine serum albumin (BSA), glomerulonephritis and arteritis develop coincident with the appearance of immune complexes. Arteritis occurs predominantly at branching points in the aorta and in pulmonary arteries. Rabbit IgG, C3 and BSA are detectable adjacent to the elastic lamina of arteries. The inflammatory infiltrate predominates, the arterial necrosis being less important. Once arteritis has developed, it is difficult to identify the host IgG, C3 and BSA, as they are promptly degraded by the polymorphonuclear (PMN) inflammatory response. This important feature of acute serum sickness has relevance to immunofluorescent studies in acute human vasculitis in which frequently neither immunoglobulin nor complement are iden-

Table 1. Experimental models contributing to the understanding
of the pathogenesis of vasculitis.

---

1. - Classic rabbit serum sickness and modified models.

2. - Arthus reaction (included Arthus nephritis).

3. - Virus-induced vasculitis.

   - Lymphochoriomeningitis virus in mice.
   - Aleutian disease of mink.
   - Lactic dehydrogenase virus in mice.
   - Equine viral arteritis.

4. - Mycoplasma gallisepticum-induced vasculitis in turkeys.

---

tified. A contrast between acute and chronic serum sickness is the
absence of arteritis in the chronic model produced in rabbits by
daily, single injections of BSA (9). However Brentjens et al. (7)
have shown that rabbits displaying a hyperactive antibody response
to multiple injections each day of high doses of BSA, developed
glomerulonephritis and accumulation of inflammatory cells and immune
deposits in the walls of renal peritubular capillaries. It is probable that the large amount of IC formation and/or the longer persistence of critical levels of IC in the circulation could explain
that difference. In the same way, in an accelerated serum sickness
in the rabbit a striking angiitis occurred predominantly in the
pulmonary artery. Unlike the vasculitis seen in acute serum sickness,
it was characterized by a prominent accumulation of monocytes in
the subendothelial space (21). Recently in a new experimental model
of chronic serum sickness in rats, mulitple immune deposits were
seen in the vessel walls of many tissues (2).

The classical Arthus reaction, in which locally injected
antigens bind with precipitating antibodies to form complexes that
activate the complement system, has represented an excellent model
of neutrophil and complement-mediated injury (10). Furthermore an
accumulation of platelets, possibly via IgE antibodies and mast
cells, has been observed at sites of antigen-antibody mediated
injury in the Arthus reaction (23). Recently segmental necrotic

lesions, segmental crescent and granuloma formation, similar to those found in human granulomatous vasculitis, were seen in one Arthus-type nephritis model induced by the infusion of BSA into the left renal artery of rabbits that had been sensitized with BSA (36).

Other experimental models such as the virus induced vasculitis have probably a greater relevance to arteritis in man. Certain strains of mice injected with lymphocytic choriomengitis virus, either in utero or shortly after birth, have persistence of the virus in blood, in large part as IC. Granular deposits of immuno-globulins and complement are seen in the glomerular basement membrane and in the vessel walls, histologically showing arteritis similar to that seen in acute immune complex diseases. The general picture may be analogous to HBs associated arteritis in man. Other models of virus-induced vasculitis are the Aleutian disease in minks and the immune complex disease associated with lactic dehydrogenase virus in mice. Evidence exists that virus-carrying animals are somehow deficient in their ability to make an immune response (see review in reference 9).

Apart from these well-known cases of immune complex-induced vasculitis there are some observations suggesting that occasionally virus (as in equine viral arteritis) or mycoplasma (as in turkeys injected with mycoplasma gallisepticum) may initiate arteritis shortly before the immune response.

Mechanisms Involved in the Vascular Deposition of Immune Complexes

The appearance of systemic immune complex disease in experi-mental animals and in humans depends, among other factors, on the presence of circulating IC in sufficient quantity, on the size of the IC, and on the status of the mononuclear phagocyte system. Complexes of large size are rapidly removed from the circulation by the Kupffer cells while small IC persisted longer. The saturation of the mononuclear-phagocyte system by carbon particles or large IC results in prolonged circulation of these IC and thereby increases the risk of tissue deposition (28). These observations raise the possibility that malfunction of the mononuclear-phagocyte system may prolong the circulation of large size IC, as has been demon-strated in some patients with vasculitis (17). The mechanisms of

small IC removal from the circulation have not been clearly identified.

The existence of circulating immune complexes is not sufficient
to explain the appearance of either glomerular or vascular disease,
or both. In the acute serum sickness of rabbits the deposition of
immune complexes in the blood-vessel walls involves an increase in
vascular permeability induced by the release of vasoactive amines
from platelets (9). In these animals the occurrence of the glomerular
lesions is particularly associated with antigen-dependent histamine
release from leukocytes (10). On addition of the antigen, basophils
sensitized with IgE degranulate and release also a platelet activating factor (PAF) which aggregates platelets and releases their
histamine (5). We have recently shown that a diminution of circulating basophils precedes the onset of proteinuria in acute experimental serum sickness; and those rabbits whose basophil count
remain normal showed no proteinuria, although they had circulating
immune complexes bigger than 19 S in size (4). The administration
of antihistamines and antiserotonin drugs markedly diminished the
presence of renal immune deposits and protected some animals from
glomerulonephritis (22). Likewise, disodium cromoglycate, a specific
inhibitor of mastocyte degranulation, also prevented the appearance
of nephritis in some rabbits (12).

However large differences exist between animal species with
regard to various types of nephritis. In rats, in which histamine
and serotonin seem to be important mediators of the inflammatory
response, a similar anaphylactic mechanism does not seem to exist,
at least in the Heymann nephritis model. In this disease, considered
as a chronic immune complex nephritis, specific homocytotropic
antibodies (IgE and IgG classes) and anti-brush border antigens
were systematically absent (1, 14). Accordingly treatment with
antihistamines and/or disodium cromoglycate did not produce any
beneficial effect upon the course of the disease (1, 14). Probably
the type of nephritis could also be important. In this sense we
have just observed the existence of specific IgG antibodies in the
recently described model of serum sickness in rats (2). The fall of
precipitating antibodies was followed by the appearance of proteinuria while the titer of IgG antibodies remained persistently high
(33). In addition to the increase in vascular permeability other
factors, such as blood flow turbulence at vessel bifurcations and
hydrostatic forces, predispose to the preferential localization of

IC in certain sites of vascular tree.

Mediator Systems in Immunologic Injury of the Vessels.

Abundant evidence indicates that immune complexes injure tissue through the mediation of cellular and humoral factors. Among these factors, an important role is played by the PMN cells, that intervene either in the Arthus vasculitis or in vasculitis of acute serum sickness of rabbits. When the IC are deposited in the vascular wall, or are formed in situ, a large accumulation of PMNs occurs within the first few hours. Several serum factors might attract PMNs toward the site of tissue injury, the complement being the most important. In animals depleted of serum complement, prior to the Arthus reaction or to the induction of acute serum sickness in rabbits, the PMN infiltration is abolished. The specific removal of PMNs by treatment with nitrogen mustard or heterologous anti-PMN sera has been shown to inhibit the appearance of the necrotizing arteritis. In the same experimental condition intimal proliferation was markedly diminished or absent and neither destruction of the internal elastic lamina nor fibrinoid necrosis in the arterial walls occurred (9). On the other hand, the glomerulonephritis normally seen in rabbit serum sickness is not affected by PMN removal; this is in accordance with the known paucity of PMNs in glomeruli in acute immune complex disease of rabbits (a review can be found in reference 9).

Once the PMN accumulation has occurred, their intracytoplasmic lysosomal enzymes, particularly collagenase and elastase, are released, causing damage and necrosis of the vessel walls. The specific structures damaged depend on the size of the vessels. In the arteritis of serum sickness, the internal elastic lamina is disrupted. In the arterioles the vascular basement membrane is apparently the critical target structure. Sometimes, depending on the size of the immune precipitates, these lesions may be accompanied by thrombosis, occlusion, hemorrhage, and ischemic changes in the surrounding tissue.

The list of potentially injurious constituents of PMNs is enormous. Recently, it has been demonstrated that platelet activating factor (PAF), a mediator of anaphylaxis found initially in basophils (3) and later in mouse and rat macrophages, may be released by PMNs of several species including man (32).

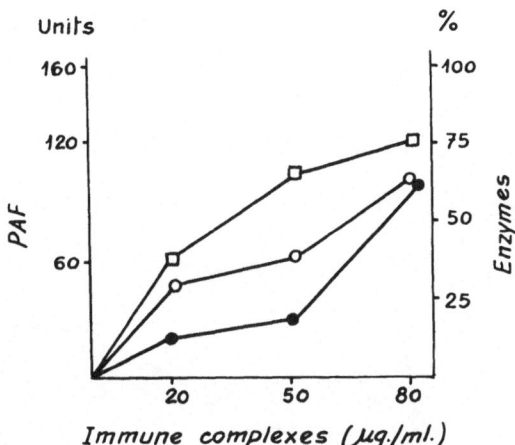

Fig. 1.   Human buffy coat cells were incubated for 45 min with
          "in vitro" preformed heterologous immune complex. The
          release of PAF (□), B-glucuronidase (○) and acid
          phosphatase (●) was measured. (Taken from Sanchez
          Crespo et al., Immunology).

Phagocytic leukocytes (monocytes and PMNs) released PAF physico-
chemically analogous to the PAF obtained by anaphylactic reactions in
rabbits, when challenged with Zymosan, Zymosan coated with complement,
immunocomplexes and immunoglobin aggregates (21) (fig. 1). Liberation of PAF and that of the lysosomal content follow different
mechanisms as they have different kinetics (PAF is liberated earlier
following membrane activation) and are modified in an opposite way
by drugs acting on the cytoskeleton (21) (Fig. 2). Although the
role of this mediator in inflammation remains to be established,
it can represent a link between phagocytosis of IC and platelet
aggregation and even intravascular coagulation. Theoretically their
release could also explain, among other factors, the high frequency
of the presence of platelets at the sites of immune complex injury

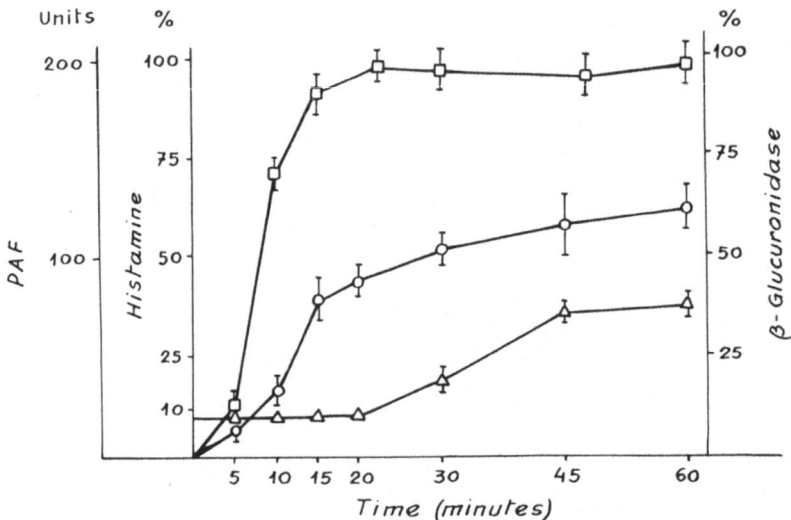

Fig. 2. Human buffy coat cells were incubated with 1.25 mg
         Zymosan at 37°C. Reactions were stopped at different
         times. As shown in the figure, the liberation of PAF
         (□) precedes B-glucuronidase (○) and histamine (△)
         releases, which agrees with our data of a different
         cellular origin for histamine and PAF. (Taken from
         Sanchez Crespo et al, Immunology).

(23). Furthermore we have recently observed that a desensitization
of platelets to PAF occurred in the serum sickness of rats coin-
cidentally with the appearance of proteinuria. Interestingly the
maximal quantities of PAF released were induced by immune complexes
in slight excess of antigen, considered to be the most injurious
to the vessels, at least in some models of experimental arteritis.

     The immune complexes formed in blood vessel walls, either in
Arthus arteritis or in the arteritis of acute serum sickness of
rabbits, are short-lived. Ingestion by neutrophils of immune com-

plexes and complement results in their rapid degradation within 8 hr and in complete or almost complete removal in 24-48 hr. Then the mononuclear cells accumulate (monocytes, macrophages, lymphocytes, fibroblasts and endothelial cells). The relative numbers of each cell type vary with the duration and rate of healing. This mononuclear cell response, unless it is very severe, regresses and the vessel heals (9).

Besides classic immune complex-mediated vasculitis, other types of immunopathogenic mechanism may be involved in vascular damage (8, 16). One of these is cell-mediated immune reactivity. The sequence of events is not clearly understood. Probably sensitized lymphocytes are triggered by circulating antigen to release a variety of lymphokines. Influx and accumulation of monocytes-macrophages occurs in and around vessels. The latter cells release their lysosomal content, causing effects similar to PMN-mediated vascular damage. In addition, these cells may further become transformed into epithelioid cells and ultimately participate in granuloma formation, giving place to the granulomatous vasculitis similar to that found in Wegener and Churg-Strauss granulomatosis.

This cell-mediated vasculitis may, under certain circumstances, be related to the immune complex-mediated vasculitis. In the Arthus nephritis model poorly soluble or insoluble IC produce granulomatous lesions at the capillaries of the glomerulus similar to those described in cell-mediated injury (36).

Furthermore the role of mononuclear cells has recently been investigated in an accelerated form of nephrotoxic nephritis in rats and in acute serum sickness of rabbits which could explain, at least in part, the PMN-independent proteinuria observed in experimental animals. The same explanation can apply to the mononuclear cell infiltration seen in certain types of vasculitis. In this situation we have found that the quantity of PAF liberated from the monocytes is approximately ten times higher than that found in PMNs upon similar stimulation (32).

Theoretically other mechanisms of tissue injury could be implicated in the production of vasculitis. Actually there is little evidence to support the contribution of antibodies specifically directed against the vessel itself, or via cytotoxic effector cells in antibody-dependent cellular cytoxicity, in the patho-

genesis of vasculitis. The role of Hageman factor pathways, that
represent together with the complement the major plasma protein
system potentially capable of mediating inflammatory reaction, is
not clearly established either in glomerular or in vascular injury.

2. IMMUNOLOGICAL ASPECTS OF HUMAN VASCULITIS

Since the initial description of necrotizing vasculitis by
Kussmaul and Maier over 100 years ago, many attempts have been made
to classify the broad spectrum of these disorders. Only the aspects
concerning the immunopathogenetic mechanisms of human vasculitis
will be considered in the following discussion.

Several data are in favour of the hypothesis that the human
vasculitides have an immunological origin. Among these data, the
detection of immunoglobulin and of complement in a granular pattern
in various forms of vasculitis is one of the most firmly established.
The proportion of patients in whom a positive immunofluorescence is
found depends on many factors. The severity of the reaction and
the age of the lesion are probably the most important. In this
aspect the detection of immunoglobulin and complement in the vessel
walls cannot be used to predict the presence or absence of circulating immune complexes in patients with vasculitis (27). This is
not suprising, taking into account that the immune reactants are
undetectable 24-48 hr after injection of antigen-antibody complexes
in experimental animals (9, 36).

Several studies have reported about the existence of circulating
immune complexes in these patients (8, 24, 27). A recently published
collaborative study, comparing multiple methods for immune complex
detection, shows that the IC can be detected in 50-75% of the cases,
depending on the technique employed (24). By using density-gradient
ultracentrifugation and gel-filtration chromatography most sera,
showing fixation of the C1q and anticomplementary properties,
eluted in high molecular weight fractions, a size consistent with
antigen-antibody complexes (27).

The existence of circulating immune complexes is not sufficient
to explain the appearance of vascular damage. Vasculitis occurs
only when complexes, greater than a critical size and in sufficient
amount, become deposited beneath the endothelium. The need of an

increased vascular permeability for the deposition of immune complexes in man has received little attention. In SLE we have suggested the existence of an immediate hypersensitivity mechanism similar to that found in acute serum sickness of rabbits (15). Circulating basophil counts in lupus patients are significantly lower than in controls, with a close correlation with some parameters considered as indicators of immunological activity (C3, C4 and DNA antibodies and immune complexes). The basophils of the lupus patients showed a much greater content of surface IgE than controls, and also a positive basophil degranulation test and release of histamine when challenged with variable amounts of native-DNA (15, 31).

Similar basophil sensitization has been found towards some bacterial antigens in acute postinfectious glomerulonephritis. The participation of IgE dependent vasopermeability in the onset of vasculitis is difficult to establish clearly. However a simple procedure, such as the intradermal injection of histamine which induces a locally increased vascular permeability, could yield some information. An immune complex deposition induced by histamine has been observed in patients with leucocytoclastic vasculitis (6). Furthermore, in a recently observed patient with cutaneous necrotizing vasculitis, urticaria and serum hypocomplementaemia, the study of the skin biopsy after cold stimulation showed an early massive degranulation of the mast cells followed by sequential infiltration of neutrophilic, eosinophilic and basophilic polymorphonuclears, together with the development of venular endothelial cell necrosis and the deposition of fibrin (34).

The antigens involved in the immunocomplexes of the human vasculitis are generally unknown. Probably exogenous and endogenous antigens may be implicated. Recent reports on the high incidence of HBs Ag and HBs Ag-anti HBs immune complexes, in the sera (18, 37) and in the vascular lesions of patients with periarteritis nodosa (29), support the hypothesis that the lesions of some forms of vasculitis can result from antigen-antibody interaction. Additional antigens have seldom been identified. However other viruses are the most probable antigens due to their capacity of replication inside the cells giving place to a situation of chronic viraemia similar to that found in experimental viral vasculitis. Bacteria are another potential source of antigen (Fig. 3).

Endogenous antigens may also be implicated. This is the case
in several diseases, notably rheumatoid arthritis, systemic lupus
erythematosus (SLE) and essential cryoglobulinemia, that are usually accompanied by striking serologic abnormalities, supporting
the concept that active immune phenomena are taking place. In these
diseases active vasculitis is often accompanied by hypocomplementaemia, the formation of antibodies or rheumatoid factors, and by the
presence of serum cryoglobulins.

The possible contribution of rheumatoid factors (antiglobulins
for denatured or aggregated IgG) in the induction of vasculitis is
now considered highly likely (30). Their presence by sensitive
techniques has been demonstrated in over 90% of patients with
vasculitis. Rheumatoid factors occur in a wide range of disorders,
particularly in those with chronic circulating immune complexes.
The binding of rheumatoid antiglobulin to immune complexes may
enhance the pathogenetic properties of complexes, changing soluble
complexes into larger less soluble aggregates, increasing the size
of immune complexes and increasing the amount of complement fixed.
Also some vasculitis lesions contain rheumatoid antiglobulin factors,
possibly locally formed, bound to aggregated IgG. It is probable
that these antiglobulins could form complexes perpetuating the
inflammation (30).

A variety of host cellular antigens, particularly deoxyribonucleic acid (DNA), have been found in immunocomplexes in systemic

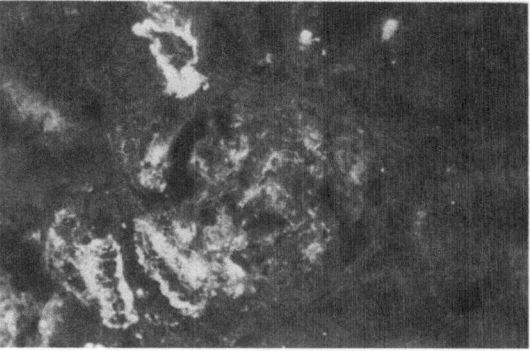

Fig. 3.   Photomicrograph of immunochemical staining of vessels
          and glomerulus showing granular deposits of IgG in a pat-
          ient with staphylococcal osteomyelitis and endocarditis.

lupus erithematosus (SLE). This disease affects multiple systems but the most significant lesions, which determine the course of the disease, are found in renal glomeruli and blood vessels. The deposits of immunoglobulins and complement demonstrable by immuno- fluorescence in a granular pattern in the glomerulus, the decrease in serum complement and the presence of fluctuating levels of anti DNA antibody in relation to clinical exacerbations of the disease, make SLE a typical prototype of immune complex disease. Treatment of glomeruli with acid buffer elute antibodies reactive with both native and single-stranded DNA and other nuclear antigens. Locali- zation of immune complexes in SLE is not limited to renal glomeruli. Immunofluorescent studies have demonstrated immunoglobulins, com- plement and DNA deposits in medium-sized and small blood vessels of the kidney, spleen and other tissues (20). Whereas deposits in vascular locations most probably derive from circulating complexes, the importance of in situ passive formation of the immune complexes, at least at the glomerular basement membrane level, has recently been stressed. In this sense Izui et al (19) have demonstrated, in vitro, a particular affinity of glomerular basement membrane and collagen for DNA. Probably this new theory could explain in part the failure to demonstrate circulating immune complexes by different techniques in some acute vasculitis. The occurrence of an immune complex disease of vessels without deposits in the glomeruli in SLE is very rare. We have recently seen such a patient with acute renal failure and necrotizing vasculitis with large deposits of immunoglobulins and complement, thrombi and slight mesangial glomerulonephritis (Fig. 4). The demonstration of large sized, poorly soluble immune complexes could probably explain this picture.

Cryoglobulins may be found in a variety of situations with or without vasculitis, fundamentally in the broad spectrum of connective tissue disorders (especially in SLE and rheumatoid arteritis). However, a distinct clinical syndrome, essential mixed cryoglobuli- nemia, has been described, in which, for the most part, there is no identifiable underlying disease, although Levo et al (25) have found the presence of HBs antigen or its antibody in the sera and cryoprecipitates in 74% of patients with mixed cryoglobulinemia (Fig. 5).

Henoch-Schonlein purpura is a typical leukocytoclastic vascu- litis with a characteristic clinicopathological picture. IgA is the antibody class most often seen in the vessel walls, even without

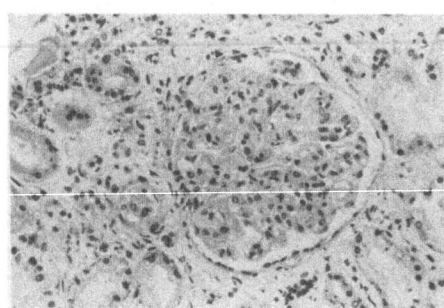 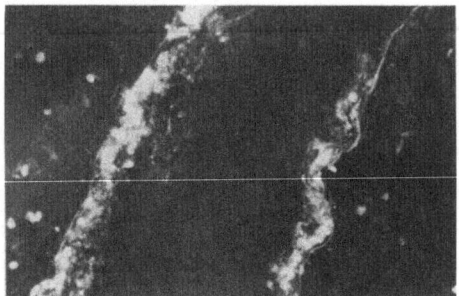

Fig. 4.   Left. Photomicrograph of immunochemical staining of
          a small artery showing granular deposits of IgG.
          Right. Glomerulus showing moderate mesangial glomerulo-
          nephritis (Hematoxylin). Autopsy kidney study in a
          patient with SLE with acute renal failure and gener-
          alized necrotizing vasculitis.

signs of inflammation, and in the glomerular mesangium. The existence of serum cryoglobulins and circulating immune complexes has
been demonstrated. Recently we have suggested that this entity may
be pathogenetically related with the "primitive" mesangial IgA
glomerulonephritis (Berger's disease) through the existence of high
levels of polymeric IgA (in part as immune complexes) in serum and
in the mesangium of both groups of patients (13, 26).

     The Wegener and Churg-Strauss granulomatosis show a distinctive
clinicopathologic picture with necrotizing granulomatous vasculitis
of the respiratory tracts, glomerulonephritis and variable degrees
of disseminated small-vessel vasculitis. The etiology of these
entities is unclear, although a hypersensitivity reaction to an as
yet unidentified antigen is highly suspected. Although circulating
immune complexes have been found in patients with active Wegener's
granulomatosis, the presence of immunoglobulins and complement in
the vessel walls are rarely seen. Thereby a clear immunopathologic
evidence for immunocomplex deposition in granulomatous vasculitis
is lacking. In contrast, a delayed hypersensitivity response to
some hypothetical antigen may elicit the granulomatous response
(8, 11, 35).

     Furthermore, as we have cited above, a typical vascular

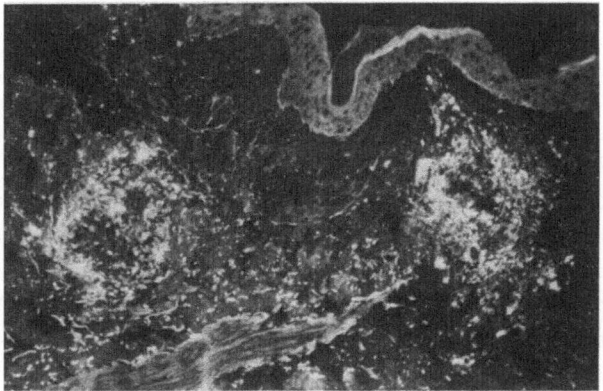

Fig. 5. Photomicrograph of immunochemical staining of dermis
        showing vessels with necrotizing vasculitis and mul-
        tiple deposits of IgG. (IgM, C1q, C3 and HBsAg were
        also seen). The same proteins were demonstrated in
        glomeruli and other vessel walls. HBsAG was found in
        serum and cryoprecipitates (Patient with a mixed
        cryoglobulinemia).

granuloma may be seen in the Arthus nephritis or in nephritis
obtained with poorly soluble or insoluble immunocomplexes. The
significance of the high levels of serum IgE and/or eosinophilia in
some cases of these entities is unclear (8). In any case the high
sensitivity to cyclophosphamide therapy of Wegener's granulomatosis
suggests the involvement of some immunological participation in
these diseases.

ACKNOWLEDGMENTS

    This work was supported in part by a grant from the Instituto
Nacional de Salud (INSALUD).

    We would like to express our gratitude to Dr. Barat and Dr.
Mampaso for the photomicrographs and Miss Isabel Navajos for secretarial assistance.

REFERENCES

1. Alonso, F., Egido, J., Sanchez Crespo, M. and Hernando L.:
    Heymann nephritis of rats: is a vasopermeability mechanism
    really needed for immune complex deposition? Proc. EDTA. 16:
    706-708, (1979).

2. Arisz, L., Noble, B., Milgrom, M., Brentjens, J.R. and Andres,
    G.A.: Experimental chronic serum sickness in rats. A model of
    immune complex glomerulonephritis and systemic immune complex
    deposition. Int. Arch. Allergy Appl. Immun. 60: 80-88, (1979).

3. Benveniste, J.: Platelet activating factor: A new mediator of
    anaphylaxis and immune complex deposition from rabbit and human
    basophils. Nature 249: 581-582, (1974).

4. Benveniste, J., Egido, J. and Gutierrez Millet, V.: Evidence
    for the involvement of the IgE basophils system in acute serum
    sickness of rabbits. Clin. Exp. Immunol. 26: 449-456, (1976).

5. Benveniste, J., Henson, P.M. and Cochrane, C.G.: Leukocyte
    dependent histamine release from rabbit platelets: the role of
    IgE basophils and platelet activating factor. J. Exp. Med.
    136: 1356-1377, (1972).

6. Braverman, I.M. and Yen, A.: Demonstration of immune complexes
    in spontaneous and histamine-induced lesions and in normal skin
    of patients with leukocytoclastic angiitis. J. Invest. Dermatol.
    64: 105-112, (1975).

7. Brentjens, J.R., O' Connell, D.W., Pawloski, I.B. and Andres,
    G.A.: Extraglomerular lesions associated with deposition of
    circulating antigen-antibody complexes in kidneys of rabbits
    with chronic serum sickness. Clin. Immunol. Immunopathol. 3:
    112-126, (1974).

8. Conn, D.L., McDuffie, F.C., Holley, K.E. and Schroeter, A.L.:
    Immunologic mechanism in systemic vasculitis. Mayo Clin. Proc.
    51: 511-518, (1976).

9. Cochrane, C.G. and Koffler, D.: Immune complex disease in
    experimental animals and man, in "Advances in Immunology"
    (F. Dixon and H. Kunkel Eds.) 16: 185-264, Academic Press,
    New York, 1973.

10. Cochrane, C.G. and Janoff, A.: The Arthus Reaction. A model
    of neutrophil and complement-mediated injury. In "The Inflam-
    matory Process" 2nd ed. vol. 3: 85-162. Edited by Zweifach,
    Grant and McCluskey. Academic Press, New York, 1974.

11. Christian, C.L. and Sergent, J.S.: Vasculitis syndromes. Clini-

cal and experimental models. Am. J. Med. 61: 385-392, (1976).

12. Egido, J., Lopez Trascasa, M., Sanchez Crespo, M., Barat, A., Garcia Sanchez, M., Hernando, L. and Casado, S.: Effects of disodium cromoglycate and antihistamines in acute serum sickness of rabbits. Proc. EDTA 15: 581-583, (1977).

13. Egido, J., Sancho, J., Mampaso, F., Lopez Trascasa, M., Sanchez Crespo, M. and Hernando, L.: A possible common pathogenesis of the mesangial IgA glomerulonephritis in patients with Berger's disease and Schönlein-Henoch syndrome. Proc. EDTA, 17: 660-666, (1980).

14. Egido, J., Alonso, F., Sanchez Crespo, M., Barat, A. and Hernando, L.: Absence of an anaphylactic vasopermeability mechanism for the immune complexes deposition in the Heymann nephritis of rats. Clin. Exp. Immunol. 42: 96-106, (1980).

15. Egido, J., Sanchez Crespo, M., Lahoz, C., Garcia, R., Lopez Trascasa, M. and Hernando, L.: Evidence of an immediate hyper-sensitivity mechanism in systemic lupus erythematosus. Ann. Rheum. Dis. 39: 312-317, (1980).

16. Fauci, A.S.: The spectrum of vasculitis. Clinical, pathologic, immunologic and therapeutic considerations. Ann. Intern. Med. 89: 660-676, (1978).

17. Frank, M.M., Hamburger, M.I., Lawley, T.J., Kimberly, R.P. and Plotz, P.H.: Defective reticuloendothelial system Fc-receptor function in systemic lupus erythematosus. N. Engl. J. Med. 300: 518-523, (1979).

18. Gocke, D.J., Hsu, K., Morgan, C., Bombardieri, S., Lockshim, M. and Christiam, C.L.: Association between polyarteritis and Australia antigen. Lancet. 2: 1149-1153, (1970).

19. Izui, S., Lambert, P.H. and Miescher, P.A.: In vitro demon-stration of a particular affinity of glomerular basement membrane and collagen for DNA. A possible basis for local formation of DNA-anti DNA complexes in systemic lupus erythema-tosus. J. Exp. Med. 144: 428-443, (1976).

20. Koffler, D., Agnello, V., Thoburn, R. and Kunkel, H.G.: Systemic lupus erithematosus. Prototype of immune complex nephritis in man. J. Exp. Med. 134: 169-179s, (1971).

21. Kondo, Y., Niwa, Y., Takizawa, J., Akikusa, B., Sano, M. and Shigematsu, H.: Accelerated serum sickness in the rabbit. IV Characteristic endoarteritis in the pulmonary artery. Lab. Invest. 41: 119-127, (1979).

22. Kniker, W.T. and Wagner, D.F.: Amelioration of immune complex nephritis by antihistaminic drugs. Proceeding of the VIIth

International Congress of Nephrology Abs 418, Firenze, 1975.

23. Kravis, T.C. and Henson, P.M.: Accumulation of platelets at
    sites of antigen antibody mediated injury: a possible role
    for IgE antibody and mast cells. J. Immunol. 118: 1569-1573,
    (1977).

24. Lambert, P.H., Dixon, F.J., Zubler, R.H., Agnello, V., Cambiaso,
    C., Casali, P., Clarke, J., Cowdery, J.S., McDuffie, F.C.,
    Hay, F.C., Maclennan, I.C.M., Masson, P., Müller-Eberhard,
    H.J., Penttinen, K., Smith, M., Tappeiner, G., Theofilopoulos,
    A.N. and Verroust, P.A.: WHO collaborative study for the
    evaluation of eighteen methods for detecting immune complexes
    in serum. J. Clin. Lab. Immunol. 1: 1-15, (1978).

25. Levo, Y., Gorevic, P.D., Kassab, H.J., Zucker-Franklin, D.
    and Franklin, E.C.: Association between hepatitis B virus and
    essential mixed cryoglobulinemia. N. Engl. J. Med. 296: 1501-
    1504, (1977).

26. Lopez Trascasa, M., Egido, J., Sancho, J. and Hernando, L.:
    IgA glomerulonephritis (Berger's disease): evidence of high
    serum levels of polymeric IgA. Clin. Exp. Immunol. 42: 247-254,
    (1980).

27. Mackel, S.E., Tappeiner, G., Brumfield, H. and Jordon, R.E.:
    Circulating immune complexes in cutaneous vasculitis. J. Clin.
    Invest. 64: 1652-1660, (1979).

28. Mannik, M.: Clearance and glomerular deposition on circulating
    immune complexes. In Protides of the biological fluids. 26:
    265-270. Edited by H. Peeters, Pergamon Press, Oxford, 1979.

29. Michalak, T.: Immune complexes of hepatitis B surface antigen
    in the pathogenesis of periarteritis nodosa. Am. J. Pathol.
    90: 619-628, (1978).

30. Parish, W.E.: Hypersensitivity vasculitis: The acute phase of
    leukocytoclastic (necrotizing) lesions. In "Comprehensive
    Immunology" 6: 537-567. Edited by S. Gupta and R.A. Good.
    Plenum Medical Book Company, New York, 1979.

31. Sanchez Crespo, M., Alonso, F., Ubeda, I. and Egido, J.: Par-
    ticipation of two inflammatory mediators: histamine and plate-
    let - activating factor in the pathogenesis of systemic lupus
    erythematosus.(submitted to publication)

32. Sanchez Crespo, M., Alonso, F. and Egido, J.: Platelet acti-
    vating factor in anaphylaxis and phagocytosis. I. Release from
    human peripheral polymorphonuclears and monocytes during the
    stimulation by ionophore A23187 and phagocytosis and not from
    degranulating basophils. Immunology 40: 645-655, (1980).

33. Sanchez Crespo, M., Alonso, F., Egido, J. and Hernando, L.: Release of inflammatory mediators in rat immune complex disease. Fourth international congress of immunology, Paris, 1980.
34. Soter, N.A., Mihm, C.C., Dvorak, H.F. and Austen, K.F.: Cutaneous necrotizing venulitis: a sequential analysis of the morphological alterations occurring after mast cell degranulation in a patient with a unique syndrome. Clin. Exp. Immunol. 32: 46-58, (1978).
35. Shillitoe, E.J., Lehner, T., Lessof, M.H. and Harrison, D.F.N.: Immunological features of Wegener's granulomatosis. Lancet. 1: 281-284, (1974).
36. Shigematsu, H., Niwa, Y., Takizawa, I. and Akikusa, B.: Arthus type nephritis. I. Characterization of glomerular lesions induced by insoluble and poorly soluble immune complexes. Lab. Invest. 40: 492-502, (1979).
37. Trepo, C.G., Zuckerman, A.J., Bird, R.C. and Prince, A.M.: The role of circulating hepatitis B antigen/antibody immune complexes in the pathogenesis of vascular and hepatic manifestations in polyarteritis nodosa. J. Clin. Pathol. 27: 863-868, (1974).

THE IMMUNOLOGY OF GLOMERULAR NEPHROPATHIES

B. Van Damme and V.J. Desmet

Laboratorium voor Histochemie en Cytochemie
Departement Medische Navorsing
Katholieke Universiteit Leuven
Leuven (Belgium)

ABSTRACT

Immunological damage is a frequent cause of glomerulonephritis. This can be related to antibodies against glomerular basement membrane antigens or to the deposition or the in situ formation of immune complexes in the glomerulus. Depending on the localisation and the severity of the aggression different degrees of inflammation occur. Adaptation of the mesangium or of the visceral epithelium results in intracapillary proliferation and membranous transformation respectively, while breaks in the glomerular capillary wall result in crescent formation.

A discussion on the immunology of glomerulonephritis should cover different aspects: the type of aggression, the inflammatory response evoked by this aggression, and the reaction patterns of the glomerular cells and structures triggered either by the aggression itself, or by the inflammatory process.

1. The immunological aggression in glomerulonephritis

Classically two types of immunological aggression are considered. One is the rather rare formation of antibodies against components of the glomerular basement membrane (GBM), the other the more common glomerular damage due to the presence of immune complex-related deposits in the glomerulus.

1.Antibodies against the glomerular basement membrane. The glomerular basement membrane is a highly complex structure mainly secreted by the visceral epithelial cells (17), and composed of collagen and noncollagen molecules. The major nephritogenic antigens reside in the noncollagenous glycoproteins (33), the nature of which has not yet been completely identified. These nephritogenic antigens seem to be lacking in patients with Alport's syndrome (18), and in normal neonates (1). High avidity antibodies are more harmful at low doses than less avid antibodies (38). Also fixation of small quantities of heterologous antiglomerular antibody causes glomerular damage only in the autologous phase (16) (i.e. when antibodies against the heterologous antibodies are formed), while fixation of large amounts of heterologous antibodies cause immediate glomerular damage (5, 45) suggesting that the speed of fixation, or the density of fixation or both are of importance in the generation of glomerular damage.

2.Immune complexes in the glomerulus. Damage of the glomerulus is caused most frequently by the presence of immune complexes. The origin of these complexes is a matter of debate. Theoretically different possibilities exist, and the fact that one possibility is operative does not exclude other mechanisms.

A.Deposition of immune complexes from the circulation. This hypothesis was formulated as a result of many series of experiments in the fifties and early sixties, and had until recently not been challenged seriously (18). It was proposed that immune complexes are formed in the circulation in different circumstances, and that these complexes deposit in the vessel walls and in the glomeruli according to their sizes and composition (24). The complexes are divided in three groups: large immune complexes formed in antibody excess, or near equivalence, which are cleared rapidly by the mononuclear phagocyte system (M.P.S.) and are not nephrotoxic. Medium sized immune complexes in small antigen excess are trapped in the mesangium and lead to inflammation. Small immune complexes formed in great antigen excess pass through the basement membrane and locate at the epithelial side of the basement membrane.

Recent studies on the kinetics of immune complexes of widely varying sizes injected in different animals have shown that the amount of immune complexes trapped in the glomerulus is very small (less than 1% of the amount injected) (3) and that blocking the

uptake by the M.P.S. by different means increases the amount deposited
in the glomerulus (20, 25). No instances have been found in which injected immune complexes locate in the subepithelial zone (26).Therefore
probably other mechanisms must exist, which are responsible for the
location of immune complexes in the different sites of the glomerulus.

B. In situ formation of immune complexes. Apart from the deposition
of circulating complexes the in situ assembly of immune complexes has
been demonstrated.

1) The heterologous immune complex glomerulonephritis (passive
Heymann nephritis) is caused by the fixation of the heterologous
antibody against tubular brush border antigens, on related antigens
present in the subepithelial zone of rat glomeruli (8, 47, 48).

2) Antibodies eluted from diseased kidneys in autologous immune
complex glomerulonephritis (active Heymann nephritis) and perfused
in an isolated normal kidney form immune complexes in the subepithelial zone (14).

3) In chronic mercury intoxication of rats, subepithelial immune
complexes are formed in situ by an antibody against nuclear antigens
crossreacting with antigens in the subepithelial zone (51).

4) In situ formation of mesangial and subendothelial immune
complexes seems to be far more pathogenic for the glomerulus than
the deposition of immune complexes from the circulation could account
for (47).

The conclusion from these experiments are probably valid in
human disease as well. In acute human glomerulonephritis (44) as well
as in acute experimental serum sickness (37) free antigen was demonstrated in the glomerulus early in the course of disease or before
antibodies could be demonstrated. In chronic membranous glomerulonephritis, antibodies against tubular brush border antigens were
shown to be present in the serum of patients in Japan (30, 31),
although this was not confirmed in Europe or in the U.S. (52).

It is likely that both mechanisms, deposition from the circulation, and in situ formation are operative in glomerular disease. In
order to induce glomerulonephritis more complexes are required in

the circulation than the amount of antigen and antibody necessary
for the formation of complexes in situ,provided the proper conditions
are present (47).

Which are now the conditions for the in situ formation of immune
complexes?

1. Formation of antibodies against constituents of the glomerulus. This was shown to be operative in different forms of immune
complex glomerulonephritis in the rat (48, 51).

2. Formation of antibodies against antigens "planted" in
different areas of the glomerulus. Different experimental models
with immune complexes inside (27, 47) or outside (19) the basement membrane used this mechanism. The size of the antibody is important
since large IgM antibodies are not able to cross a (normal) capillary
wall, and even IgG antibodies have only a limited access to the subepithelial area (34).

The antigens are planted in the glomerulus according to different factors:

a) The size. It is obvious that very large antigenic molecules cannot
pass the glomerular basement membrane (34), but may be trapped in
the mesangium (29).

b) The charge. Different layers of the capillary wall possess strong
negative charges (29) which will attract or temporally immobilize positi-
vely charged antigens, conversely negative antigens may be arrested
in the less negatively charged zones of the lamina densa.

c) Affinity for glomerular structures. Plant lectins are shown to
have affinities for the subepithelial zones (6). Other products of
plant or bacterial origin may as well fix to different areas in the
glomerulus (19). Antibodies against these molecules fix on these
trapped antigens.

d) Alternating antigen/antibody excess. This system was shown to be
operative in an ex vivo system, where alternating perfusion of an
isolated kidney with antigen, buffer, antibody, and buffer resulted
in the formation of immune complexes in the subepithelial zone (15).
An analogous daily alternation of antigen-antibody excess is created

in chronic serum sickness, by the daily injection of antigen exceeding the amount of circulating antibody (9) in an immunized animal.
This excess of antigen would bind to the already deposited immune
complexes, which, in the presence of free circulating antibody, had
free antigen combining sites. These complexes would turn into complexes with free antigenic determinants. These again are covered by
newly synthetised antibody, and turned into "antibody excess" complexes (47).

e) The presence of complexes as an antigen. In the autologous phase
of the (Heymann) immune complex glomerulonephritis the heterologous
antibody present in the subepithelial complexes, acts as an antigen
upon which new autologous antibody is deposited (11, 13).

f) The simultaneous presence of low affinity antibodies and antigen
in the serum. In such a situation free antigen and antibody could be
present together in the serum. Due to the differential permeability
of the glomerular capillary wall small antigens and antibody could
diffuse at different speeds changing the relative concentrations of
antigen and antibody, resulting in in situ complex formation (12).

Any combination of the mentioned mechanisms could also be operative.

2. Inflammation: A non specific response to aggression

Inflammation has been discussed earlier in this symposium, and
has been reviewed recently (35). Only a few points will be stressed.

1. The pathophysiology of inflammation is extremely complex and
involves activation of complement, coagulation, fibrinolysis and the
kinin generating systems. These systems are interrelated and are controlled by positive and negative feedback mechanisms.

2. Activation of complement (7), especially C_1, plays a central
role in the induction of inflammation in the anti-GBM and immune
complex-mediated glomerulonephritis.

3. The intensity of inflammation will depend on the speed of
activation of complement. Rapid activation of complement will lead
to a high rate of production of leukotactic factors and anaphylotoxins, and recruitment or activation of coagulation, fibrinolysis

and kinin generation systems. Rapid activation of complement is only possible in sites easily accessible to antibodies (anti-GBM or others), immune complexes and complement, i.e. the sites located inside the basement membrane (subendothelial and mesangial). Large molecules such as antibodies and many of the complement factors do not cross the basement membrane easily. Therefore deposition or formation of immune complexes in the subepithelial zone is slow and does not lead to inflammation. Moreover, complement split products liberated in that site will be eliminated in the urine and their back diffusion into the blood stream is hard to conceive.

4. Grossly two levels of intensity of inflammation can be distinguished. The milder form consists of attraction of leukocytes which adhere to the deposited material in the capillary wall and in the mesangium. These cells try to clear the offensive material by means of phagocytosis and degranulation. The more severe form of inflammation includes the activation of the coagulation system, and results in necrosis of part of the glomerulus with breaks in the basement membrane and crescent formation. All possible variants between the milder ant the more severe forms of inflammation exist, and not all glomeruli are of necessity equally affected.

3. Glomerular adaptation to injury, and repair.

The presence of injurious agents such as bulky immune complexes or related macromolecules (e.g. cryoglobilins) leads to adaptations of the glomerular structures. Moreover the inflammation may damage glomerular structures resulting in anatomical and functional alterations and reparative process.

1) The subendothelial-mesangial complex.

a) Immune material deposited in the subendothelial space is shifted towards the mesangium (29). Molecules may be taken up and degraded by the mesangial cells, or they may be directed towards the vascular pole from where they are eliminated by mechanisms until now poorly understood (29). The mesangial cells proliferate and may be assisted temporarily in this clearing function by infiltrating monocytes, which actively phagocytose the deposited immune complexes. The clearing capacity of the mesangium is limited and depends among other factors on the quality of the material stored. Recently it was shown that IgA-containing immune complexes are cleared less afficiently than equimolar amounts of those containing IgG (50). These-mechanisms are operating in the different forms of intracapillary proliferative glomerulonephritis.

b) The presence of mesangial immune complexes blocks the disposal of newly deposited complexes (22). Chronic overload of the mesangium by large amounts of immune material, as e.g. in severely active Lupus Erythematosus, leads to flooding of the mesangium and overflow into the subendothelium (32). Besides inflammatory reactions mesangial cell extensions also migrate between the endothelium and the basement membrane (2). Being derivatives of smooth muscle cells they secrete fragments of basal lamina (36), as do the endothelial cells, deprived of their normal support by the glomerular basement membrane (53). The end result is a severely distorted capillary wall composed of different layers of cells and basal laminae intermingled with immune complexes, typical of membranoproliferative glomerulopathies.

2) The crescent: a destructive reparation.

The most severe reaction in the glomerulus consists in the formation of extracapillary crescents: the urinary space is obliterated by partly epithelial, partly monocyte-derived cells and fibrin (4, 23). Crescent formation is related to major breaks in the glomerular basement membrane induced by the release of proteolytic enzymes from leukocytes or other severe inflammatory processes. The activation of the coagulation system plays a central role in the formation of crescents since anticoagulants given in addition to the administration of anti-GBM antiserum to rabbits inhibit the formation of crescents, but not the fixation of the antibody or the inflammatory reaction (49). Analogous findings were reported in chronic serum sickness (41).

Not all gaps in the basement membrane lead to crescent formation. Gaps responsible for microscopic hematuria in glomerulonephritis may occur in single capillary loops and be closed quickly by protrusion of endothelial or mesangial cells, with little or no proliferation of cells in the urinary space (39).

Ultimately these gaps are closed by synthesis of new basement membrane. Also not all gaps are the result of severe inflammation: the deposition of immune complexes inside or outside the basement membrane may decrease the resistance of the GBM to the high pulsatile intracapillary pressure (39, 40).

3) The subepithelial space.

Immune complexes formed or deposited in the subepithelial space can only be degraded slowly either by endocytosis or extra-cellular digestion by visceral epithelial cells (43). In acute glomerulonephritis or acute serum sickness, which are examples of a one shot immune complex disease with subendothelial and sub-epithelial depositions, the subepithelial immune complexes can be found even weeks after their deposition, when the clinical disease and the inflammation have already subsided (43). The presence of subepithelial immune complexes seems to disturb locally the normal secretion and turnover of the basement membrane. At this site the normal negative charge of the lamina rara externa due to the pres-ence of glycosaminoglycans (21) (mainly heparin sulfate) is abol-ished (16). If the deposition is chronic, the visceral epithelial cells secrete increased amounts of basement membrane material around the deposited material, resulting in the light microscopic picture of "spikes" on the epithelial side of the basement membrane (10). The deposits become progressively encircled and covered by the newly formed basement membrane and may be pushed downwards slowly towards the endothelial side of the GBM (42). This mechanism is exemplified in membranous glomerulopathy.

The immunological events summarized in this outline may account for the great variability of the expression of glomerular damage between different forms of diseases with common pathogenetic mech-anisms, and for the great variability of the glomerular damage within a single disease entity. Superimposed on the above mentioned processes a dynamic fourth (temporal) dimension should be consid-ered.

REFERENCES

1. Anand, S.K., Landing, B.K., Heuser, E.T., Olson, D.L., Grushkin,
    C.M. and Lieberman, E.: Changes in glomerular basement membrane
    antigen(s) with age. J. Pediat. 92: 952, (1978).
2. Arakawa, M. and Kimmelstiel, P.: Circumferential mesangial
    interposition. Lab. Invest. 21: 276, (1969).
3. Arend, W.P. and Mannik, M.: Studies on antigen-antibody complexes.
    II. Quantification of tissue uptake of soluble complexes in
    normal and complement depleted rabbits. J. Immunol 107: 63,
    (1971).
4. Atkins, R.C., Holdsworth, S.R., Glasgow, E.F. and Matthews, F.E.:
    The macrophage in human rapidly progressive glomerulonephritis.
    Lancet 1: 830, (1976).
5. Blantz, R.C., Tucker, B.J. and Wilson, C.B.: The acute effects
    of anti-glomerular basement membrane antibody upon glomerular
    filtration in the rat. The influence of dose and complement
    depletion. J. Clin. Invest. 61: 910, (1978).
6. Bretton, R. and Bariety, J.: A comparative ultrastructural local-
    ization of concanavalin A, wheat germ and ricinus communîs on
    glomeruli of normal rat kidney. J. Histochem. Cytochem. 24: 1093,
    (1976).
7. Cochrane, Ch., Unanue, E.R. and Dixon, F.J.: A role of polymor-
    phonuclear leukocytes and complement in nephrotoxic nephritis.
    J. Exp. Med. 122: 99, (1965).
8. Couser, W.G., Steinmuller, D.R., Stilplant, M.M., Salant, D.J.
    and Lowenstein, L.M.: Experimental glomerulonephritis in the
    isolated perfused rat kidney. J. Clin. Invest. 62: 1275, (1978).
9. Dixon, F.J., Feldman, J.D. and Vazquez, J.J.: Experimental
    glomerulonephritis. The pathogenesis of a laboratory model re-
    sembling the spectrum of human glomerulonephritis. J. Exp. Med.
    113: 899, (1961).
10. Ehrenreich, T. and Churg, J.: Pathology of membranous nephropathy.
    Path. Annual 3: 145, (1968).
11. v. Es, L.A., Blok, A.P.R., Schoenfeld, L. and Glassock, R.J.:
    Chronic nephritis induced by antibodies reacting with glomerular
    bound immune complexes. Kidney Int. 11: 106, (1977).
12. Evans, D.J.: Pathogenesis of membranous glomerulonephritis.
    Lancet 1: 1143, (1974).
13. Feenstra, K., Lee, R.V.D., Greben, H.A., Arends, A. and Hoede-
    maeker, Ph.J.: Experimental glomerulonephritis in the rat induced
    by antibodies directed against tubular antigens. I. The natural

history: a histologic and immunohistologic study at the light
microscopic and the ultrastructural level. Lab. Invest. 32: 235,
(1975).

14. Fleuren, G.J., Grond, J. and Hoedemaeker, Ph. J.: The pathogenetic role of recirculating antibody in autologous immune complex
glomerulonephritis. Clin. Exp. Immunol. 41: in press (1980).

15. Fleuren, G.J., Grond, J. and Hoedemaeker, Ph. J.: In situ
formation of subepithelial glomerular immune complexes in passive
serum sickness. Kidney Int. 17: in press (1980).

16. Fleuren, G.J.: Personal communication (1980).

17. Foidart-Willems, J., Dechenne, Ch. and Mahieu, Ph.: Biosynthesis
of basement membrane collagen in cultures of renal glomerular
and tubular epithelial cells. Diabète et Métabolisme 1: 227,
(1975).

18. Germuth, F.G. and Rodriguez, E.: Immunopathology of the renal
glomerulus. Immune complex deposit and antibasement membrane
disease (Little & Brown, Boston 1973).

19. Golbus, S.M. and Wilson, C.B.: Experimental glomerulonephritis
induced by in situ formation of immune complexes in glomerular
capillary wall. Kidney Int. 16: 148, (1979).

20. Hakenstaad, A.O. and Mannik, M.: Saturation of the reticulo-
endothelial system with soluble immune complexes. J. Immunol.
112: 1939, (1974).

21. Kanwar, Y.S. and Farquhar, M.G.: Anionic sites in the glomerular
basement membrane. In vivo and in vitro localization to the
laminae rarae by cationic probes. J. Cell Biol. 81: 137, (1979).

22. Keane, W.F.: Impaired mesangial cleaning of macromolecules in
chronic mesangial immune complex injury. Kidney Int. 16: 784,
(1979).

23. Kondo, Y., Shigematsu, H. and Kobayashi, Y.: Cellular aspects
of rabbit Masugi nephritis. II. Progressive glomerular injuries
with crescent formation. Lab. Invest. 27: 620, (1972).

24. Koyama, A., Niwa, Y., Shigematsu, H., Taniguchi, M. and Tada, T.:
Studies on passive serum sickness. II. Factors determining the
localization of antigen-antibody complexes in the murine renal
glomerulus. Lab. Invest. 38: 253, (1978).

25. Kylstra, A., v.d. Lely, A., Knutson, D.W., Fleuren, G.J. and
v. Es, L.A.: The influence of phagocyte function on glomerular
localization of aggregated IgM in rats. Clin. Exp. Immunol. 32:
207, (1978).

26. Mannik, M. and Haakenstad, A.O.: Circulation and glomerular deposition of immune complexes. Arthritis Rheum. 20: 148, (1977).

27. Mauer, S.M., Sutherland, D.E.R., Howard, R.J., Fish, A.J., Najarian, J.S. and Michael, A.F.: The glomerular mesangium. III. Acute immune mesangial injury: a new model of glomerulonephritis. J. Exp. Med. 137: 553, (1973).

28. McCoy, R.C., Johnson, H.K., Stone, W.J. and Wilson, C.B.: Variation in glomerular basement membrane antigens in heridity nephritis. Lab. Invest. 34: 325, (1976).

29. Michael, A.F., Nevins, T.E., Raij, L., Keane, W.F. and Scheinman, J.I.: Macromolecular transport in the glomerulus: studies of the mesangium and epithelium in vivo and in vitro. In: Immunologic mechanisms of renal disease. C.B. Wilson, B.M. Brenner, J.H. Stein (eds.). (Churchill-Livingstone, London 1979).

30. Miyakawa, Y., Kitamura, K., Shibata, S. and Naruse, T.: Demonstration of human nephritogenic tubular antigen in the serum and organs by radioimmunoassay. J. Immunol. 117: 1203, (1976).

31. Naruse, T., Kitamura, K., Miyakawa, Y. and Shibata, S.: Deposition of renal tubular epithelial antigen along the glomerular capillary walls of patients with membranous glomerulonephritis. J. Immunol. 110: 1163, (1973).

32. Pirani, C.L. and Silva, F.G.: The kidneys in systemic lupus erythematosus and other collagen diseases: recent progress. In: Kidney disease: present status. International Academy of Pathology Monograph. J. Churg, B.H. Spargo, F.K. Mostofi, M.R. Abell (eds.) Williams & Wilkins Co., Baltimore (1979).

33. Rothbard, S. and Watson, R.F.: Renal glomerular lesions induced by rabbit anti rat collagen serum in rats prepared with adjuvant. J. Exp. Med. 109: 633, (1959).

34. Ryan, G.B., Hein, S.J. and Karnovsky, M.J.: Glomerular permeability to proteins. Effects of hemodynamic factors on the distribution of endogenous immunoglobulin G and exogenous catalase in the rat glomerulus. Lab. Invest. 34: 415, (1976).

35. Ryan, G.B. and Mayno, G.: Acute inflammation, a review. Am. J. Path. 86: 185 (1977).

36. Scheinman, J.I., Brown, D. and Michael, A.F.: Collagen synthesis by human glomerular cells in culture. Biochim. Biophys. Acta 542: 128, (1978).

37. Segoloni, G., Camussi, G., Ragni, R., Stratta, P., Piccoli, G. and Vercellone, A.: Kinetics of glomerular deposition of immune complexes in rabbit acute serum nephritis. Kidney Int. 11: 148, (1977).

38. Shimizu, F., Mossman, H., Takamiya, H. and Vogt, A. Effect of antibody avidity on the induction of renal injury in anti-

glomerular basement membrane nephritis. Brit. J. Exp. Path. 59:
624, (1978).

39. Stejskal, J., Pirani, C.L., Okada, M., Mandalenakis, N. and
    Pollak, V.E.: Discontinuities (gaps) of the glomerular capillary
    wall and basement membrane in renal disease. Lab. Invest. 28:
    149, (1973).

40. Tallqvist, G., Pasternack, A. and Törnroth, T.: Indentations of
    the glomerular basement membrane in renal diseases: a light and
    electron microscopic study on ultrathin serial sections. Lab.
    Invest. 35: 327, (1976).

41. Thomson, N.M., Simpson, I.J., Evans, D.J. and Peters, D.K.:
    Defibrination with ancrod in experimental chronic immune complex
    nephritis. Clin. Exp. Immunol. 20: 527, (1975).

42. Törnroth, T. and Skrifvars, B.: The development and resolution
    of glomerular basement membrane changes associated with subepi-
    thelial immune deposits. Am. J. Path. 79: 219, (1975).

43. Törnroth, T.: The fate of subepithelial deposits in acute post-
    streptococcal glomerulonephritis. Lab. Invest. 35: 461, (1976).

44. Treser, G., Semar, M., McVicar, M., Franklin, M., Ty, A., Sagel,
    I. and Lange, K.: Antigenic streptococcal components in acute
    glomerulonephritis. Science 163; 676, (1969).

45. Unanue, E.R. and Dixon, F.J.: Experimental glomerulonephritis.
    V. Studies on the interaction of nephrotoxic antibodies with
    tissues of the rat. J. Exp. Med. 121: 697, (1965).

46. Unanue, E.R. and Dixon, F.J.: Experimental glomerulonephritis.
    VI. The autologous phase of nephrotoxic serum nephritis. J.
    Exp. Med. 121: 715, (1965).

47. Van Damme, B.: Studies on mechanisms of deposition of immune
    complexes in experimental glomerulonephritis. Academic Thesis,
    Groningen. (Arscia Uitgaven, Brussels 1977).

48. Van Damme, B.J.C., Fleuren, G.J., Bakker, W.W., Hoedemaeker,
    P.J. and Vernier, R.L.: Experimental glomerulonephritis in the
    rat induced by antibodies directed against tubular antigens.
    V. Fixed glomerular antigens in the pathogenesis of heter-
    ologous immune complex glomerulonephritis. Lab. Invest. 38:
    502, (1978).

49. Vassalli, P. and McCluskey, R.T.: The pathogenic role of the
    coagulation process in rabbit Masugi nephritis. Am. J. Path.
    45: 653, (1964).

50. Ward, D.M., Spiegelberg, H.L. and Wilson, C.B.: Persistence of
    IgA aggregates in the glomerular mesangium in mice. Kidney Int.
    16: 801, (1979).

51. Weening, J.J.: Mercury induced immune complex glomerulopathy.
     Academic Thesis, Groningen (1980).
52. Witworth, J.A., Leibowitz, S., Kennedy, M.C., Cameron, J.G.,
     Evans, D.J., Glassock, R.J. and Schoenfeld, L.S.: Absence of
     glomerular renal tubular epithelial antigen in membranous
     glomerulonephritis. Clin. Nephrol. 5: 159, (1976).
53. Zollinger, H.U. and Mihatsch, M.J.: Renal pathology in biopsy.
     (Springer, Berlin 1978).

IMMUNOPATHOLOGY OF TUBULO-INTERSTITIAL NEPHRITIS IN MAN AND IN
EXPERIMENTAL ANIMALS*

T. Sugisaki and G. Andres

Departments of Microbiology, Pathology and Medicine
State University of New York at Buffalo
Buffalo, New York 14214 (U.S.A.)

ABSTRACT

It is generally acknowledged that many forms of glomerular
diseases are antibody-mediated, either as a result of deposition of
circulating immune complexes or due to anti-glomerular basement
membrane antibodies. Until recently little consideration was given
to the possibility that there might be comparable forms of diseases
involving other parts of the kidney. However, there is now clear
evidence, obtained both in experimental models and human disease,
that tubular and interstitial lesions can result from deposition or
local formation of immune complexes or because of antibodies directed
against tubular basement membranes or against the brush border of
proximal convoluted tubules. Moreover, it appears that the renal
interstitium is a site where cell-mediated reactions may occur,
accounting for some forms of interstitial nephritis.

We shall discuss in sequence experimental models of immunologically-mediated tubular and interstitial renal diseases and the
evidence concerning the occurrence and probable significance of
corresponding lesions in man.

---

* This work was supported by the Deparment of Health, Education and
  Welfare, United States Public Health Service, H.I.C. Grant AI-10334.

91

I. Antibody-Mediated Tubular and Interstitial Renal Diseases

    A. Tubular and interstitial immune complex diseases

        1. Experimental models
           a. Autologous immune complex disease
           b. Exogenous immune complex disease

        2. Human disease
           a. Renal transplants, Sjögren's syndrome,
              systemic lupus erythematosus

    B. Anti-tubular basement membrane antibody disease (anti-TBM
       disease)

        1. Experimental models

        2. Human disease

    C. Cytotoxic damage to tubular cells induced by antibody to the
       brush border of proximal convoluted tubules.

        1. Experimental models

        2. Human disease (renal transplant ?)

    D. Tubular and interstitial diseases induced by interaction of
       antigen with IgE antibody (immediate hypersensitivity).

II. Cell-Mediated Tubular and Interstitial Renal Diseases

    A. Autologous renal antigens (tubular cells, Tamm-Horsfall protein)

        1. Experimental models

        2. Human disease

    B. Exogenous antigens (bacterial, viral)

        1. Experimental models

        2. Human disease

INTRODUCTION

    Human renal interstitial tissue, composed of tubular basement
membrane (TBM), vessels and various mesenchymal cells, is known as
one of the areas most susceptible to inflammatory reactions. The
term "interstitial nephritis" was generally used to indicate an
inflammatory reaction in the renal interstitium. More recently,
however, it has been claimed that interstitial nephritis is usually
associated with tubular cell damage and, thus, the term "tubulo-
interstitial nephritis" (TIN) appears more appropriate to define
this lesion.

    There are many cases of glomerular disease which are frequently
associated with interstitial inflammation and tubular damage. These
tubulo-interstitial changes are sometimes considered as secondary
to the glomerular lesion. Other, unexplained, TIN without concomitant
glomerular pathology may be due to toxic damage, ischemia or viral
infection. In other TIN of unknown origin, the diagnosis of "pyelonephritis" is made without evidence of bacterial infection and/or
obstruction (22).

    Recently, immunopathological studies have been focused on the
pathogenesis of TIN. These studies now allow the recognition of a
group of clinically important, though relatively rare, renal diseases
(4, 38, 43, 73). The study of renal tissue obtained by renal biopsy
and, especially, the widespread use of immunofluorescence, electronmicroscopy and of immunoserological tests have made it possible to
recognize and to elucidate the pathogenesis of anti-TBM- or immune
complex-mediated TIN. In addition to anti-TBM and immune complex-
mediated TIN, it appears that cell-mediated (delayed hypersensitivity) and atopic (immediate hypersensitivity) mechanisms may
induce lesions in the tubules and the interstitium, although convincing evidence indicating a role for these reactions is not
available at the present time.

    A variety of studies on TIN induced by immunological procedures
in experimental animals has greatly contributed to this progress.
These studies comprise TIN characterized by the deposition of antibodies against various renal tissue antigens such as those present
in TBM, brush border (BB), Henle's loop (HL) or distal convoluted
tubule (DCT) as well as TIN characterized by deposition of immune
complexes along the TBM and in the interstitium. Other studies have

Table 1. Comparative classification of immunologically-mediated TIN in experimental animal and in man.

Classification of TIN	Animal	Man
Anti-TBM TIN		
nephrotoxic nephritis autoimmune anti-TBM TIN	various animals guinea pig, rat	Goodpasture's disease after APGN(a), after Tx(b), methicillin-induced TIN
alloimmune anti-TBM TIN	rat	after Tx
IC TIN		
autologous IC TIN	rabbit, rat	SLE, Tx, MPGN(c)
autologous IC TIN due to anti-Henle's loop antibody	rat	pyelonephritis? Sjögren's syndrome?
IgE-mediated TIN	unknown	drug allergy?
Cell-mediated TIN		
autoimmune cell-mediated TIN	rat	Tx?, TIN of unknown origin?
cell-mediated TIN due to heterologous antigen	guinea pig, rat	renal tubercolosis?, bacterial or parasitic infections?
Autoimmune anti-brush border antibody disease	rat	unknown
Heterologous anti-brush border antibody disease	rat	unknown

(a) acute post-streptococcal glomerulonephritis, (b) transplantation, (c) membranoproliferative glomerulonephritis.

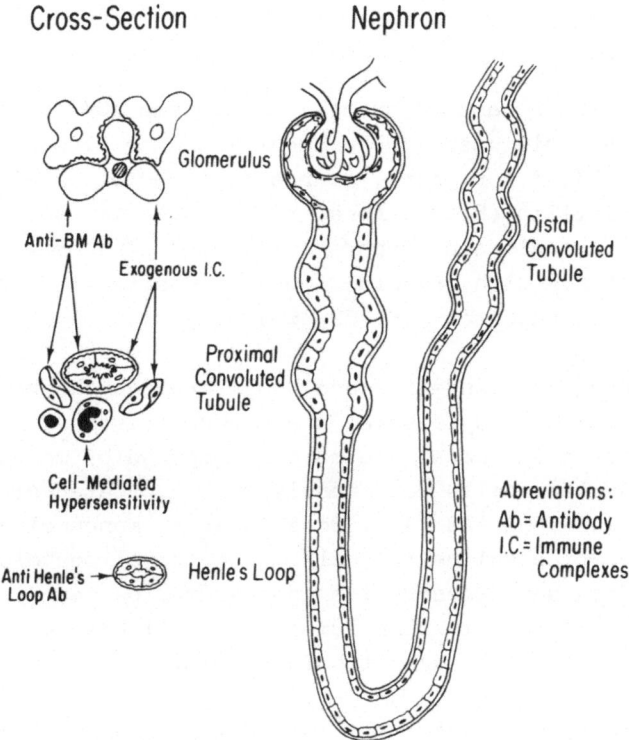

Fig. 1.   Glomerular and tubulo-interstitial architecture with
          the immunologically-mediated tubulo-interstitial
          nephritis.

indicated that cell-mediated TIN characterized by reactivity to
autologous renal antigen or heterologous antigen can also be induced
in experimental animals.

    In this article, we will review the present knowledge concerning
immunologically-mediated TIN in man. In addition, we will discuss
the immunologically-mediated experimental animal models which mimic
human TIN. A comparative classification of human and experimental
TIN is presented in table 1. Schematic representation of glomerular
and tubulo-interstitial architecture with the immunologically-
mediated TIN is shown in Fig. 1.

I. ANTI-TBM TIN

A. Human Disease

Anti-TBM TIN has been frequently (more than 70%) recognized in association with antiglomerular basement membrane glomerulonephritis (anti-GBM GN) (71). A more rare, sporadic, association with administration of methicillin (5, 7), renal transplantation (6, 30, 39, 72), immune complex glomerulonephritis (IC GN) (46) or with Fanconi's syndrome in patients with granular immune deposits along GBM and TBM, has also been reported (38, 39, 68).

Light microscopy. Exudative and proliferative changes in and around the peritubular capillaries are characteristic features of anti-TBM TIN. The infiltrates are mainly composed of mononuclear and polymorphonuclear cells (3), rarely multinucleated giant cells (only one case was reported (4)). Occasionally, mononuclear cells are found between the epithelial cells of proximal convoluted tubules. Cell proliferation and degenerative changes may be seen in the tubular epithelium. Some tubular lumina are obliterated by casts formed by components of damaged tubular cells.

The appearance of giant cells may be important from the pathogenetic viewpoint because giant cells are occasionally found in tuberculoid, a lesion related to delayed hypersensitivity. It would be interesting to determine which cells are transformed into giant cells. It has been shown that interstitial giant cells result from fusion of monocytes and epithelioid cells (67). These cells, infiltrated in the interstitium, are similar to the epithelioid and giant cells occasionally observed in glomerular crescents of patients with rapidly progressive glomerulonephritis (28, 41, 55).

In diseased areas the TBM may be thin and occasionally disrupted. In late stage disease, widespread thickening and splitting of TBM can be seen in association with tubular atrophy and interstitial fibrosis.

Immunofluorescence. Linear deposits of IgG and C3 are generally seen in the TBM of the majority of tubules. The deposits may occasionally have a focal distribution (75). These linear TBM deposits are most often found in patients with anti-GBM GN who have linear deposits of IgG and C3 along GBM (3, 38).

Serum and renal eluates from patients with anti-GBM GN react
in vitro with GBM and TBM, and sometimes with alveolar basement
membrane (ABM) as well. Absorption of the eluates or serum with
solubilized GBM antigen was shown to remove anti-GBM, anti-TBM and
also anti-TBM antibody (45), indicating that the antibodies are
directed against shared antigenic determinants of GBM, TBM and ABM.

Whereas the observation of idiopathic anti-TBM TIN is a very
rare event, several cases of anti-TBM TIN lacking significant deposits in glomeruli have been reported, for example, after renal
transplantation (30, 54, 72), after administration of drugs (7, 25,
51) or in association with IC GN (21, 40, 46). The IgG antibody
present in the serum and/or in renal eluate reacted with TBM or
normal human, rabbit and rat (although there were some refractory
strains) kidneys.

Electronmicroscopy. In studies of biopsy specimens, it is
generally difficult to obtain precise information, because the
lesion is usually focal in distribution. However, severe interstitial
infiltrates, mainly composed of monocytes and lymphocytes, have been
observed. Polymorphonuclear leukocytes, lymphocytes and monocytes
sometimes penetrate between tubular epithelial cells. The giant
cells do not appear in contact with TBM and phagocytosis of TBM
fragments is usually not observed. Epithelial cells of proximal
convoluted tubules occasionally show proliferation resulting in
partial or complete obliteration of tubular lumina. The TBM may
appear split and thickened and occasional gaps are also seen.

Immunopathogenesis. Relationship between anti-TBM antibody
and interstitial tissue damage. It is rather difficult to evaluate
the contribution of anti-TBM antibody to the development of histologic damage, because cellular infiltration is seen frequently in
renal allograft rejection or in severe glomerulonephritis. The
most compelling evidence that anti-TBM antibody is directly responsible for TIN obviously lies in the few reports of idiopathic
anti-TBM TIN. These few reports suggest that anti-TBM antibody may
induce complement activation and accumulation of inflammatory cells,
a series of events similar to those responsible for anti-GBM GN.
On the other hand, one should consider the possibility that a superimposed delayed-hypersensitivity mechanism may be operative.

These immunological reactions in the interstitium can initiate an inflammation leading to tissue destruction of phagocytic cells, synthesis and polymerization of collagen components, and finally, to severe interstitial fibrosis and tubular atrophy.

Formation of anti-TBM antibody. Although the precise mechanism of formation of anti-TBM antibody is still unknown, two hypotheses have been proposed, one considering increased synthesis or denaturation of TBM or GBM components and another involving an immune response against TBM alloantigen(s).

According to the first hypothesis, GBM or TBM damage may liberate the antigen into the circulation and stimulate formation of autoantibodies reacting with the TBM. Furthermore, the TBM may develop a new antigenicity by binding chemicals, such as dimethoxyphenyl-penicilloyl haptenic group, a derivative of the penicillin, which is largely secreted by PCT (14). This compound could act as a hapten protein conjugated to TBM thereby leading to a production of anti-TBM antibody (5, 7, 14).

The second hypothesis is concerned with the production of anti-TBM antibody after renal transplantation (30). The allotypic anti-TBM antibodies found in a patient with a renal allograft reacted with the TBM of the allograft and all other human kidneys tested, but not with the patient's own kidney (72). Therefore, it was suggested that the immune response was generated by a reaction to a foreign TBM antigen introduced in the host "via" transplantation. This interpretation does not exclude the possibility that in some recipients of renal grafts the anti-TBM response is of autoimmune nature. In fact, it has been reported that some renal allografts may stimulate formation of antibodies against autologous and isologous TBM antigen (30).

B. Animal Model

(a) Heteroimmune anti-TBM TIN. Similar to anti-GBM in man, anti-GBM antibody produced by immunization of animals with heterologous GBM is reactive not only with GBM, but also with TBM and ABM. The glomerular lesions are usually severe, in contrast to tubulo-interstitial changes which are mild or absent (for review, see 74).

(b) Autoimmune anti-TBM TIN. In the first experimental model, described by Stebley et al. (62), guinea pigs, immunized with rabbit cortical basement membrane in complete Freund's adjuvant, developed severe TIN in association with linear deposits of guinea pig IgG, and sometimes C3, along TBM and GBM. A similar lesion occurred in guinea pigs immunized with bovine TBM. In the second experimental model, described by Sugisaki et al. (65), both Brown Norway (BN) and/or BN/Lewis hybrid rats which were immunized with isologous renal antigen developed severe TIN in association with linear deposits of rat IgG and C3 along the TBM of proximal convoluted tubules, only. A similar lesion was later described in the same strain of rats immunized with bovine TBM (37). The animals with anti-TBM TIN developed glycosuria and, sometimes, aminoaciduria, indicating tubular dysfunction.

The histological features of this severe TIN are characterized by polymorphonuclear and mononuclear cell infiltration, tubular cell destruction and giant cell formation. In the guinea pigs the infiltration of mononuclear cells predominates whereas in the rat infiltration of polymorphonuclear leukocytes may also been present. High titers of anti-TBM antibodies are detected by indirect immunofluorescence in the sera or in the eluates obtained from diseased kidneys.

Both types of TIN have been passively transferred to normal guinea pig or to BN or BN/Lewis rats by means of circulating anti-TBM antibodies obtained from nephritis donors (61, 65, 69). Thus, it is evident that anti-TBM antibodies initiate TIN. However, the precise mechanism by which anti-TBM antibodies and complement bring about tubular damage is not known. Although guinea pigs depleted of C3 by administration of cobra venom are resistant to the passive transfer of the disease (58), C3 deposits are sometimes absent in TBM, even in animals with severe disease (35, 62, 69). In addition, TIN develops in guinea pigs genetically deficient in C4 (57). Therefore, the combination of antigen (TBM), antibody (anti-TBM antibody) and complement does not appear a likely cause of tubulo-interstitial damage.

It has been stressed that the infiltrating mononuclear cells are important for the development of TIN in guinea pigs (69). The nature of the infiltrating cells in the tissue has been studied by rosetting technique using IgG and/or C3 coated red blood cells (EA

and EAC, respectively). The EA bound on many of the monocytes/
macrophages bearing Fc receptor (Fc receptors on B cells do not
bind in this assay). No EAC bound-B lymphocytes were detected, indicating that many of the infiltrating lymphocytes are T cells or
null cells. However, considering the late appearance of plasma cells
in this disease, the presence of B cells may also have pathogenic
significance. Thus, it is conceivable that a cell-mediated immune
response may cooperate with anti-TBM antibody in the development
of progressive, chronic, lesions. It must be noted, however, that
attempts to induce TIN in guinea pig by transfer of lymphoid cells
have been unsuccessful (36, 69).

(c) Alloimmune anti-TBM TIN. An alloantibody response to TBM
antigens can occur in certain strains of rat after renal transplantation when recipient rats lacking TBM alloantigen (such as
Lewis, Wistar-Furth and MAXX strains) are grafted with a kidney
bearing the TBM alloantigen (such as kidneys from Fischer 344, BN,
Sprague-Dawley, etc.) (34, 65). The allotypic specificity seems to
be unrelated to the major histocompatibility locus (RT1) in the
rats, since Lewis and Fischer 344 or MAXX and BN are RT1 identical.

(d) Genetic influence of immune response to TBM antigen. Recently, the genetic control of the immune response to TBM antigen
has been recognized (26, 27). When strain XIII and strain II guinea
pigs are immunized with rabbit TBM, strain XIII develop more severe
anti-TBM TIN and higher titer of anti-TBM antibody than strain II.
However, the kidney from strain II possesses the TBM antigen since
it binds anti-TBM antibodies from strain XIII animals. These observations indicate that the cause of less severe TIN in strain II
is not due to a lack of pertinent TBM antigens but probably to
failure to synthetize anti-TBM antibodies. Furthermore, it has been
reported that a dominant immune response (Ir) gene(s) linked to the
major histocompatibility locus probably control the autoimmune
response to TBM in guinea pig (26).

II. IMMUNE COMPLEX MEDIATED TIN

A. Human Disease

TIN associated with granular deposits of immunoglobulin and

Table 2. Extraglomerular granular deposits of IgG and C3 in
          kidney of patients with active SLE (2).

Mery, Morel-Maroger, and Boelaert (1974)	8/37	(22%)
Lehman, Wilson and Dixon (1975)	22/32	(69%)
Brentjens et al. (1975)	24/45	(53%)

complement, presumably immune complexes localized in tubulo-interstitial structures, is observed infrequently, generally in association with glomerular diseases (29).

The condition in which granular tubulo-interstitial immune
deposits are most frequently found is in patients with diffuse
proliferative or exudative lupus glomerulonephritis (about 50% of
these patients) (9, 29, 38). Table 2 summarizes the results of three
studies concerning the frequency of extraglomerular immune deposits
in active SLE. Other forms of glomerulonephritis in which tubulo-
interstitial deposits have been found include mixed cryoglobulinemia,
membranoproliferative, rapidly progressive (29, 38) and post-
streptococcal (Sugisaki, unpublished) glomerulonephritis.

In patients with renal failure due to lupus nephritis, granular
immune deposits may be more abundant in tubulo-interstitial structures than in glomeruli (29, 38). Furthermore, TBM deposits in
absence of glomerular changes have been found in renal allografts
(1). Thus, the development of immune complex deposits in TBM does
not appear due to a primary saturation of glomerular sites with
secondary glomerular "spillover" but rather the consequence of
pathogenic factors which are still unknown.

It is difficult to assess the role of immune complexes in the
pathogenesis of TIN. The association of tubulo-interstitial change
with local immune deposits, especially in patients without prominent
glomerular changes, is consistent with the interpretation that the
immune deposits have pathogenic significance. However, it is usually
difficult to reach this conclusion because similar histologic changes
may be seen in the interstitium of patients without local immune
deposits. Furthermore, in most patients with lupus nephritis, the

severity of tubulo-interstitial damage parallels that of glomerular
disease.

Light microscopy. Local infiltration of inflammatory cells,
mainly mononuclear cells, tubular cell damage or tubular atrophy,
thickening and/or duplication of the TBM and interstitial fibrosis
may be seen.

Immunofluorescence. The granular or ribbon-like deposits of
IgG and C3 (rarely IgM and/or IgA, as in renal transplantation)
appear located either along or within the TBM, around peritubular
capillaries, or in the interstitium (6, 9, 29, 38). The distribution
of the deposits is often local but occasionally may be widespread.

Electronmicroscopy. Dense deposits are found along or within
tubular and capillary basement membranes or in the interstitium.
These findings usually parallel those obtained by immunofluorescence
microscopy. However, because of the focal nature of the deposits,
electronmicroscopy is of limited use in these studies.

Immunopathogenesis. Direct evidence that the antigen is present
in the immune deposits, and thus, that these deposits indeed contain
immune complexes is lacking, with the exception of lupus nephritis
where denatured DNA has been demonstrated as a constituent of the
immune deposits in TBM (9), and membranoproliferative glomerulonephritis in a patient with candida endocrinopathy syndrome, where
candida antigens were shown in tubular deposits (12).

The precise mechanism by which immune complexes localize along
the TBM or in the interstitium is still unknown. One possibility is
that immune complex may be formed "in situ". Antigens liberated
from tubular cells, presumably DNA or cytoplasmic component, may
react with corresponding antibodies at the level of the TBM and
precipitate, thus forming immune deposits. This reaction has been
well documented in the experimental animal (rabbit) given repeated
renal transplant or repeated injection of renal tissue antigen
(the precise mechanism will be described later) (31). Another possibility is that exogenous immune complexes formed in the circulation
may be "trapped" in tubulo-interstitial tissue. This mechanism can
have a role in bacterial endocarditis, acute post-streptococcal
glomerulonephritis and some cases of SLE. In addition, deposition
of circulating immune complex along TBM can occasionally be associ-

ated with deposition of immune complexes in the walls of peritubular capillaries. In this condition, it is possible that an enhanced permeability of peritubular capillaries may be the primary event which allows deposition of immune complexes along the TBM.

B. Animal Model

The demonstration of immune complex TIN in experimental animals led to a similar discovery in man. Immune complex TIN has been induced by two different immunization procedures; one, by using homologous renal antigen, the other, by using heterologous antigen (bovine serum albumin (BSA)).

(a) Autologous immune complex TIN. Experimental TIN associated with immune complex deposits along TBM has been induced in rabbits by repeated injections of homologous renal extracts in complete Freund's adjuvant (31). The TIN was characterized by severe interstitial fibrosis, mononuclear cell infiltration and tubular damage. The glomeruli were generally normal. By immunofluorescence, granular deposits of rabbit IgG and C3 were seen mainly along the TBM of proximal convoluted tubules. Electronmicroscopy study revealed that the foreign deposits were localized inside the TBM or in subepithelial portion. A similar lesion was transferred by means of large amount of the immune sera. Since eluates obtained from the kidneys or the serum of nephritic animals reacted by indirect immunofluorescence with the proximal tubular cells, it was proposed that the immune complexes are formed at the level of the TBM by antigen released from the tubular cells and the corresponding antibody coming from the circulation. This study was the first to show that immune complexes may be formed in the kidney by a mechanism similar to Arthus reaction. Immunopathologic lesions comparable to those described above were then induced in rabbits given repeated renal allografts (32).

Another example of immune complex formation along the TBM of proximal convoluted tubules is in Sprague-Dawley rats immunized with homologous renal antigen in adjuvant. These rats develop autologous immune complex glomerulonephritis characterized by production of antibody to the brush border of proximal convoluted tubules (Heymann nephritis) (23). Recently, it has been suggested that this type of glomerular injury results from a reaction between antigen(s)

present at the surface of the podocytes and circulating anti-brush
border antibodies (16). The nephritis is occasionally associated
with granular deposits of rat IgG and C3 along the TBM, particularly
in rats with severe glomerulonephritis (33). Since brush border
antigen(s) have been shown in these tubular deposits, it is conceivable that they contain brush border antigen(s) released from
tubular cells and corresponding antibodies coming from the circulation or filtered through the glomerulus and reabsorbed by tubular
cells.

Rats immunized with homologous renal tissue (33) or crude
extract of portions of Henle's loop (HL) (64) may develop broad
deposits of rat IgG, rarely C3, along the basal portions of HL and
also distal convoluted tubules (DCT). Mild local inflammatory
reaction (mainly composed of mononuclear cells) occurs around the
HL. The serum and the renal eluates obtained from these rats react
with the tubular cells of HL in normal rats. A similar reactivity
was detected in normal rats by transfer of the immune serum (64).
It is conceivable that local formation of the complexes can occur
when cellular antigens are released into the extracellular space.

More recently, autologous IC TIN of HL was developed in rats
immunized with Tamm-Horsfall protein (24), a glycoprotein, which
is a major constituent of casts in HL and DCT. Immune deposits are
seen along the ascending thick limb of the HL and also DCT; these
segments of the nephron are the sites of Tamm-Horsfall protein
production. Mild interstitial mononuclear infiltration is seen in
the vicinity of the immune deposits.

(b) Heterologous immune complex (serum sickness-type)TIN.
Rabbits receiving repeated i.v. daily injection of BSA develop
chronic serum sickness glomerulonephritis which may be associated
with TIN (8). Immunofluorescence reveals deposits of BSA and IgG
(anti-BSA antibody) along renal basement membranes, including GBM,
TBM and peritubular capillary. Electronmicroscopy confirms the
presence of foreign deposits along these basement membranes. The
deposits are seen not only in the renal tissue, but also in the
lung, the liver, the muscle and the intestine. These findings may
result from an overload of the reticulo-endothelial system which
cannot clear the large amount of immune complexes formed in the
circulation. In addition, some other factors, such as involvement
of IgE antibodies (13), "in situ" immune complex formation (18) and

presence of Fc receptor in the renal interstitium (19) may account
for the development of this systemic pathology.

III. CELL-MEDIATED TIN

A. Human Disease

To what extent the interstitial cellular infiltration that
characterizes many forms of glomerulo- or tubulo-interstitial-
nephritis results from cell-mediated hypersensitivity (delayed
hypersensitivity), is still unknown. The renal interstitium appears
as an ideal site where sensitized lymphocytes can react with either
heterologous or autologous antigens. A few in vitro observations
suggest that cell-mediated immunity to heterologous antigens, such
as bacterial, viral antigens (15) and certain drugs (aspirin) (44)
can lead to a mononuclear cell infiltration in the interstitium.
In addition, the compound resulting from the conjugation of a drug
with a renal antigen may be important for the development of cell-
mediated TIN. Renal allograft rejection obviously provides the most
characteristic example of TIN due to cell-mediated immunity against
histocompatibility antigens or organ-specific alloantigen(s). In
conclusion, a convincing proof that cell-mediated hypersensitivity
plays a role in the pathogenesis of human TIN is still lacking. It
is apparent that new approaches for the identification of this
mechanism of renal injury are needed.

B. Animal Model

Two models of cell-mediated TIN have been developed in experimental animals, one characterized by reactivity to exogenous antigens
localized in the renal tissue (70) (cell-mediated TIN induced by
heterologous antigen), and the other, characterized by reactivity
to autologous renal antigen (autoimmune cell-mediated TIN) (66).

(a) Cell-mediated TIN induced by heterologous antigen. Guinea
pigs and rats were sensitized to bovine gamma globulin (BGG) in a
manner devised to elicit delayed hypersensitivity and then received
direct injection of aggregated or soluble BGG into the renal cortex.
The histology of the kidneys injected with aggregated BGG, but not

with soluble BGG, showed mononuclear cell infiltration with focal
tubular destruction. Similar lesion could be transferred with lymph-
node cells, but not with serum (70).

Another experimental study showed that rats with streptococcal
pyelonephritis developed cellular reactivity to the bacterial
antigen, as indicated by the lymphocyte stimulation test (15). It
was not proved, however, that streptococci elicited hypersensitivity
at the level of the renal interstitium. This mechanism could have
importance in the pathogenesis of human pyelonephritis.

(b) Autoimmune cell-mediated TIN. It has been suggested that
delayed hypersensitivity directed against autologous antigen(s) may
participate in the development of TIN in rats (33). Lewis rats
injected with renal tissue in complete Freund's adjuvant, plus
pertussis vaccine, developed severe TIN at 10-14 days, usually
before evidence of antibody production (56). The lesions were
characterized by irregular mononuclear cell infiltration and tubular
cell damage, prior to apparent deposits of rat IgG and complement
in the glomeruli and in the interstitium. Passive transfer of
lymphoid cells resulted in focal interstitial infiltrates and tubular
cell damage. Furthermore, the migration of peritoneal macrophages
from kidney-injected rats was inhibited by kidney antigens. There
are a few in vitro studies (based on the migration inhibition (20),
lymphocyte blast formation and skin test (42)), which confirm the
hypothesis that cellular immunity may be responsible for tissue
injury in rats injected with renal antigen (Heymann nephritis). All
these tests, however, have been focused on the glomerular lesions.
Considering that the renal antigen used for these tests was of nonglomerular origin and that the cell-mediated TIN was induced by
nonglomerular antigen, it seems appropriate to use the same tests
for in vitro studies which could contribute to the elucidation of
the pathogenesis of cell-mediated TIN.

IV. TUBULAR CELL DAMAGE DUE TO ANTI-BRUSH BORDER ANTIBODY

A. Human Disease

Anti-brush border antibodies have been implicated in the
pathogenesis of human glomerular disease, such as "idiopathic"

membranous glomerulonephritis (48, 59), membranous glomerulonephritis associated with sickle cell anemia (63), renal vein thrombosis (53) or renal allograft rejection (29). However, at variance with rat Heymann nephritis, in human renal diseases there is no evidence indicating that anti-brush border antibodies may react "in vivo" with the tubular brush border. Lack of this evidence suggests that human and other species, such as rabbit, guinea pig and mouse, are not capable of producing anti-brush border antibody, except in very rare instances. Therefore, it is difficult to assess the possible role of anti-brush border antibody in the pathogenesis of tubular damage.

B. Animal Model

In vivo binding of IgG to the brush-border of proximal convoluted tubules is frequently seen in rats with Heymann nephritis, in rats injected with anti-brush border antibody when glomerular permeability is artificially increased by administration of aminonucleoside (Sugisaki, unpublishied) or by the induction of chronic serum sickness glomerulonephritis (50) and, rarely, in normal rats injected with the antibody. These findings indicate that glomerular injury is necessary in order to allow an in vivo reaction of anti-brush border antibody with the corresponding tubular antigen. This reaction may involve complement fixation and may result in loss of microvilli of proximal convoluted tubules, increased clasmatosis and evidence of autophagocytosis.

V. IgE-MEDIATED TIN

A. Human Disease

It has been suggested that IgE-mediated hypersensitivity may participate in the development of some forms of acute TIN. The evidence that favors this interpretation may be summarized as follows:

1) Drugs, such as penicillin or its derivatives, and phenobarbitone are agents which frequently induce immediate hypersensitivity characterized by skin rash, elevated IgE levels in the serum (52) and increased numbers of eosinophils in the peripheral blood.

2) End-products of these drugs are largely excreted by the kidney.

3) Eosinophil and basophil infiltration is frequently observed in the renal interstitium (14), together with accumulation of IgE-containing plasma cells (17).

In spite of sometimes impressive histological, clinical and laboratory findings, direct evidence that acute TIN may be induced by IgE-type hypersensitivity is still lacking.

B. Animal model

At present, there is no experimental model of IgE-mediated TIN.

VI. CLINICAL MANIFESTATION OF HUMAN TIN

As described earlier, since immunologically-mediated TIN without glomerulonephritis is a very rare clinical and histological entity, it is difficult to analyze the characteristic symptoms.

On the basis of the immunopathological findings, it should be expected that the symptoms of immune complex TIN are mild because of the scattered distribution of the immune deposits which usually induce only a mild inflammatory reaction. In contrast, the symptoms of anti-TBM TIN (with or without anti-GBM antibody-mediated glomerulonephritis) are severe and sometimes rapidly progressive because of generalized and diffuse binding of antibody to TBM and widespread inflammatory reaction. If the proximal convoluted tubules are the site of cell damage, proteinuria (particularly low molecular weight proteins), glucosuria, aminoaciduria and sometimes Fanconi's syndrome (59) may develop.

Drug-induced acute TIN (probably generated by an immediate hypersensitivity reaction) may result in acute renal failure. The symptoms of a TIN mediated by anti-Henle's loop antibody or by cell-mediated hypersensitivity mechanisms, are still unknown, because well-controlled clinicopathological observations have not yet been reported.

VII. LABORATORY FINDINGS

A. Urine Examination

Besides proteinuria (small molecular weight proteins), aminoaciduria and glucosuria, ß2-microglobulinemia, all reflecting a
dysfunction of tubular cells, may be found. Urinary casts of polymorphonuclear leukocytes may be seen in patients with severe exudative TIN without evidence of renal bacterial infection (11).

B. General Laboratory Examination

Serum sodium chloride, potassium, calcium, phosphate, creatinine,
urea and uric acid should be routinely measured. In the case of
glucosuria, a glucose tolerance test should be performed, to rule
out a primary diabetes mellitus.

C. Immunological Examination

If anti-basement membrane (anti-TBM or anti-GBM), anti-brush
border or anti-Henle's loop antibodies, are detected by indirect
immunofluorescence in the sera, the findings are strongly indicative
of the presence of antibody-mediated TIN. A positive in vitro lymphocyte stimulation test using drugs or purified tissue antigen,
such as GBM, may be of assistance in evaluating the participation
of cell-mediated immunity in the development of TIN. Measurements
of immune complex in the serum may also provide an indication. Skin
tests are sometimes useful to prevent the occurrence of drug-induced
anaphylactoid reaction. However, they are not indicated in patients
with drug allergy because even small doses of antigen may exacerbate
the symptoms of the disease.

D. Comparative Histological and Functional Studies

It has been proposed that the data resulting from the histologic
study of renal tissue obtained by needle biopsies provide a good
correlation between glomerular abnormalities and functional changes
evaluated by measurement of the creatinine clearance (10, 47).

However, the value of this correlation is not generally accepted
(56, 60). In contrast, it has been proposed that a significant
correlation between morphologic.and functional changes exists in
TIN. These conflicting reports make it difficult to assess the
value of a correlation between renal functional data and severity
of TIN. Physiological studies in animals with immunologically-
mediated TIN not associated with glomerulonephritis are needed in
order to elucidate the functional abnormalities of the tubules.

VIII. TREATMENT

At present, it is difficult to discuss the treatment of immunologically-mediated TIN because only a small number of cases have
been reported and many questions remain to be elucidated. It is
obvious that the basic approach in the treatment of TIN should be
similar to that of glomerulonephritis because the etiology and the
pathogenesis of the two diseases are similar or identical.

A general outline concerning the treatment of the TIN should
include:

a) Elimination or avoidance of the causative antigens, although
the majority of the antigens are still undetected.

b) Immunosuppressive agents.

c) Plasmapheresis which may eliminate circulating antibodies
or immune complexes.

d) Anti-inflammatory agents and/or corticosteroid hormones
which may contribute to reduce certain aspects of the inflammatory
reaction.

CONCLUSION

In the last decade, it has been possible to identify some
immunological mechanisms which induce injury to the renal tubules
and the interstitium. These mechanisms are basically similar to
those responsible for glomerulonephritis. The study of immunologically
mediated TIN in experimental animals has greatly contributed to the

understanding of similar TIN in man. However, much remains to be done in order to get a deeper insight of the pathogenesis and the functional abnormalities of human TIN. Additional progress is necessary in order to devise a more rational and efficient therapy of these diseases.

REFERENCES

1. Andres, G. A., Accinni, L., Hsu, K.C., Penn, I., Porter, K.A., Randall, J.M., Seegal, B. C. and Starzl, T.E.: Human renal transplants. III. Immunopathologic studies. Lab. Invest. 22: 588-604, (1970).

2. Andres, G.A., Albini, B. and Ossi, E.: Pathogenesis, diagnosis, prognosis and treatment of immunologically mediated tubulo-interstitial nephritis. In: Prevention of Kidney and Urinary Tract Diseases. Fogarty International Center Series on Preventive Medicine, Coggins, C.H. and Cummings, W.B., eds. Vol. 5, pp. 251-262 (DHEW, NIH, Bethesda, 1978).

3. Andres, G.A., Brentjens, J. Kohli, R., Anthone, R., Anthone, S., Baliah, T., Montes, M., Mookerjee, B., Prezyna, A., Sepulveda, M., Venuto, T. and Elwood, C.: Histology of human tubulo-interstitial nephritis associated with antibodies to renal basement membranes. Kidney Int. 13: 480-491, (1978).

4. Andres, G.A. and McCluskey, R.T.: Tubular and interstitial renal disease due to immunologic mechanisms. Kidney Int. 7: 271-289, (1975).

5. Baldwin, D.S., Levine, B.B., McCluskey, R.T. and Gallo, G.R.: Renal failure and interstitial nephritis due to penicillin and methicillin. New Engl. J. Med. 279: 1245-1252, (1968).

6. Bergstein, J. and Litman, N.: Interstitial nephritis with anti-tubular basement membrane antibody. New Eng. J. Med. 292: 875-878, (1975).

7. Border, W., Lehman, D., Egan, J., Sass, H., Glode, J. and Wilson, C.B.: Antitubular basement membrane antibodies in methicillin-associated interstitial nephritis. New. Engl. J. Med. 291: 381-384, (1974).

8. Brentjens, J., O'Connell, D., Pawloski, I. and Andres, G.A.: Extraglomerular lesions associated with deposition of circulating antigen-antibody complexes in kidneys of rabbits with chronic serum sickness. Clin. Immunol. Immunopath. 3: 112-122, (1974).

9. Brentjens, J., Sepulveda, M., Baliah, T., Bentzel, C., Erlanger, B., Elwood, C., Montes, M., Hsu, K. and Andres, G.A.: Interstitial immune complex nephritis in patients with systemic lupus erythematosus. Kidney Int. 7: 342-350, (1975).

10. Brod, J. and Benesova, D.: A comparative study of fuctional and morphological renal changes in glomerulonephritis. Acta Med. Scand. 157: 23-32, (1957).

11. Case records of the Massachusetts General Hospital: Case 2 - 1976. New Engl. J. Med. 294: 100-105, (1976).

12. Chesvey, R., O'Regon, S., Guyda, H.Y. and Drummond, K.N.: Candida endocrinopathy syndrome with membranoproliferative glomerulonephritis: demonstration of glomerular candida antigen. Clin. Nephrol. 5: 232-238, (1976).

13. Cochrane, C.G.: Mechanisms involved in deposition of immune complexes in tissue. J. Exp. Med. 134: (Pt. 2): 75s-89s, (1971).

14. Colvin, R.B., Burton, N.E., Hyslop, N.E.Jr., Spitz, L. and Lichtenstein, N.S.: Penicillin-associated interstitial nephritis. Ann. Int. Med. 81: 404-405, (1974).

15. Cotran, R.S. and Peissens, W.F.: Pathogenesis of chronic pyelonephritis. Proc. Int. Congr. Nephrol., 6th, pp. 509-523 (Karger-Basel, 1976).

16. Couser, W. and Salant, G.: In situ immune complex formation and glomerular injury. Kidney Int. 17: 1-14, (1980).

17. Faarup, P. and Cristensen, E.: IgE containing plasma cells in acute tubulo-interstitial nephropathy. Lancet 2: 718, (1974).

18. Fleuren, G., Grond, J. and Hoedemacker, P.J.: In situ formation of subepithelial glomerular immune complexes in passive serum sickness. Kidney Int. 17: 631-637, (1980).

19. Gelfand, M.C., Frank, M.M., Green, I. and Shin, M.L.: Binding sites of immune complexes containing IgG in the renal interstitium. Clin. Immunol. Immunopath. 13: 19-29, (1979).

20. Grupe, W.E.: An "in vitro" demonstration of cellular sensitivity in experimental autoimmune nephrosis in rats. Proc. Soc. Exp. Biol. Med. 127: 1217-1222, (1968).

21. Harner, M.H., Nolte, M., Wilson, C.B., Talwalker, Y.B., Musgrave, J.E., Brooks, R.E. and Campbell, R.A.: Anti-tubular basement membrane antibody and nephrotic syndrome associated with milk hypersensitivity. Abstr. 3rd Int. Symp. Pediat. Nephrol., Washington, p. 8 (1974).

22. Heptinstall, R.H.: Interstitial nephritis; a brief review. Amer. J. Pathol. 83: 214-236, (1976).

23. Heymann, W., Hackel, D.B., Harwood, S., Wilson, S.G.F. and

Hunter, J.L.P.: Production of nephrotic syndrome in rats by Freund's adjuvant and rat kidney suspension. Proc. Soc. Exp. Biol. Med. 100: 660-664, (1959).

24. Hozer, J.R.: Tubulointerstitial immune complex nephritis in rats immunized with Tamm-Horsfall protein. Kidney Int. 17: 284-292, (1980).

25. Hyman, L.R., Ballow, M. and Knieser, M.R.: Diphenylhydantoin nephropathy: Evidence for an autoimmune pathogenesis. Kidney Int. 8: 450, (1975) (Abstr.)

26. Hyman, L.R., Colvin, R.B. and Steinberg, A.D.: Immunopathogenesis of autoimmune tubulointerstitial nephritis. I. Demonstration of differential susceptibility in strain II and strain XIII guinea pigs. J. Immunol. 116: 327-335, (1976).

27. Hyman, L.R., Steinberg, A.D., Colvin, R.B. and Bernard, E.F.: Immunopathogenesis of autoimmune tubulointerstitial nephritis. II. Role of an immune response gene linked to the major histocompatibility locus. J. Immunol. 117: 1894-1897, (1977).

28. Kawowski, S., McKay, D.G.: Howes, E.L.Jr., Csavossy, I. and Wolfson, M.: Multinucleated giant cells in antiglomerular basement membrane antibody-induced glomerulonephritis. Nephron 16: 415-426, (1976).

29. Klassen, J., Andres, G., Brennen, J. and McCluskey, R.T.: An immunologic renal tubular lesion in man. Clin. Immunol. Immunopath. 1: 69-83, (1972).

30. Klassen, J., Kano, K., Milgrom, F., Menno, A., Anthone, R., Sepulveda, M., Elwood, C. and Andres, G.A.: Tubular lesions produced by autoantibodies to tubular basement membranes in human renal allografts. Int. Arch. All. Appl. Immunol. 45: 675-689, (1973).

31. Klassen, J., McCluskey, R.T. and Milgrom, F.: Nonglomerular renal disease produced in rabbits by immunization with homologous kidney. Amer. J. Pathol. 63: 333-350, (1971).

32. Klassen, J. and Milgrom, F.: Autoimmune concomitants of renal allografts. Transplant. Proc. 1: 605-608, (1969).

33. Klassen, J., Sugisaki, T., Milgrom, F. and McCluskey, R.T.: Studies on multiple renal lesions in Heymann nephritis. Lab. Invest. 25: 571-585, (1971).

34. Lehman, D.H., Lee, S., Wilson, C.B. and Dixon, F.J.: Induction of antitubular basement membrane antibodies in rats by renal transplantation. Transplantation 17: 429-431, (1976).

35. Lehman, D.H.; Marquardt, H., Wilson, C.B. and Dixon, F.J.: Specificity of autoantibodies to tubular and glomerular basement

membranes induced in guinea pigs. J. Immunol. 112: 241-248, (1974).

36. Lehman, D.H. and Wilson, C.B.: Role of sensitized cells in antitubular basement membrane interstitial nephritis. Int. Arch. All. Appl. Immunol. 51: 168-176, (1976).

37. Lehman, D.H., Wilson, C.B. and Dixon, F.J.: Interstitial nephritis in rats immunized with heterologous tubular basement membrane. Kidney Int. 5: 187-195, (1974).

38. Lehman, D.H., Wilson, C.B. and Dixon, F.J.: Extraglomerular immunoglobulin deposits in human nephritis. Amer. J. Med. 58: 765:786, (1975).

39. Levy, M., Gagnadoux, M.F., Beziall, A. and Habib, R.: Membranous glomerulonephritis associated with antitubular and antialveolar basement membrane antibodies. Clin. Nephrol. 10: 158-165, (1978).

40. Levy, M., Gagnadoux, M.F. and Habib, R.: An immunologic Fanconi syndrome. Int. Symp. Pediat. Nephrol., 3rd, Washington, D.C. (1974).

41. Lewis, E.J., Cavallo, J.A., Harrington, J.T. and Cotran, R.S.: An immunopathologic study on rapidly progressive glomerulo-nephritis in the adult. Human Pathol. 2: 185-208, (1971).

42. Letwin, A., Adams, L.E., Levy, R., Cline, S. and Hess, E.B.: Cellular immunity in experimental glomerulonephritis of rats: I. Delayed hypersensitivity and lymphocyte stimulation studies with renal tubular antigen. Immunology 20: 755-766, (1971).

43. McCluskey, R.T. and Klassen, J.: Immunologically mediated glomerular tubular and interstitial renal disease. New. Engl. J. Med. 288: 564-570, (1973).

44. McLeish, K. Senitzer, D. and Gohara, A.: Acute interstitial nephritis in a patient with aspirin hypersensitivity. Clin. Immunol. Immunopath. 14: 64-69, (1979).

45. McPhaul, J.J. and Dixon, F.J.: Characterization of human anti-glomerular basement membrane antibodies eluated from glomerulo-nephritic kidneys. Clin. Invest. 49: 308-317, (1970).

46. Morel—Maroger, L., Kourilsky, O., Mignon, F. and Richet, G.: Antitubular basement membrane antibodies in rapidly progressive poststreptococcal glomerulonephritis. Clin. Immunol. Immuno-path. 2: 185-194, (1974).

47. Muehrcke, C., Kark, R.M., Pirani, C.L. and Pollak, V.E.: Lupus nephritis, a clinical and pathologic study based on renal biopsies. Medicine (Baltimore) 36: 1-145, (1957).

48. Naruse, F., Kitamura, K., Miyakawa, Y. and Shibata, S.: Depo-

sition of renal tubular antigen along the glomerular capillary walls of patients with membranous glomerulonephritis. J. Immunol. 110: 1163-1166, (1973).

49. Nicastri, A.D., Chen, C.K., Rao, T.K.S., Ginzler, E.M., Kaplan, D. and Friedman, E.A.: Renal disease with tubular immunofluorescence deposits. Kidney Int. 8: 452, (1975) (Abstr:).

50. Noble, B., Mendrick, D., Brentjens, J.R. and Andres, G.A.: Damage of renal proximal tubules by passive transfer of homologous brush border antibodies. Fed. Proc. 39, I: 815A, (1980).

51. Olsen, S. and Asklund, M.: Interstitial nephritis with acute renal failure following cardiac surgery and treatment with methicillin. Acta Med. Scand. 99: 305-310, (1976).

52. Ooi, B.S., First, M.R., Pesce, A.J., Pollak, V.E., Bernstein, I.L. and Iag, W.: IgE levels in interstitial nephritis. Lancet 1: 1254-1256, (1974).

53. Ozawa, T., Boedecker, E.A., Scharr, W., Guggenheim, S. and McIntosh, R.M.: Immune-complex disease with unilateral renal vein thrombosis. Arch. Path. 100: 279-282, (1976).

54. Paul, L., Van Es, L., Stuffers-Heiman, M., Brutel, de la, Riviere, G. and Kalff, M.: Antibodies directed against tubular basement membranes in human renal allograft recipients. Clin. Immunol. Immunopath. 14: 231-237, (1979).

55. Proskey, A.J., Weatherbee, L., Esterling, R.E., Greene, J.A. and Weller, J.M.: Goodpasture's syndrome. A report of five cases and review of the literature. Amer. J. Med. 48: 162-173, (1970).

56. Risdon, R.A., Sloper, J.C. and de Wardener, H.E.: Relationship between renal function and histological changes found in renal biopsy specimens from patients with persistent glomerular nephritis. Lancet. 2: 263-266, (1968).

57. Rudofsky, U.H., McMaster, P.R.B., Ma, W., Steblay, R.W. and Pollara, B.: Experimental autoimmune renal cortical tubulointerstitial disease in guin a pigs lacking the fourth component of complement (C4). J. Immunol. 112: 1387-1393, (1974).

58. Rudofsky, U.H., Steblay, R.W. and Pollara, B.: Inhibition of experimental autoimmune renal tubulointerstitial disease in guinea pigs by depletion of complement with cobra venom factor. Clin. Immunol. Immunopath. 3: 396-402, (1975).

59. Shwayder, M., Ozawa, T., Boedecker, E.: Guggenheim, S. and McIntosh, R.H.: Nephrotic syndrome associated with Fanconi syndrome. Immunopathogenic studies of tubulointerstitial nephritis with autologous immune-complex glomerulonephritis.

Ann. Intern. Med. 84: 432-437, (1976).

60. Sloper, J.C., de Wardener, H. and Woodrow, D.F.: Relationship
between renal structure and function deduced from renal biopsies.
In: Renal Pathophysiology. Leaf, A, Giebisch, G., Bolis, L.
and Gorini, S. eds. pp. 109-120 (Raven Press, New York, 1980).

61. Steblay, R.W. and Rudofsky, U.H.: Transfer of experimental
autoimmune renal cortical tubular and interstitial disease in
guinea pigs by serum. Science 180: 966-968, (1973).

62. Steblay, R.W. and Rudofsky, U.H.: Renal tubular disease and
autoantibodies against tubular basement membrane induced in
guinea pigs. J. Immunol. 107: 589-594, (1971).

63. Strauss, J., Pardo, V., Koss, M.N., Griswold, W. and McIntosh,
M.: Nephropathy associated with sickle cell anemia: An autologous immune complex nephritis. Amer. J. Med. 58: 382-387,
(1975).

64. Sugisaki, T., Klassen, J., Andres, G.A., Milgrom, F. and McCluskey,
R.T.: Species specific renal lesions in different rat strains.
VI Intern. Cong. Nephrol. (Abstr.), Firenze, Italy, p. 304,
(1975).

65. Sugisaki, T., Klassen, J., Milgrom, G., Andres, G.A. and
McCluskey, R.T.: Immunopathologic study of an autoimmune tubular
and interstitial renal disease in Brown Norway rats. Lab.
Invest. 28: 658-671, (1973).

66. Sugisaki, T., Yoshida, T., McCluskey, R.T., Andres, G.A. and
Klassen, J.: Autoimmune cell-mediated tubulointerstitial
nephritis induced in Lewis rats by renal antigens. Clin. Immunol. Immunopath. 15: 33-43, (1980).

67. Sutton, J. and Weiss, L.: Transformation of monocytes in tissue
culture into macrophages, epithelial cells and multinucleated
giant cells. An electron microscopic study. J. Cell. Biol. 28:
303-332, (1966).

68. Tung. K. and Black, W.: Association of renal glomerular and
tubular immune complex disease and antitubular basement membrane
antibody. Lab. Invest. 32: 696-700, (1975).

69. Van Zwieten, M.J., Bhan, A.K., McCluskey, R.T. and Collins,
A.B.: Studies on the pathogenesis of experimental anti-tubular
basement membrane nephritis in the guinea pig. Amer. J. Pathol.
83: 531-546, (1976).

70. Van Zwieten, M.J., Leber, P.D., Bhan, A.K. and McCluskey, R.T.:
Experimental cell mediated interstitial nephritis induced with
exogenous antigens. J. Immunol. 118: 589-593, (1977).

71. Wilson, C.B. and Dixon, F.L.: Antiglomerular basement membrane

antibody-induced glomerulonephritis. Kidney Int. 3: 74-89,
(1973).

72. Wilson, C.B., Lehman, D., McCoy, R., Gunnels, J. and Stubel,
     D.: Antitubular basement membrane antibodies after renal
     transplantation. Transplantation, 18: 447-452, (1974).

73. Wilson, C.B. and Dixon, F.J.: Diagnosis of immunopathologic
     renal disease. Editorial, Kidney Int. 5: 389-401, (1974).

74. Wilson, C.B. and Dixon, F.J.: The renal response to immuno-
     logical injury. In: The Kidney. Brenner, B.M. and Rector, F.C.
     Jr., eds, Vol. II, pp. 838-940 (W.B. Saunders Co., Philadelphia,
     1976).

75. Wilson, C.B. and Dixon, F.J.: Renal injury from immune reactions
     involving antigens in or of the kidney. In: Immunologic Mecha-
     nisms of Renal Disease. Wilson, C.B., Brenner, B.M. and Stein,
     J.H., eds., pp. 35-66 (Churchill Livingstone, New York, 1979).

MONITORING THE ALLOTRANSPLANT RECIPIENT

H. Kreis and J. Crosnier

Department of Nephrology
Hôpital Necker
Paris (France)

Since the first long term successful transplantation in the human in 1959, our group at the Necker Hospital has been concerned that a range of immunological tests, inspired by basic research into graft immunity, be introduced into clinical transplantation. We were thus the first in 1959 to concentrate our efforts on the immunologic selection of the donor and on discovering the means of assessing the degree of immunologic similarity between donor and recipient.

During the past ten years, our group, as well as many others, has proposed a number of immunologic tests for routine study of the immunologic response of the organ allograft recipient. It was tempting to extrapolate clinical tests from experimental models in order to follow the kinetics of the allograft recipient response. This would supplement clinical follow-up of human allograft recipients by immunologic surveillance of the various factors involved in rejection and would lead to more rational management of these patients.

However, it is not yet possible to follow the kinetics of all immunologic events responsible for rejection (25).

There are several reasons for the lack of total success in this area, and they are easily understandable:

a) The differing results observed by various groups using the same test are probably due to variations in techniques (differences in incubation time, the culture media, etc.) and to a lack of standardization of in vitro tests, which limits interpretations and comparison.

b) Episodes of renal failure following kidney transplantation are too often systematically thought to be of immunologic origin when no evident cause of renal failure is detected. Based on the results of renal biopsy, they are considered to be acute rejection. We know, however, that even in uncomplicated cases with normal renal function tests, allograft biopsy reveals latent changes in more than 50% of cases (23). This means that such alterations, particularly oedema and focal cellular infiltration into the interstitial tissue, may be found in patients with renal failure without being the cause of the failure. On the other hand, analyzing retrospectively 91 episodes of renal failure thought to be rejection crises, it was possible to suspect, a posteriori, a non-immunological factor responsible for or contributing to the occurrence of renal failure in 45 percent of these episodes (36). Undetected, these factors would strongly bias attempts to use immunological monitoring to forecast and even recognize rejection episodes.

c) Even when renal failure is actually the result of immunologic rejection, the highly complex events playing a role in the recipient response are intricately and, above all, variously interlocked. Thus, results of immunologic tests in vitro are not consistent in an overall population of patients, whereas their interpretation would perhaps be feasible in well selected patients (5).

d) While rejection processes occur within the graft itself, the great majority of immunological tests proposed are performed on blood samples. It is possible that the blood imperfectly reflects what actually occurs within the renal parenchyma.

e) Finally, the results of immunologic tests may be modified because they are performed on blood samples taken from patients treated by immunosuppressive drugs.

Despite these difficulties, every transplant team is concerned to carry out the necessary "Immunologic follow-up of the renal allograft recipient" (22), in order to permit:

a) Early detection and even prediction of rejection processes,

thus avoiding (by rapid institution of treatment) irreversible
alterations of the graft.

b) Constant adjustment of immunosuppressive therapy to the
immunological response of the recipient, allowing prescription of
the optimum dose rather than of the maximum tolerated dose.

c) Detection of patients with a state of tolerance, with specific non reactivity to graft antigens, in whom immunosuppression could
be withdrawn or drastically reduced. We have been able to discontinue
azathioprine in 14 patients with no deleterious effects but seven
others in the same condition experienced a rejection episode (8).
Similarly, in most of our HLA-identical recipients, although not
in all, we have been able to interrupt corticosteroids after the
second post-transplantation year (12).

RECIPIENT PRESENSITIZATION TO DONOR ANTIGENS

Presensitization to HLA-A, B and C antigens. First, Terasaki (60)
and Kissmeyer-Nielsen (34) clearly showed that the presence of preformed circulating antibodies to donor antigens in serum of an
allograft recipient, demonstrated by a positive cross-match, inevitably results in an immediate and irreversible rejection of the graft.
This remains an important fact in transplantation today, and a
negative cross-match, generally based on the microcytotoxicity technique (41), now is an important prerequisite for organ transplantation. In fact, very little is known today about the specificities
of these antibodies. They are usually thought to have anti-HLA-A, B
and C specificity, but it is almost certain that other systems,
whether a part of the HLA or not, are also involved. In 1976 and
1978 Ettenger et al. (18, 19) demonstrated that it was possible to
find patients with preformed antibodies against donor B lymphocytes
and not against T lymphocytes. When a graft was performed in such
cases, no hyperacute rejection was observed and the incidence of
graft failure was not higher than in cross-match negative donor
recipient pairs. Similar findings by other groups (39, 42, 47) could
suggest that when a positive cross match is found to be positive
against only donor B lymphocytes and not T lymphocytes, a renal
allograft might be safely carried out in most instances. The present situation however, is more complex:

1) Because the above hypothesis is far from being confirmed or

universally accepted, and many authors (1, 10, 58) have reported
either acute rejection or poorer graft survival in B cell cross-
match positive transplants, while the lack of such presensitization
is associated with stable graft function.

2) Because B cell antibodies are certainly not of a unique type.
Terasaki's group (30, 31) was the first to show that sera of prospective kidney transplant recipients (29) contain not only antibody
to HLA antigens but also antibody to non-HLA antigens. Thus, patients
who have antibodies reactive to B lymphocytes in the cold have a
higher graft survival rate than patients with either no preformed
antibody or with B or T warm antibodies. The graft survival rate in
the latter case is lower than that of patients with no cytotoxins.
The opinion of these authors is that warm B-cell cytotoxins are
probably not only directed against HLA-DR, but in many instances
against HLA-A, B and C specificities. In fact, HLA-A, B and C
antibodies that are too weak to react against T lymphocytes can
react against B lymphocytes only (31). Simple sera dilution can
discriminate between these two types of antibodies. It also appears
that antibodies which react against B lymphocytes in the cold do
not have HLA specificities and are not demonstrated after dilution
of the HLA antibodies. Cold antibodies reacting against B lymphocytes
are probably autoreactive antibodies.

Thus, antibodies responsible for a positive cross-match can be
characterized as being:

- T (or T + B) cell reactive and thus directed against
allogeneic HLA-A, B and C specificities, although T cell autoreactive
antibody can be found.

- B-cell reactive. They can then be directed at HLA-DR antigens
or at HLA-A, B and C antigens when they are weak, because of the
greater density of the HLA-A, B and C antigens on B cells than on T
cells. They can as well be present in the cold and be directed at
non-HLA antigens found specifically on B cells.

From all these rather confusing data two facts emerge:

1) that warm antibodies directed at donor T lymphocytes are
associated with rapid rejection of the graft, and 2) that a positive cross-match reaction due to autoreactive or cold anti-B-cell

antibodies is likely not to affect graft outcome. The influence of
other varieties of preformed anti-donor antibodies on graft survival,
especially warm anti-B-cell antibody, is still a matter of debate.

For practical purposes, it should be remembered that a strongly
positive cross-match reaction, even with 50 or 60% cell death, does
not always indicate the presence of anti-HLA-A, B or C antibody,
particularly when spleen cells are tested. Because of the present
policy of transfusing patients awaiting kidney transplantation, more
and more potential recipients have high concentrations of preformed
anti-B-cell antibody only. In these cases, a positive cross-match
reaction probably is not a definite contraindication to grafting.
In any case, the need to identify the precise type of anti-donor
antibodies in the serum of a potential recipient requires that cross-
match reaction be performed at least on T cells and on B cells, both
at 37° and in the cold.

Antibody Dependent Cell Mediated Cytotoxicity (ADCC). ADCC positive
cross-matches against a potential donor are frequent. Whether it is
a contraindication to transplantation is still debated. In fact,
the main question involved has to do with the specificity of the ADCC.
ADCC can be directed at HLA-A, B and C antigens or at B-cell antigens.
Adsorption of the serum with platelets from the donor thus appears
to be an essential step for precisely determining the specificity
against HLA-A, B and C antigens or against anti-Ia antigens. This
may partly explain the great discrepancy in results reported by
various groups. The Richmond group (42) reported a statistically
significant association between a positive pretransplant ADCC and
chronic rejection or early severe acute rejection. On the other
hand, the Johannesburg group (44) found no correlation between the
ADCC cross-match and graft rejection, and in Dossetor's experience,
platelet nonadsorbable ADCC was not detrimental to graft survival
(52). It thus appears that present data on ADCC are so conflicting
that the ADCC cross-match should not be taken into account in donor
selection.

Lymphocyte Mediated Cytotoxicity (LMC). The pretransplant detection
of LMC and its correlation with kidney transplant survival has been
studied by many groups. Almost all studies report a significant
correlation between a positive LMC cross-match and kidney graft loss
from rejection (44, 52). Results obtained by the various groups with
this assay were so similar that the conclusion of a session espe-

cially devoted to this matter at the Sixth Congress of The Transplantation Society was to recommend that the LMC cross-match be
included as a part of routine cross-matching (52). As this will
make the cross-match procedure much more complicated, further data
are still required before including this test in everyday practice.

Inhibition of the mixed lymphocyte reaction by recipient sera
was shown to correlate well with donor B-lymphocytotoxic reactivity
by Ettenger et al. (18). This suggests that antibodies against
B-lymphocytes thus detected were actually antibodies against LD
determinants. Unfortunately although these authors were able to
correlate the presence of MLR inhibition and favorable graft outcome,
it was found to be associated with a poor graft survival by Suciu-
Foca et al. (57) when assessed on a non-donor-specific panel.
Inhibition of MLR thus appears to be a complex phenomenon, often
involving non-antibody factors.

OTHER DONOR-SPECIFIC ASSAYS

The sera of potential recipients may contain IgG antibodies
directed against antigens on the endothelium of donor kidney arteries.
Cerilli et al. (6) reported a threefold increase in the number of
accelerated rejections in such cases over that in recipients without
endothelial antibodies. On a panel of endothelial cells isolated
from umbilical veins they showed that the presence of pre-existing
endothelial antibodies prior to grafting correlates more closely to
the occurrence of accelerated transplant rejection than does the
presence of anti-lymphocyte antibodies. Blood transfusion can induce
endothelial antibodies in the serum of potential recipients. They
are associated with accelerated acute rejection only when they are
directed against the donor kidney (48). These endothelial antibodies, also present in monocytes, are possibly involved in the
vascular lesions which are known to play the major role in kidney
allograft failure. Thus, cross-matching recipient serum with donor
monocytes may become a necessity.

Recipient Presensitisation to Non-Donor-Specific HLA Antigens. The
question of whether a previous immunization of the recipient against
non-donor-specific HLA antigens has any influence on graft survival
is still without a clear answer. Terasaki et al. (61) were the first
to show an unfavorable influence of recipient presensitization to
HLA antigens. A similar poor graft prognosis in recipients with

preformed antibodies was not observed by other groups except when
additional factors were considered. A retrospective analysis of 191
cadaveric renal transplantation performed in our group (13) shows
virtually no difference in the 2-year graft survival rate between
121 patients without pre-existing anti-HLA-A and B antibodies (58%)
and 70 preimmunized recipients (60%). However when compatible and
incompatible recipients are distinguished in each group, a striking difference appears between the two groups, i.e., presensitized
patients have a significantly lower graft survival rate when they
receive kidneys from incompatible donors (42% at 2 years) than from
compatible donors (79% at 2 years). This difference is not seen
among nonsensitized patients, where graft survival at 2 years is
similar in compatible (56%) and incompatible (61%) pairs. Similar
results observed by other authors (9, 45, 46, 69, 70) clearly
indicate that close HLA compatibility is of great importance in
patients having anti-HLA preformed antibodies but appears to be
unnecessary in nonpresensitized patients.

However, in retrospective studies the terms "presensitized
recipient" (or responder) and "nonpresensitized recipient" (or nonresponder) are based on the demonstration of the absence of cytotoxic
antibodies at any moment prior to transplantation. It is well known
today, however, that antibodies may develop after antigenic stimulation, such as blood transfusion, and then disappear, giving a false
negative result. Sera must thus be screened for antibodies at frequent intervals and after every blood transfusion. On the other
hand, in a retrospective study potential responders may be classed
as "nonresponders" if they have not received antigenic stimulation.
A "nonsensitized" recipient must not be confused with a "nonresponder", for the latter implies that adequate antigenic stimulation
has been provided. Thus, the "antibody negative" group of patients
reported in retrospective studies includes both nonsensitized and
true nonresponder patients. In order to clarify this situation, we
have undertaken a prospective study (still in progress) in which
all potential allograft recipients receive up to 10 units of blood,
unless they develop antibodies with less. Only those who do not
produce antibody after 10 units of blood are labelled "nonresponders".
Our preliminary results are in complete agreement with those of the
retrospective study. They re-emphasize the fact that matching for
HLA-A and B antigens does not influence posttransplantation outcome,
and they confirm that HLA compatible "responder" recipients have a
much better outcome than "nonresponder" recipients, whatever the

degree of compatibility.

POST TRANSPLANT IMMUNOLOGIC RESPONSE

Antibody mediated responses - Complement dependent cytotoxicity (CDC).
For all authors (11, 20, 55), CDC in the recipient's serum against
donor peripheral blood lymphocytes is associated with rejection.
However, there are many false positive and false negative results
(16), and the clinical usefulness of CDC for predicting rejection
is lessened by the high variability of its temporal relationship with
rejection (53).

Complement-dependent cytotoxicity against donor B cells. Whereas a
positive pretransplant B cell crossmatch does not appear to indicate
poor graft survival (17, 51), the advent of antidonor B cell antibodies in the post transplant period is frequently associated with,
but is not predictive of rejection (67).

Antibody-dependent cell-mediated cytotoxicity (ADCC). There is total
disagreement on the meaning of the presence of ADCC and its association or not with rejection (63) or tolerance (11). These discrep-
ances may be due to technical differences in the assays (16). The
Brigham group pointed out the favorable meaning of disappearance of
ADCC associated with treatment of rejection episodes by high doses
of corticosteroids and the unfavorable meaning of its persistence
in this case (20).

Anti-Fc receptor antibody. The Brigham group here observed a close
correlation between rejection and the occurrence of antidonor Fc
receptor antibodies detected by Ea rosette inhibition (59), but these
results have not been confirmed. The same group has found anti-Ia
antibodies by erythrocyte antibody rosette inhibition in the eluate
of transplanted kidneys nephrectomized after rejection (21).

CELL-MEDIATED IMMUNITY

Macrophage inhibition factor. The leukocyte migration inhibition
test, proposed by various authors in 1970 as a test of cell-mediated
immunity, was extensively used by our group. Using spleen extracts
from the cadaver donor (or even less specific antigens), we found
that these antigens inhibit the recipient leukocyte migration in
practically all rejection episodes when the test is performed before

the corticosteroid treatment (15). The test becomes positive before
any detectable clinical manifestations and is predictive of rejection
(15). Unfortunately, there are many false negative results (16),
perhaps due to lack of standardization of the technique.

Detection of transformed lymphocytes in vivo. Response to mitogens.
Morris, in the sheep (49) and our group in man (24) showed that
transformed lymphocytes, similar to those observed in vitro in mixed
lymphocyte cultures, may be seen in lymph collected from the renal
graft when manifestations of rejection are observed. It has been
suggested (28) that transformed lymphocytes can be detected in the
circulating blood by measuring tritiated thymidine incorporation
into mononuclear blood cells. Unfortunately, elevated levels of
incorporation are inconsistently associated with rejection (27, 28).
In addition , Dimitriu et al. (14), using autoradiography, found
that incorporating cells were not transformed lymphocytes but immature cells of the myeloid series and that the influx of these cells
might result from several factors, e.g. infection, or even the
administration of high doses of corticosteroids in treating rejection.

    Similarly, highly variable results are observed in the response
of recipient peripheral blood lymphocytes to various mitogens (16).

Rosette-forming cells. There is considerable disagreement on the
correlation between the number of T rosettes observed and the immunological responses of the recipient. For some authors, changes in
T cell number are associated with rejection and may even be predictive (62), but others have not found this association (4). Our
group (68) has shown that the number of rosette-forming cells is
generally low in rejection but that in some cases it may rise. This
inconsistency may perhaps reflect the reactions of various cell
populations (32).

Mixed lymphocyte culture (MLC) and cell-mediated lymphocytotoxicity
(CML). All studies, in animals and in man, conclude that after
transplantation an MLC nonreactivity consistently occurs, both
during rejection and during quiescence (3, 26, 71). During rejection, this nonreactivity has been explained by a possible sequestration of responder cells in the graft (26) and during quiescence
it has been considered to reflect immunological tolerance (3).
However, when the mixed lymphocyte reaction is studied after a 48-

hour interruption of immunosuppressive treatment, specific depression
of recipient cell reactivity is less consistent.

More recently, several groups have observed that the CML production specific to donor cells was depressed except during rejection
(71).

Finally suppressor cells have been reported in association with
good graft function by two groups (37, 66) who observed that suppression by recipient lymphocytes of third-party cell CML production
specific to the donor is seen in long-term graft survival.

Lymphocyte-mediated cytotoxicity (LMC). For groups that have experience with this test (16, 20, 35, 40, 55), there is a very good
correlation between positive LMC and acute rejection. They report
that in 70% of cases LMC is positive prior to rejection and thus has
a predictive value (55). Moreover, the effect of rejection therapy
on the results of these tests would have a prognostic value because
in those instances in which the LMC became negative post treatment,
graft survival was better than in those patients who remained LMC
positive.

Technical problems (incubation time, types of cells used as
targets) can be responsible for "false positive" results.

Assessment of immunosuppressive effectiveness. In order to avoid
excessive depression of mechanisms responsible for defence against
infection, it is necessary to adjust the immunosuppressive dose to
the needs of each individual recipient. To accomplish this, reliable
parameters are required.

For most groups (5, 7, 65) the determination of E rosette
numbers, first proposed by J. F. Bach (2), is the most informative
test for adjusting azathioprine and ATG dosage. Stinnet (56) proposed
a monitoring protocol including T cell number determination, PHA
responsiveness, study of bactericidal activity by neutrophils for
various bacteria and serum opsonic activity, to predict and then
perhaps to avoid infectious complications in the high risk patients
treated by immunosuppressive drugs.

There is no way of measuring the immunosuppressive effect of
a specific dose of steroids, and the immunosuppressive activity of

corticosteroids is not related to the level of the drug in the blood
(50). The biological significance of serum levels must be considered
in the light of steroid binding to carrier proteins, of in vivo half-
life of the steroid used and of modifications in its catabolism. In
the future, however, the various aspects of individual variations
for a given dosage may be covered by using individual steroid blood
level determination with the available radioimmunoassays and the
evaluation of the density of lymphocyte receptors for steroids (38).

Detection of specific nonreactivity to graft antigens. We have seen
above that, in a limited number of patients, we did not observe any
alteration of renal function in 66% of patients after prolonged
withdrawal of azathioprine, but that we were not able to predict
which patients would tolerate this withdrawal and which would undergo
rejection (8). Several tests have been proposed to detect tolerance –
the search for ADCC (11) and for complement-dependent anti-B cell
antibodies (43), inhibition of CML (3, 37, 66), T rosette number
(33) – but none of these assays has yet proved to be sufficiently
reliable to allow innocous withdrawal of all immunosuppressive therapy
in patients with good clinical allograft tolerance.

REFERENCES

1. Albrechtsen, D., Arnesen, E., Solheim, B.G. and Thorsby, E.:
    Significance of HLA-DR matching of B cell cross-match tests in
    vitro and in cadaver renal transplantation. Transplant. Proç.
    9: 743, (1979).
2. Bach, J.F., Dardenne, M. and Fournier, C.: In vitro evaluation
    of immunosuppressive drugs. Nature. 222: 998, (1969).
3. Bach, M.L., Engstrom, M.A., Bach, F.M., Etheredge, E.E. and
    Najariam, J.G.: Specific tolerance in human kidney allograft
    recipients. Cell. Immunol. 3: 161, (1972).
4. Buckingham, J.M., Ritts, R.E., Woods, J.E. and Ilstrup, D.M.:
    An assessment of cell-mediated immunity in acute allograft
    rejection in man. A prospective study. Mayo Clin. Proc. 52:
    101, (1977).
5. Byfield, P.E., Barth, C.L., Brodie, J.A., Cosimi, A.B., Dienst,
    S.G., Elberg, A.J., Elfing, G.L., Hardy, M.A., Hayashi, H.,
    Kountz, S.L., Lamborn, K.R., Loughman, B.E., Sakai A., Satoh,
    P.S., Zielinski, C.M. and Wechter, W.J.: The use of circulating
    rosette forming cells as a guide to determination of ATG dose
    (Dose by rosette). Transplant. Proc. 10: 627, (1978).
6. Cerilli, J., Holliday, J.E., Dawne, B.S, Fesperman, B.A. et al.:
    Antivascular endothelial cell antibody – Its role in trans-
    plantation. Surgery. 81: 132, (1977).
7. Cosimi, A.B., Delmarico, F.L., Burdick, J.F. and Russel, P.S.:
    Individualized management of immunosuppression according to
    serial monitoring of immunocompetence. Transplant. Proc. 10:
    647, (1978).
8. Dandavino, R., Trunet, P., Descamps, B. and Kreis, H.: Prolonged
    withdrawal of Azathioprine in kidney transplantation. Transplant.
    Proc. 10: 655,(1978).
9. Dausset, J., Hors, J., Busson, M., Festenstein, H., Oliver,
    R.T.D., Paris, A.M.I. and Sachs, J.A.: Serologically defined
    HL-A antigens and long term survival of cadaver kidney trans-
    plants.New Engl. J. Med. 18: 979, (1974).
10. Dejelo, C.L. and Williams, T.C.: B. Cell cross-match in renal
    transplantation. Lancet. 2: 241, (1977).
11. Descamps, B., Gagnon,R., Debray-Sachs, M., Barbanel, C. and
    Crosnier, J.: Lymphocyte dependent and complement dependent
    antibodies in human renal allograft recipients. Transplant.
    Proc. 7: 635, (1975).
12. Descamps, B., Hinglais, M. and Crosnier, J.: Renal transplant-

ation between 33 HL-A identical siblings. Transplant. Proc. 5: 231, (1973).

13. Descamps, B. N'Guyen, A.T. and Kreis, H.: New insights into immunologic selection of human cadaver renal allograft recipients based on immune response capacity criteria. Transplant. Proc. 10: 497, (1978).

14. Dimitriu, A., Debray-Sachs, M., Descamps, B., Sultan, C. and Hamburger, J.: Tritiated thymidine incorporation in blood leukocytes of renal allograft recipients. Transplant. Proc. 3: 1577, (1971).

15. Dormont, J., Sobel, A., Galanaud, P., Crevon, M.C. and Colombani, J.: Leukocyte migration inhibition with spleen extracts and other antigens in patients with renal allografts. Transplant. Proc. 4: 265, (1972).

16. Dossetor, J.B. and Myburgh, J.A.: Posttransplant immunologic monitoring summation. Transplant. Proc. 10: 661, (1978).

17. Ettenger, R.B., Opelz, G., Walker, J., Terasaki, P.I., Malekzadeh, M.K., Pennisi, A.J., Uittenbogaart, C.M. and Fine, R.N.: Donor-specific preformed lymphocyte antibodies and mixed lymphocyte culture (MLC). Blocking in cadaver renal allograft recipients. Transplant. Proc. 10: 479, (1978).

18. Ettenger, R.B., Opelz, G., Walker, J., Terasaki, P.I., Uittenbogaart, C., Pennisi, A.J., Malekzadeh, M.H. and Fine, R.N.: Antibodies to donor B lymphocytes and mixed lymphocyte culture blocking in cadaveric renal transplantation. Transplantation. 25: 169, (1978).

19. Ettenger, R.B., Terasaki, P.I., Opelz, G., Malekzadeh, M., Pennisi, A.J., Uittenbogaart, C. and Fine, R.: Successful renal allograft across a positive cross-match for donor B-lymphocyte alloantigens. Lancet. 2: 56, (1976).

20. Gailiunas, P., Suthanthiran, M., Person, A., Strom, T.B., Carpenter, C.B. and Garovoy, M.R.: Posttransplant immunologic monitoring of the renal allograft recipient. Transplant. Proc. 10: 609, (1978).

21. Garovoy, M.R., Suthanthiran,M., Gailiunas, P., Carpenter, C.B., Graves, M., Busch, G. and Tilney, N.L.: Anti-Ia antibody eluted from rejected human renal allograft. Transplant. Proc. 10: 613, (1978).

22. Hamburger, J.: Immunologic follow-up of renal allograft recipients. Transplant. Proc. 4: 669, (1972).

23. Hamburger, J., Crosnier, J., Dormont, J. and Bach, J.F.: Renal Transplantation Theory and Practice. Williams and Wilkins,

Baltimore. 1972, p. 26.

24. Hamburger, J. Dimitriu, A., Bankir, L., Debray-Sachs, M. and Auvert, J.: Collection of lymph from kidneys homotransplanted in man: cell transformation in vivo. Nature. 232: 633, (1971).

25. Hamburger, J., Vaysse, J., Crosnier, J., Tubiana, M., Lalanne, C.M., Antoine, B., Auvert, J., Soulier, J., Dormont, J., Salomon, C., Maisonnet, M. and Amiel, J.L.: Transplantation d'un rein entre jumeaux non monozygotes après irradiation du receveur. Bon fonctionnement au quatrième mois. Presse. Med. 67: 1771, (1959).

26. Hattler, B.G. and Miller, J.: Changes in human mixed lymphocyte culture reactivity as an indicator of kidney rejection. Transplant. Proc. 4: 655, (1972).

27. Hersh, E.M., Butler, W.T., Rossen, R.D. and Morgan, R.O.: Lymphocyte activation: a rapid test to predict allograft rejection. Nature. 226: 757, (1970).

28. Hersh, E.M., Butler, W.T., Rossen, R.D., Morgan, R.O. and Saki, W.: In vitro studies of the human response to organ allografts: appearance and detection of circulating activation lymphocytes. J. Immunol. 107: 571, (1971).

29. Iwaki, Y., Tarasaki, P.I., Park, M.S. and Billing, R.: Enhancement of human kidney allograft by cold B lymphocyte cytotoxins. Lancet. 1: 1228, (1978).

30. Iwaki, Y., Terasaki, P.I., Park, M.S., Hientz, R., Silberman, H. and Berne, T.: Dilutions and specificity analysis of pretransplant sera. Transplant. Proc. 11: 944, (1979).

31. Iwaki, Y., Terasaki, P.I., Weil, R., Koep, L. and Starzl, T.: Retrospective tests of B cold lymphocytotoxins and transplant survival at a single center. Transplant. Proc. 11: 941, (1979).

32. Kerman, R.H. and Geis, W.P.: T-RFC monitoring of CMI events in renal allograft recipients. Transplant. Proc. 10: 633, (1978).

33. Kerman, R.H., Ing, T.S., Hano, J.F. and Geis, W.P.: Predictive value of active T-RFC in renal allograft survival. Transplant. Proc. 10: 637, (1978).

34. Kissmeyer-Nielsen F., Olsens, S., Petersen, V.P. and Fjelborg, O.: Hyperacute rejection of kidney allografts associated with pre-existing humoral antibodies against donor cells. Lancet. 2: 662, (1966).

35. Kovithavongs, T., Schlaut, J., Pazderka, V., Lao,V., Pazderka, F., Bettcher, K.B. and Dossetor, J.B.: Posttransplant immunologic monitoring with special consideration of technique and interpretation of LMC. Transplant. Proc. 10: 547, (1978).

36. Kreis, H., Noel, L.H., Chailley, J., Lacombe, M., Descamp, J.M.
     and Crosnier, J.: Kidney graft rejection: Has the need for
     steroids to be re-evaluated.Lancet. 2: 1169, (1978).

37. Liburd, E.M., Pazderka, V., Kovithavongs, T. and Dossetor, J.B.:
     Evidence for suppressor cells and reduced CML induction by the
     donor in transplant patients. Transplant. Proc. 10: 557, (1978).

38. Lippman, M.E., Halterman, R.H., Leventhal, B.G., Perry, S. and
     Thompson, E.B.: Glucocorticoid-binding proteins in human acute
     lymphoblastic leukemic blast cells. J. Clin. Invest. 52: 1715,
     (1973).

39. Lobo, P.I., Westervelt, F.B. and Rudolf, L.E.: Kidney trans-
     plantability across a positive cross-match. Cross match assays
     and distribution of B lymphocytes in donor tissues. Lancet. 1:
     225, (1977).

40. McConnachie, P.R., Finch, W.T. and Birtch, A.G.: Monitoring and
     predicting rejection. Transplant. Proc. 10: 543, (1978).

41. Mittal, K.K., Mickey, M.R., Singal, D.P. and Terasaki, P.I.:
     Refinement of microdroplet lymphocyte cytotoxicity test. Trans-
     plantation. 6: 913, (1968).

42. Morris, J.P.: Histocompability antigens in human organ trans-
     plantation. Surgical Clinics of North America. 58: 233, (1978).

43. Myburgh, J.A. and Smit, J.A.: B-cell antibodies and clinical
     kidney transplantation. Transplant. Proc. 10: 529, (1978).

44. Myburgh, J.A. and Smit, J.A.: Pre-transplant lymphocyte mediated
     cytotoxicity (LMC) and antibody dependent cell-mediated
     cytotoxicity (ADCC) in kidney transplantation. Transplant.
     Proc. 10: 425, (1978).

45. Oliver, R.T.D., Sachs, J.A., Festenstein, H., Pegrum, G. and
     Moorhead, J.F.: The influence of HL-A matching, antigenic
     strenght and immune responsiveness on the outcome of 349 ca-
     daver renal grafts. Lancet. 2: 1381, (1972).

46. Opelz, G., Mickey, M.R. and Terasaki, P.I.: HLA matching and
     cadaver kidney transplant survival in North America. Influence
     of center variation and presensitisation. Transplantation. 23:
     490, (1977).

47. Parks, M.S., Terasaki, P.I. and Bernoco, D.: Autoantibody
     against B lymphocytes. Lancet. 2: 465, (1977).

48. Paul, L.: Antibodies directed against donor antigens in human
     renal allograft, Thesis, Drukkeris, J.H., Pasmans, Gravenhage,
     1979, p. 140.

49. Pedersen, N. and Morris, B.: The role of the lymphatic system
     in the rejection of homografts: a study of lymph from renal

transplants, J. Exp. Med. 131: 936, (1970).

50. Sells, R.A., Brookes, L., Baser, P. and Whitemore, D.: Methyl-
     prednisolone blood levels in cadaveric renal allografts recip-
     ient. Transplant. Proc. 10: 651, (1978).

51. Soulillou, J.P., Peyrat, M.A. and Guenel, J.: Studies of the
     antibodies against HLA, Ia-like, FC and/or C3 receptors present
     in retransplant sera: Anti B cell antibodies not associated
     with accelerated graft loss. Transplant. Proc. 10: 475, (1978).

52. Stiller, C.R., Dossetor, J.B., Carpenter, C.B. and Myburgh, J.A.:
     Immunologic monitoring of the transplant recipient. Transplant.
     Proc. 9: 1245, (1977).

53. Stiller, C.R. and Sinclair, N.R. Stc.: Monitoring of rejection.
     Transplant. Proc. 11: 343, (1979).

54. Stiller, C.R., Sinclair, N.R. Stc., Abrahams, S., McGirr, D.,
     Singh, H., Howson, W.T. and Ulan, R.A.: Anti-donor immune
     responses in prediction of transplant rejection. New Engl. J.
     Med. 294: 978, (1976).

55. Stiller, C.R., Sinclair, N.R. Stc, McGirr, D., Jevnikar, A.
     and Lkan, R.A.: Diagnostic and prognostic value of donor spe-
     cific posttransplant immune response clinical correlates and
     in vitro variables. Transplant. Proc. 10: 525, (1978).

56. Stinnett,J.D., Alexander, J.W., Ogle, C.K., McClellan, M.A.,
     Brody, D. and First, R.: Immune function of kidney transplant
     patients in relation to infection. Transplant. Proc. 10: 639,
     (1978).

57. Suciu-Foca, N., Hardy, M., Kiamar, M., Weiner, J., Susinno, E.,
     Molinaro, A.P., Jacob, J. and Reemtsma, K.: MLC inhibitory
     alloantibodies and outcome of kidney allografts. Transplant.
     Proc. Vol. 10, 2: 493, (1978).

58. Suthanthiran, M., Gailiunas, P., ST Louis, G., Fagan, G.,
     Carpenter, C.B. and Garovoy, M.R.: Presensitization to donor
     B-cell ("Ia") antigens associated with early allograft failure.
     Transplant. Proc. 9: 1807, (1977).

59. Suthanthiran, M., Gailiunas P., Fagan, G., Strom, T.B.,
     Carpenter, C.B. and Garovoy, M.R.: Detection of anti-donor "Ia"
     antibodies: A strong correlate of rejection. Transplant. Proc.
     10: 639, (1978).

60. Terasaki, P.I., Marchioro, TKL and Starzl, T.E.: First finding
     of preformed cytotoxic antibodies against donor cells. In:
     Histocompatibility Testing, 1965, Washington, D.C.: National
     Academy of Sciences, 1965, p. 83.

61. Terasaki, P.I., Mickey, M.R. and Kreisler, M.: Presensitization

and kidney transplant failures. Postgrad. Med. J. 47: 89, (1971).

62. Thomas, F., Lee, H.M., Wolf, J.S., Mendez-Picon, G. and Thomas, J.: Monitoring and modulation of immune activity in human transplant recipients. Surgery. 79: 408, (1976).

63. Thomas, F., Mendez-Picon, G., Thomas, J., Lee, H.M. and Lower, R.: Effective monitoring and modulation of recipient immune reactivity to prevent rejection in early posttransplant period. Transplant. Proc. 10: 537, (1978).

64. Thomas, F., Thomas, J., Mendez, G. and Lee, H.M.: Pretransplant immune monitoring of donor-recipient compatibility. Transplant. Proc. 10: 429, (1978).

65. Thomas, F., Thomas, J., Mendez, G., Lee, H.M. and Lower, R.: Individualization of recipient immunosuppression by use of in vitro monitoring parameters. Transplant. Proc. 10: 621, (1978).

66. Thomas, J., Thomas, F., Johns, C. and Lee, H.M.: Consideration in immunologic monitoring of long-term transplant recipients. Transplant. Proc. 10: 569, (1978).

67. Ting, A. and Morris, P.J.: Pre and Posttransplant B-cell antibodies in renal transplantation. Transplant. Proc. 11: 393, (1979).

68. Tursz, T., Fournier,C., Kreis, H., Crosnier, J. and Bach, J.F.: T-lymphocytes in kidney allograft recipients. Br. Med. J. 1: 799, (1976).

69. Van Hoof, J.F., Shippers, H.M.A., Van Der Steen, G.J. and Van Rood, J.J.: Efficacy of HL-A matching in Eurotransplant. Lancet. 2: 1385, (1972).

70. Van Hoof, J.P., Hendriks, G.F.J., Schippers, H.M.A. and Van Hoof, J.J.: Influence of possible HL-A haploidentity on renal graft survival in Eurotransplant. Lancet. 1: 1130, (1974).

71. Wonigeit, K. and Pichlmayr, R.: Posttransplant monitoring of donor-specific T-cell reactivity at the precursor cell level. Transplant. Proc. 10: 563, (1978).

THE SELECTION AND INVESTIGATION OF CANDIDATES FOR RENAL

TRANSPLANTATION

V. Bonomini and A. Vangelista

Nephrology and Dialysis Department
S. Orsola University Hospital
Via Massarenti, 9
40100 Bologna (Italy)

ABSTRACT

Though renal transplantation is steadily increasing and is often considered the ideal therapy for chronic renal failure, there are many side effects, particularly when the operation is performed using cadaver donors.

A rational clinical and laboratory approach and an accurate selection of the candidates seem to greatly reduce the risk of early and later complications.

In our experience immunologic, clinical and dialytic criteria are the most important parameters in evaluating patient candidacy for inclusion in kidney transplantation programme.

Immunological

The donor-recipient histocompatibility plays a significant role only in living donor transplant, while conflicting results on kidney survival rates were obtained in the cadaver transplant, with a negative cross-match.

Clinical

An accurate clinical evaluation significantly affects the
results of either cadaver or living transplantation, regardless of
histocompatibility.

In our experience the main parameters of importance in patient
selection are: age, the renal ailment leading to functional failure,
the presence of active infection or systemic disease, abnormality
of urinary tract, gastric or duodenal ulcer, and hepatic insufficiency.

Dialytic

A careful, adequate and thorough approach to dialysis, i.e.
early dialysis, prevention of complications, such as anaemia,
peripheral neuropathy, osteodystrophy, alterations of glucose and
lipid metabolism, vascular calcification, and cardiovascular disease,
dramatically improve the patient's clinical status and social rehabilitation. It also affects the survival rate of transplanted
patients.

———

Renal transplantation is considered the ideal therapy for
chronic renal failure and its employment is becoming more and more
frequent. The number of clinical complications, however, particularly in cadaver transplantation, is still too high to consider renal
transplantation the final solution to chronic uremia.

A more appropriate clinical and laboratory evaluation of patients and more objective criteria for selection of recipients
appear to be of the utmost importance in reducing the risk of
early and late complications and in obtaining the best clinical
results in terms of survival and rehabilitation.

Criteria for selection, however, continue to vary and some
factors which were considered of decisive importance up to a few
years ago, have been modified, making satisfactory clinical results
achievable today even in patients who whould once have been excluded
from a kidney transplantation programme.

Table 1. "St. Orsola" pre-transplant check-list.

Hystocompatibility typing

Complete clinical and laboratory examination including:

- assessment of renal status;
- urological evaluation;
- bacteriological work-up;
- cardiovascular status;
- gastrointestinal evaluation;
- psychological assessment;
- dialysis status.

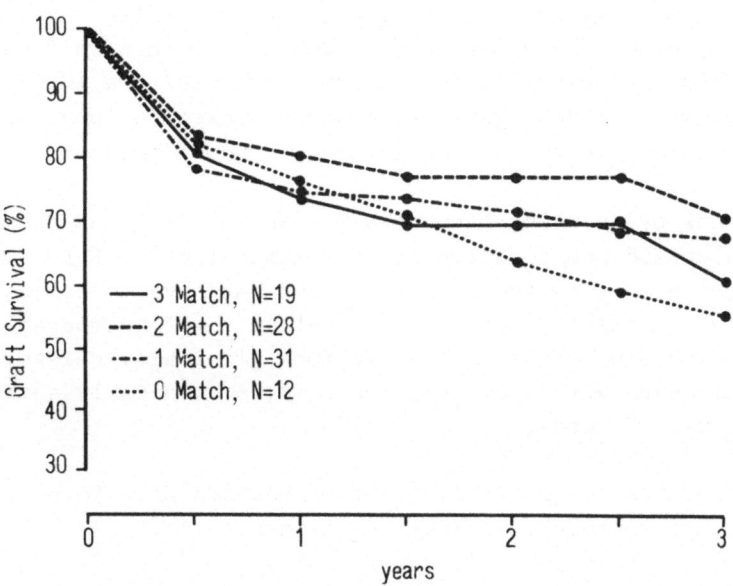

Fig. 1. HLA matching and graft survival in cadaver renal
transplantation.

From our personal experience (137 renal transplantations, 47 from living and 90 from cadaver donor) as well as from reports in the literature, some parameters continue to be considered of decisive importance for patient selection:

- Absence of cytotoxic antibodies.
- Age (patients less than 5 and more than 60 are generally excluded).
- Irreversible renal failure.
- Normal lower urinary tract.
- Absence of extrarenal complications, malignancy, active infection, severe malnutrition, pancytopenia.

Table 1 shows the main parameters we follow in evaluating patient candidacy for inclusion in Kidney Transplantation programme.

Even today opinions still differ as to the influence of histocompatibility on long-term results, especially in cadaver renal transplantation, at least as far as present laboratory techniques (HLA matching) are concerned. Some Authors have found a good correlation between HLA matching and graft outcome (5); others deny its importance (6). In our experience, in cadaver renal transplantation there is no correlation between graft survival after 3 years (Fig.1) and the number of HLA antigens mismatched. A more significant correlation seems to derive from mixed lymphocyte culture (MLC) results (locus D): these investigations, however, take too long (about 5 days) to be routinely employed in cadaver renal transplantation.

Judging by the long-term results, we can say that HLA matching cannot be considered decisive for patient selection for, even where good compatibility is found, a recipient may have to be excluded as clinically unsuitable; again, among patients with comparable histocompatibility, one might choose the more clinically reliable recipient; while even with poor compatibility, a clinically satisfactory candidate may be accepted.

Besides histocompatibility, other factors have to be considered in an attempt to improve transplantation results:

- Immunological manipulation.
- Nutrition.
- Surgery.

- Dialysis strategy.
- Regular assessment.

IMMUNOLOGICAL MANIPULATION

The measures employed today are essentially the formation of
a thoracic duct fistula and the recipient pre-treatment. These
techniques, however, can be adopted only in living donor transplantation, and the use of more specific immunological manipulation
(i.e. specific cellular suppression) is today mainly experimental
and not yet clinically applicable.

As regards immunological manipulation of patients waiting for
transplantation, mention must be made of the role which blood
transfusion has been claimed to play in the long-term results.
While its use was frowned on until a few years ago, owing to an
alleged adverse effect on graft survival, a good correlation has
recently been found between the total number of blood transfusions
performed and the graft outcome (4). However, in our experience the
number of blood transfusions prior to transplantation seems to have
no significant effect on patient and graft survival.

NUTRITION

Among the other factors to be considered prior to inclusion in
a transplantation programme, "Nutrition" seems to be of clinical
importance, both in terms of the patient's nutritional status and
of dietary measures (2), to prevent and/or to correct metabolic
alterations which may induce high-risk conditions before and after
transplantation. Hyperlipemia, impaired glucose utilization and
hyperparathyroidism, for example, may be considered as major high-
risk factors for vascular calcification and accelerated athero-
sclerosis in uremia. If present during dialysis, these conditions
may persist and even progress after transplantation.

SURGERY

Various surgical problems may exist in patient awaiting
transplantation and it often becomes essential to solve them before

transplantation takes place, in order to improve the long-term
results. In our experience the following surgical measures are
clinically important:

- Correction of lower urinary tract abnormalities (vescico-
  ureteral reflux and bladder neck obstruction).

- Bilateral nephrectomy in patients with drug-resistant
  malignant hypertension, complicated large polycystic kidneys,
  and severe and persistent urinary tract infection.

- Parathyroidectomy in the presence of severe hyperparathyroidism, which in some cases may persist even after successful transplantation.

- Correction of peptic ulcer and other surgical problems which
  may worsen during immunosuppressive therapy.

DIALYSIS STRATEGY

     The relationship between dialysis strategy and post-transplant
results are yet to be defined. However, from our experience, some
basic statements can be made. No correlation seems to exist between
duration of dialysis and development of cytotoxic antibodies and
it is likely that dialysis duration per se has no influence on the
number of rejection crises and on the final outcome of the graft.
On the contrary, the duration and, above all, the strategy of dialysis, might have other effects on the patient's clinical status,
since with prolonged standard dialysis some complications may
develop (worsening in subclinical uremic changes, infections and,
above all, accelerated vascular disease, sometimes to such a marked
degree as to raise doubts about the patient's acceptability for
transplantation. Standard Late Dialysis started with a low residual
creatinine clearance (Ccr) of less than 5 ml/min or after protracted
low protein diet may be associated with a significant survival
rate; in survivors, however, systemic uremic changes (anemia,
peripheral neuropathy, osteodystrophy, alterations of glucose and
lipid metabolism, and vascular calcifications) are not reversed,
but progress year after year, resulting in an "intrisic high-risk
condition" both in dialysis and in transplantation.

An Early Dialysis approach (1), i.e. dialysis started with a
relatively high residual Ccr of 10-15 ml/min, is associated with
moderate or minimal uremic changes. Better clinical and metabolic
rehabilitation occurs and persists with time, even after the
residual renal function has reached very low levels and more satisfactory results are achievable, as far as the patient's clinical
condition is concerned, both in dialysis and after transplantation.

REGULAR ASSESSMENT

Before putting the patient on a waiting list for transplantation,
it is of decisive importance to perform repeated clinical and laboratory checks for any clinical and subclinical alteration. Investigations include:

- Hematological parameters (white cell count, hemoglobin, platelet
count, reticulocytes);

- Immunological parameters (serum immunoglobulins and complement
fractions, circulating immunocomplexes, anti-GBM antibodies, antinuclear antibodies, erythrosedimentation rate, ASO titer);

- Renal function indexes (BUN, serum creatinine and uric acid,
electrolyte balance);

- Hepatic function investigations (SGOT, SGPT, alkaline phosphatase, gamma GT enzyme, bilirubin, hepatitis antigens and antibodies);

- Metabolic studies (arterial and venous glucose, aminoacids,
lipid prophyle, Ca/P equilibrium);

- Hormonal equilibrium including renal and extrarenal hormones;

- Radiological evaluation of gastrointestinal tract, lower urinary
tract, bones, cardiovascular and respiratory apparatus;

- Bacteriological and virological investigations in order to
ascertain the presence of clinical and subclinical infection which
often becomes evident and/or worsens after transplantation, because
of immunosuppressive therapy.

Among the various clinical factors to be considered, <u>the nature
of renal disease</u> is of great importance, even if opinions appear
to have been modified today concerning absolute or relative contraindications to transplantation. Various immunological renal diseases
may recur after transplantation; above all: membrano-proliferative
glomerulonephritis (60-100% according to the type of lesions);
anti-GBM antibody glomerulonephritis (about 50% of cases); focal
and segmental glomerulosclerosis (20%); IgA nephropathy (20-50%).
However it must be remembered that recurrence of renal lesions does
not necessarily mean graft failure and this seems to be confirmed
by the survival rate 1-2 years after surgery, which does not show
any significant difference among the various groups of renal disease,
at least in cadaver transplantation. Furthermore, some systemic
(lupus erythematosus, amyloidosis, diabetes) and congenital metabolic
disorders (gout, cystinosis, Fabry's disease) are no longer regarded
as absolute contraindications to transplantation; even considering
the more likely recurrence of the disease in the transplanted kidney.
The relatively prolonged good function of the graft in a significant
percentage of cases supports this approach at least in selected
patients. The only absolute contraindication is oxalosis, in the
presence of which both dialysis and transplantation have been
reported to be unsuccessful. Apart from oxalosis some other conditions
are generally considered as <u>absolute contraindications</u> to transplantation today:

     - active infection wich cannot be eradicated;
     - malignancy which cannot be brought under control;
     - life-threatening extrarenal disease;
     - severe psychiatric problems.

In our experience a proper consideration of the patients is of
the utmost importance in obtaining better long-term results in
transplantation. Survival and clinico-metabolic rehabilitation
prove to be influenced by a careful evaluation of the following
important factors, before accepting the patient for transplantation:
nutritional status, vascular problems, psychological assessment
and dialysis strategy.

From personal experience and the results from the literature,
the <u>ideal candidate</u> for transplantation may be considered to fulfill
several criteria. He or she should be young, with non-immunological
renal disease, a normal lower urinary tract, should have received

blood transfusion as required and have no cytotoxic antibodies.
He or she should also have a short low protein diet-time pre-dialysis,
a short waiting-time on dialysis and psychological and socio-
economic stability.

Finally "if he ends up at a center which has a low mortality
rate, he will definitely have found the winning combination" (3).

REFERENCES

1. Bonomini, V.: Early Dialysis 1979. Nephron 24: 157-160, (1979).
2. Bonomini, V., Scolari, P., Stefoni, S., Vangelista, A.: Athero-
    sclerosis in uremia: A longitudinal study. Amer. J. Clin. Nutr.
    33: 1493-1498, (1980).
3. Guttmann, R.D.: Pretransplant evaluation and treatment of
    donors and recipients. Dial. and Transpl. 7: 118-127, (1978).
4. Opelz, G. and Terasaki, P.I.: Blood transfusion in hemodyalisis
    unit: yes or not? Dial. and Transpl. 6: 46-53, (1977).
5. Opelz, G. and Terasaki, I.P.: Cadaver kidney transplantation
    in North America: Analysis 1978. Dial. and Transpl. 8: 167-
    172, (1979).
6. Salvatierra, O.: Renal transplantation in perspective. Dial.
    and Transpl. 7: 172-182, (1978).

THE IMMUNOLOGICAL BASIS OF THE TREATMENT OF GLOMERULONEPHRITIS

J.F. Bach and L. Chatenoud

Inserm U 25
Hôpital Necker
161, rue de Sèvres
75730 Paris, Cedex 15 (France)

INTRODUCTION

It is generally admitted that human idiopathic glomerulo-
nephritides are due to the pathogenetic effect of immune complexes.
The formation of the immune complexes implies that an antigen has
induced the formation of antibodies in adequate amount and with a
certain (yet undefined) quality. The participation of cell-mediated
immunity has also been recently suggested (2, 5).

Two (non-mutually exclusive) main etiopathogenic approaches
may be considered involving either the introduction of a particularly nephritogenic antigen or the abnormal immune response to a
environmental and frequently met antigen. In the latter case patients with glomerulonephritis (GN) present an abnormal and probably genetically controlled background. In fact, there are some data
in the literature indicating anomalies of the immune system that
are not directly explained by the formation of immune complexes:

- functional abnormalities of T cells in the nephrotic syndrome
  (8);

- deficiency of suppressor T cells in membranous GN and Berger's
  disease;

- production of a factor increasing vascular permeability in
  various forms of GN (9);

- abnormal level of IgA-bearing cells in Berger's disease (7);

- hereditary deficiency of some complement components in membranoproliferative GN.

Recently, using monoclonal anti-T cell antibodies, produced
by hybridomas (6), we have shown that extra membranous GN and
Berger's disease were statistically associated with a decrease in
the level of circulating suppressor T cells (4).

In view of this dual mechanism, one may think either of a
specific or of a non antigen specific therapeutic approach to GN.
In the first case, one may attempt to remove the antigen or to
deviate the immune response to this antigen. In the second case, a
non antigen specific approach may consist in the attempt to suppress
immune responses, irrespective of their targets. Another possibility
for a non-specific approach, more stimulating but even more difficult, might be focused on the attempt at correcting the basic
immunological abnormality which is at the origin of the disease.

We shall successively consider the different levels at which
the immunological therapy may apply from the peripheral lesion to
the very initial event which contributes to the triggering of the
pathologic immune response.

PERIPHERAL ACTION

The immune complexes deposit in the glomeruli and induce inflammatory lesions, in the creation of which the complement system,
various mediators of immediate hypersensitivity and perhaps coagulation factors play an important role. Several therapies are used
with the aim of interfering with the action of these molecules or
of the cells that they activate (e.g. the polymorph) which in turn
will release pharmacologically active molecules. Let us mention
here anticoagulants, anti-platelets aggregating agents (Persantin),
anti-complement agents (cobra venom factor) and anti-inflammatory
agents (such as indomethacin or aspirin). Steroids which are essentially given as immunosuppressive agents might also act at this

level.

REMOVAL OF ANTIBODIES

    Plasmapheresis has been used in the treatment of numerous
cases of GN. The idea is to remove circulating antibodies or immune
complex-induced GN. Positive results have been published both in
autoantibody mediated GN (Goodpasture syndrome) and in IC mediated
GN. One should realize however that plasmapheresis has many other
effects (removal of non immunoglobulinic factors, addition of extraneous factors). Perhaps the future of immunological depletion
lies in the use of specific immunoadsorbant (coupled with antigen,
antibody, or Clq) that would specifically remove the GN-inducing
molecules.

PREVENTING ANTIBODY FORMATION

    Numerous drugs are known to suppress immune responses in
animals. Fewer are known to have such properties in man, i.e. steroids
(although steroids are not strong immunosuppressive agents),
6-mercaptopurine or its imidazole derivate azathioprine, cyclophosphamide and chlorambucil, two alkylating agents, and metho-
trexate. Antilymphocyte sera and cyclosporin A are essentially
potent suppressors of cell-mediated immunity but they may also
depress antibody formation in certain settings. All these agents,
as well as the more recently developed method of total lymphoid
irradiation, act non specifically. Their effect on the particular
antibodies responsible for the nephritogenic immune complexes are
unknown. Their side effects, especially the promotion of various
infections are frequent and severe. It is obvious that this level
of action is not very satisfactory because it is not closely focused
on the specific pathogenetic mechanisms of GN. In fact, results
published so far with steroids, thiopurines or alkylating agents
are on the whole disappointing in idiopathic GN.

RESTORATION OF THE IMMUNOLOGICAL HOMEOSTASIS

    If there is, as suggested above, an immunological imbalance at
the origin of certain forms of GN, it is attractive to attempt to

correct it. The best approach could be to stimulate suppressor T cells. Several agents can be used nowadays - levamisole, transfer factor, thymic hormones. These treatment have not yet been attempted in human GN and only a few studies are available in experimental GN, except in lupus. The remissions obtained in murine lupus have only been very partial and sometimes paradoxical. Thymocyte grafts, factors produced by Con A-activated suppressive cells, or thymic hormones prevent some aspects of autoimmunity in NZB and B/W mice, in particular hemolytic anemia and Sjogren's syndrome (1). However the effect on GN is much less striking, often absent and one may even observe in some cases aggravation of the GN probably due to undesired stimulation of helper T cells. The future of this approach is uncertain. One will have to use agents specific for suppressor cells (their existence is not demonstrated) or to associate to the T cell-stimulating agent an anti-helper T cell agent, such as an anti-T helper cell monoclonal antibody or cyclosporin A. Perhaps, also, protocols will be found which selectively stimulate suppressor cells (e.g. high doses of thymic hormones).

Correction of complement deficiency is not yet available but could be considered in the future. Androgen treatment has been proposed in some cases.

SPECIFIC ACTION ON THE ANTIGEN OR ON THE ANTIBODY

The antigen (and the antibody) responsible for the GN is very rarely known. One may hope that the progress in the study of immune complexes will improve this knowledge. When the antigen is known, several new approaches in the treatment of GN will be available. Antigen avoidance or elimination (for example chemical treatment of a viral or parasitic disease) would be the best and most logical attitude but it would not always be feasible. Injection of the antibody (e.g. in monoclonal form after production by hybridomas) could displace the nephritogenic component of IC. Especially anti-idiotypic immunization could be considered. Antibodies bear idiotypic antigenic determinants that are characteristic for their specific antigen. One may prepare anti-idiotypic antibodies which will specifically inhibit the suppression of the corresponding antibody. Thus, in murine lupus, monoclonal anti-DNA autoantibodies have been obtained (10) and used to prepare anti-idiotypic sera specific for anti-DNA antibodies. One could inject an anti-idiotypic

antibody or even actively autoimmunize the patient against his own
idiotypes, as it has already been performed in transplantation
immunity (3). Other more conventional methods could be used to
induce tolerance to the nephritogenic antigen (with the help of
potent immunosuppressive method such as cyclosporin A, anti lymphocytic serum (ALS) or total lymphoid irradiation, but such an approach
is likely to be difficult to achieve on a practical basis.

CONCLUDING REMARKS

Many possibilities may soon be available to manipulate the
immune system of patients with GN. This diversity is in contrast
with the very small number of current methods. One may hope that
the continuous progress in immunology will permit the nephrologist
to put his hands on one of these attractive but still esoteric
methods.

REFERENCES

1. Bach, M.A. and Droz, D.: Experimental models in the search
    for new treatments of autoimmune renal diseases. Adv. Nephrol.
    9.: 187-208, (1980).
2. Bhan, A.K.; Collins, A.B., Schneeberger, E.E. and McCluskey,
    R.T.: A cell-mediated reaction against glomerular-bound immune
    complexes. J. Exp. Med. 150: 1410-1420, (1979).
3. Binz, H. and Wigzel, G.: Induction of specific transplantation
    tolerance in adult animals. Transpl. Proc. 11: 914-918, (1979).
4. Chatenoud, L. and Bach, M.A.: T-cell subsets abnormalities in
    glomerulonephritis and systemic lupus erythematosus. Kidney
    Int. (Submitted).
5. Fillit, H.M., Read, S.E., Sherman, R.L., Zabriskie, J.B. and
    Van De Rijn, I.: Cellular reactivity to altered glomerular
    basement membrane in glomerulonephritis. N. Engl. J. Med. 298:
    861-868, (1978).
6. Kung, P.C., Goldstein, G., Reinherz, E.L. and Schlossman, S.F.:
    Monoclonal antibodies defining distinctive human T cell surface
    antigens. Science 206: 347-351, (1979).
7. Nomoto, Y., Sakai, H. and Arimori, S.: Increase of IgA-bearing
    lymphocytes in peripheral blood from patients with IgA nephro-
    pathy. Amer. J. Clin. Pathol. 71: 158-162, (1979).

8. Shalhoub, R.J.: Pathogenesis of lipoid nephrosis: a disorder
    of T cell function. Lancet 2: 556-559, (1974).

9. Sobel, A., Heslan, J.M., Branellec, A. and Lagrue, G.: Actua-
    lités Néphrologiques de l'Hôpital Necker (Flammarion, Paris,
    p. 296, (1980).

10. Tron, F.: Charron, D., Bach, J.F. and Talal, M.: Establishment
    and characterization of a murine hybridoma secreting monoclonal
    anti-DNA antibody. J. Immunol. 125: 2805-2809, (1980).

TREATMENT OF GLOMERULONEPHRITIS WITH NON STEROIDAL

ANTIINFLAMMATORY DRUGS

P. Michielsen

A.Z. Sint Rafaël
Dienst Nefrologie
Kapucijnenvoer, 33
B-3000 Leuven (Belgium)

ABSTRACT

Long term experience with anti-inflammatory drugs in the treatment of glomerulonephritis is limited to Indomethacin. The design of the available controlled trials does not permit one to answer the question whether such a treatment modifies the prognosis of glomerulonephritis, but this is suggested by uncontrolled clinical data. Besides the anti-inflammatory activity, it is likely that the decrease in proteinuria and the modification of the glomerular circulation contribute to modify the response of the glomeruli to immunological injury.

———

Among the non steroidal antiinflammatory drugs available, it is Indomethacin which as been usually employed for the treatment of glomerulonephritis. Long term results are available only for this drug. In short term experiments, Lagrue (4) has demonstrated that some other non steroidal antiinflammatory drugs have similar efficiency on proteinuria. Other drugs, like Acetylsalicylic acid associated with Dipyridamole or Metiazinic acid, have only a weak and inconstant anti-proteinuric effect. Until more data are available it does not seem justified to extend indiscriminately the long term results obtained with Indomethacin in the treatment of glomerulo-

nephritis to other non steroidal antiinflammatory drugs. Besides
its inhibitory effect on prostaglandin synthesis by inhibition of
cyclo-oxygenase, Indomethacin is also a phosphodiesterase-inhibitor.
The term "anti-inflammatory drugs" thus covers only a part of the
activity of these products. In the interpretation of the clinical
results, their influence on various physiological processes must
be considered.

     Since 1965 we have treated systematically all our glomerulo-
nephritis patients with Indomethacin. Under preventive administration of antacids and taking into account both the tolerance to
the medication and the therapeutic response, the dose is increased
progressively to a maximum of 150 mg/day. This treatment is given
continuously. After disappearance of the proteinuria and nor-
malisation of the urinary sediment, a cautious progressive reduction
of the dose is attempted under control of the proteinuria and of
the sediment. If a recurrence occurs, the treatment is resumed, and
persued as long as the urinary symptoms persist. In selected non
responders cyclophosphamide in low dose (50 mg/day) is added to the
treatment for a limited period of time, preferables not exceeding
6 months (7).

     Indomethacin reduces the proteinuria in most cases. This effect
is not dependent on the small decrease in glomerular filtration rate
(GFR) usually observed at the beginning of the treatment. This fact
is well established and generally accepted. The efficiency of the
long-term treatment on the evolution of the disease has however been
met with skepticism. The main argument advanced is that the claim
for its efficiency has not been substantiated in a controlled trial.
In a German controlled trial (3) the results were evaluated, taking
into account 7 combined clinical and biochemical parameters after
only one year of treatment. In a Medical Research Council (MRC)
trial (6) a low dose of 100 mg of Indomethacin was given during two
years, followed by a final 12-month period of follow-up. In the
evaluation of these negative results it must be remembered that
glomerulonephritis is a heterogenous group with a low mortality
rate, making necessary a follow-up period of five to ten years to
give any evaluable results. Obviously the policy to adapt the treatment to the severity of the disease during long term therapy does
not lend itself to a controlled trial. A different approach is to
use the patient as his own control. In patients in whom a reduction
in GFR was observed, we have evaluated the subsequent evolution

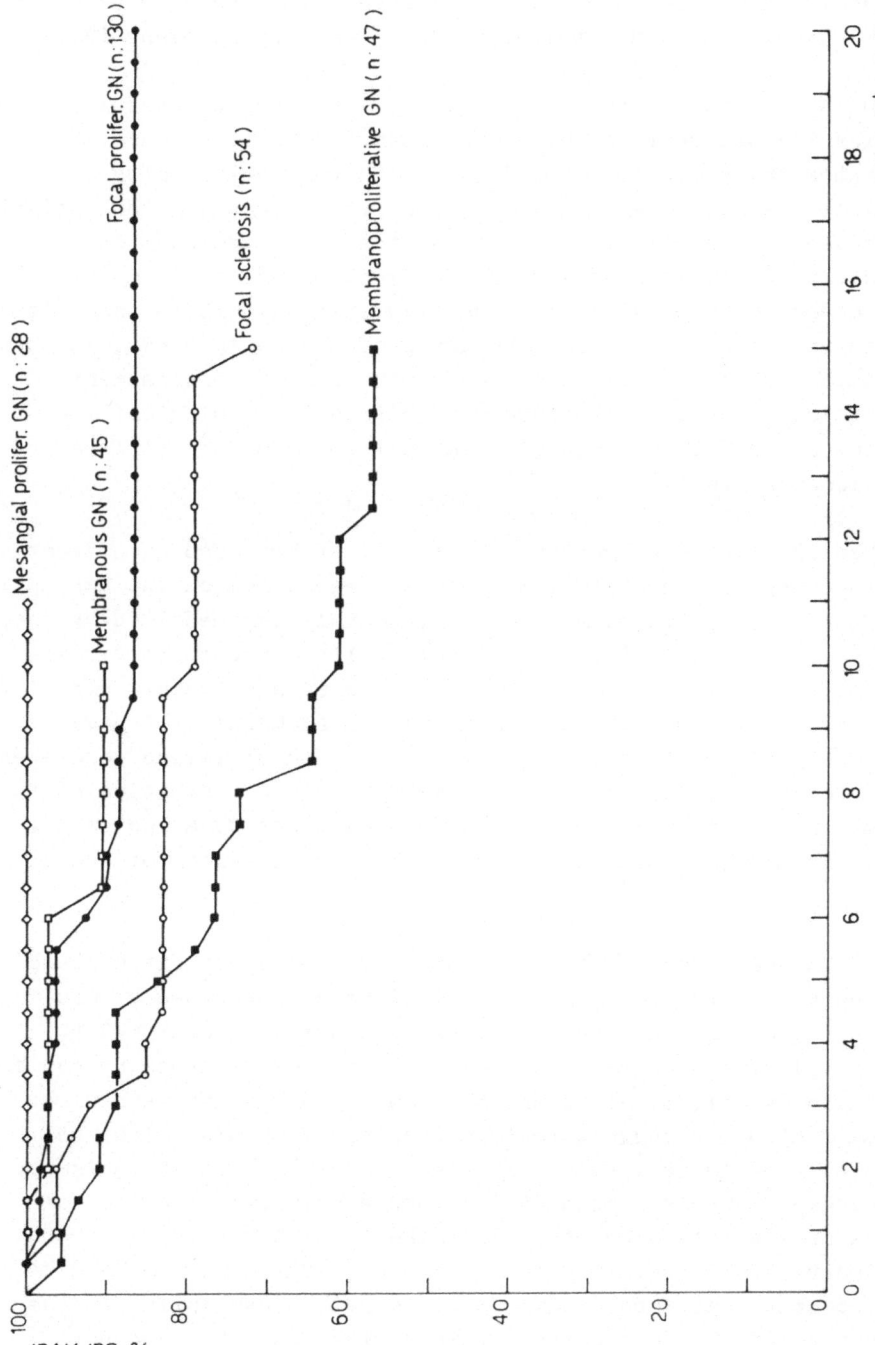

Fig. 1. Actuarial survival curve in various histological groups – Endpoint renal death.

after addition of Cyclophosphamide. Paired t-testing on measurements
made in 28 patients demonstrated a significant improvement (7).

In the various histological groups, our long term results are
significantly better than those published in the literature. Fig. 1
illustrates the evolution of patients with various histological
groups. It must be noted that these results include all the patients
who started the treatment, even if the treatment was abandoned
immediately for medical reasons or for lack of patient compliance.
Our data demonstrate an improved survival, if we compare them with
the data from the literature. The group of Lagrue in Paris, who has
followed an analogous policy of continuous long term treatment,
confirmed our results and showed a significant histological improvement after 5 to 8 years of treatment in membranoproliferative
glomerulonephritis (4).

When 15 years ago Indomethacin was first proposed as a treatment of glomerulonephritis, this approach was based on the assumption
that the immunological damage to the glomeruli was mediated by the
inflammatory reaction. A drug which could inhibit chemotactic migration of leucocytes, stabilize the lysosomes and inhibit platelet
aggregation could prevent the injury to the glomeruli, if given
continuously at least as long as the deposition of immune complexes
in the glomeruli persisted. It is likely that the inhibition of the
inflammatory reaction is important, but there are data indicating
that other factors could be involved in the progression of the
disease:

a. In the model of the spontaneous glomerulonephritis found
in ageing male rats, the glomerular lesions are preceded by proteinuria (2) consisting essentially of the sex-dependent α-2 μ
globulin (5). The occurrence of this spontaneous nephropathy can be
accelerated by unilateral nephrectomy and by a high protein diet.
These data strongly suggest that an overload mechanism might be
responsible for these lesions. Little is known of the mechanism by
which single nephrons respond to an increase in filtered load of
proteins or adapt to a decrease in number of functioning nephrons.
Preliminary results obtained in our group by Vanrenterghem and Van
Damme indicate that Indomethacin administration can influence the
proteinuria and the development of these lesions. This suggests
that the so-called antiproteinuric effect of Indomethacin could be
of more than symptomatic significance.

b.   Besides the protein intake, a high salt intake can also accelerate the occurrence of the lesions. A high salt intake increase GFR and the renal plasma flow. In humans the lesions of focal and segmental hyalinosis occur predominantly in the juxtamedullary region. These glomeruli are known to exhibit a higher glomerular filtration rate and filtration fraction than superficial glomeruli. Prostaglandins are known to play a role in the intrarenal blood distribution. A modification of the intrarenal circulation could thus also contribute to a modification of the glomerular response to an immunological aggression (1).

As long as the pathogenesis of glomerulonephritis is incompletely understood, the attemps to understand the therapeutic effect of Indomethacin will remain speculative.

REFERENCES

1.   Arisz, L., Donker, A., Bretjens, J. and Van Der Hem, G.: The effect of Indomethacin on proteinuria and kidney function in the nephrotic syndrome. Acta Med. Scand. 199: 121, (1976).
2.   Elema, J.D. and Arends, A.: Focal and segmental glomerular hyalinosis and sclerosis in the rat. Lab. Invest. 33: 554, (1975).
3.   German Glomerulonephritis Research Group: A controlled Multicenter Trial of Cyclophosphamide and Indomethacin in chronic glomerulonephritis. In: Kluthe, Vogt and Botsford. Glomerulonephritis (Georg Thieme Stuttgart 1976), p. 196.
4.   Lagrue, G.: Les antiinflammatoires et l'évolution des glomérulonéphrites primitives humaines. Gaz. Méd. France, 86: 4371, (1979).
5.   Neuhaus, O. and Flory, W.: Age-dependent changes in the excretion of urinary proteins by the rat. Nephron 22: 570, (1979).
6.   Rose, G.: Medical Research Council trials. In: Kluthe, Vogt and Botsford. Glomerulonephritis (Georg Thieme Stuttgart 1976), p. 174.
7.   Vanrenterghem, Y., Roels, L., Verberckmoes, R. and Michielsen, P.: Treatment of chronic glomerulonephritis with a combination of Indomethacin and Cyclophosphamide. Clin. Nephrol. 4: 218, (1975).

CORTICOSTEROID THERAPY IN IDIOPATHIC GLOMERULONEPHRITIS

G. Piccoli, R. Coppo, P. Stratta, M. Messina,
D. Roccatello, G.M. Bosticardo, F. Quarello, and
S. Alloatti.

Nephrology and Dialysis Units
S. Giovanni Hospital
Turin (Italy)

and Chair of Medical Nephrology
University of
Turin (Italy)

ABSTRACT

Corticosteroids have multiple effects on the movement and
functional capabilities of inflammatory or immunologically reactive
cells, on various soluble factors, on vascular and tissue responses.
There is a different sensitivity of various populations and sub-
populations of cells to the corticosteroid modulation. These mechanisms are still under discussion, but the final effects appear to
support the use of corticosteroids in certain patients with idiopathic glomerulonephritis (GN).

In minimal change GN the 10 years survival after onset was
not significantly increased by introducing corticosteroids, but
the prompt disappearance of proteinuria (80 per cent of adults by
8 weeks in our own series) supported their use. The problem of
corticosteroid treatment in focal sclerosing GN is complicated by
the probable coexistence of two histologically indistinguishable
forms (one being non-steroid sensitive). In our own series the
corticosteroid response, although transient, was present in 7 out
of 16 patients. We obtained a high number of total remissions (57

per cent) and partial remissions (14 per cent), in membranous GN,
where the conflicting data of the literature suggest differences
in the criteria of selection and admission of patients to corticosteroid treatment, calling attention to further controlled trials.
In rapidly progressive GN the combined use of corticosteroids,
immunosuppressants and heparin has elicited a stabilization or
improvement of renal function in 40 per cent of the treated patients. By the same treatment we observed a total remission in 19
per cent and a partial remission in 62 per cent of severely nephrotic
patients with the histological appearances of membranoproliferative
GN characterized by massive subendothelial deposits of the early
complement fractions (C1, C4). Although it is impossible to draw
firm conclusions either on pathogenesis of idiopathic GN or on the
biochemical, cellular and tissue effects of corticosteroid, these
drugs sometimes appear effective in clinical practice.

INTRODUCTION

    Abnormal immunological responses may be involved in various
forms of human idiopathic glomerulonephritis (GN) (36). Therefore
the use of corticosteroids in the treatment of patients with GN has
been accepted for several years (5, 11, 14, 16, 20, 21, 29, 46, 53,
68). This effect is probably both anti-inflammatory and immunosuppressive (table 1) (34).

    Decreased leukocyte and monocyte accumulation (10, 25, 30) and
inhibition of histamine-mediated reactions (22) probably represent
the main antiinflammatory effects of the corticosteroids, while
immunosuppressive actions are largely mediated by decreased lymphocyte and monocyte functions (10, 26, 27, 32, 33, 34, 71). In
addition, a decreased passage of immune complexes through the
glomerular basement membrane has been observed in experimental
models of acute glomerulonephritis whilst employing corticosteroids
(37).

    The Authors who maintain that corticosteroids are useful in
the treatment of human GN consider the possible interference of
these drugs on the immunologic mechanisms involved in GN, as the
positive blockade of damaging events.

    In GN the pathogenetic role of the polymorphonuclear leukocytes,

Table 1. Mechanism of action of corticosteroids.

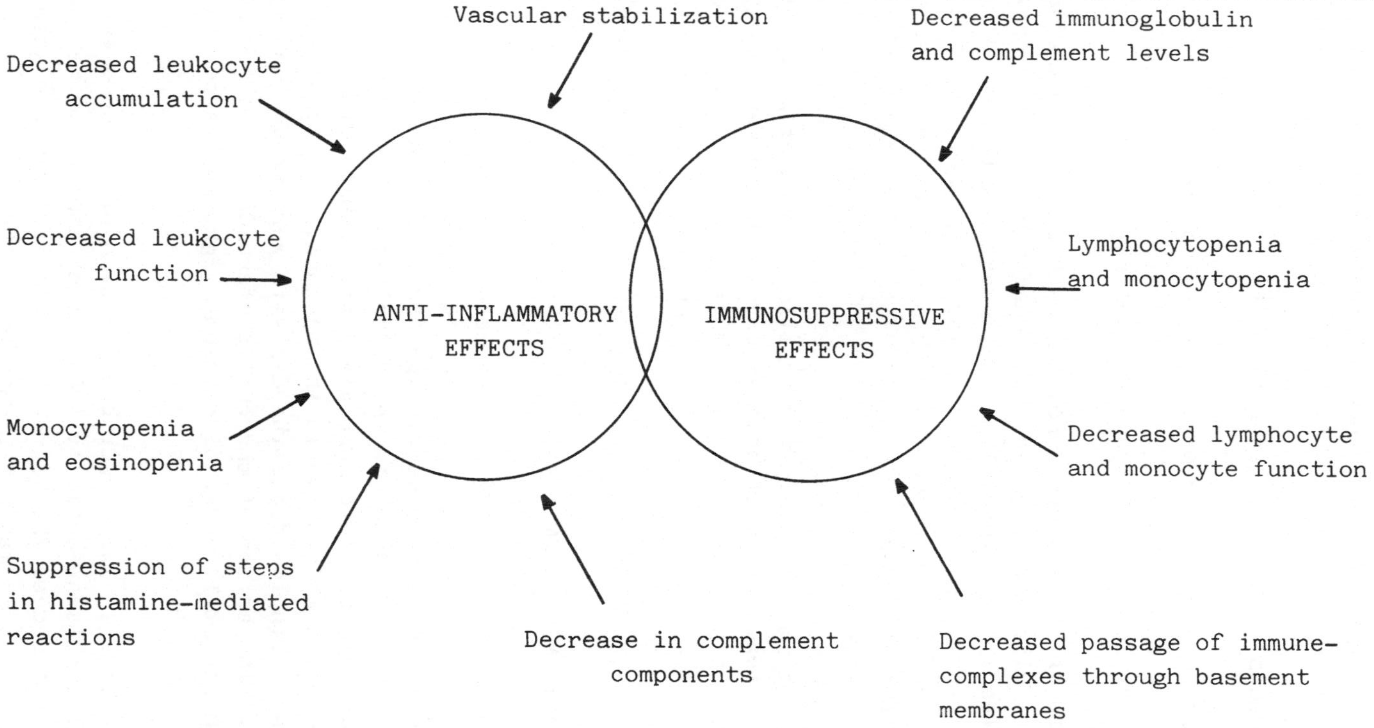

by the release of lysosomal enzymes, has been envisaged (19). This hypothesis has been taken into account in cases of acute post-infectious GN, extracapillary proliferative GN, mesangiocapillary GN, lupus nephritis, and also in Goodpasture's syndrome, in Schönlein-Henoch GN and polyarteritis (51).

Although in acute experimental immune complex GN the proteinuria is not influenced by polimorphonuclear depletion (19), the positive effect of corticosteroid blockade is theoretically valid.

The role of macrophages-monocytes in the pathogenesis of GN has not been fully investigated. Macrophages have been observed in renal biopsies in experimental and human extracapillary GN, in acute serum poststreptococcal GN, and in cryoglobulinaemic GN (23). The blockade of macrophages-monocytes in human GN may be of value in some proliferative forms.

In the acute nephritic syndrome consequent to cryoglobulinaemia, we observed (55) monocytes in the capillary lumen of glomeruli engulfed by material identifiable as cryoglobulins. As monocytes seem to play are effective role in removing cryoglobulins, the positive effect of corticosteroids is theoretically uncertain. The question is still open.

CORTICOTHERAPY: MECHANISM OF ACTION

As shown by Germuth (37) the effects of corticotherapy on the permeability of the basement membrane to immune complexes are not uniform, and are dose-dependent. High doses of these drugs modified the localization of immune complex deposition in acute serum sickness, from subepithelial to mesangial sites, while with intermediate doses the deposits were shown in the mesangial and focal subendothelial areas. With low doses the deposits were predominantly subendothelial. The action of the corticosteroids on the complement system, although still under discussion, might influence its role in pathogenesis.

Nevertheless a newly-described function of the complement, the immune complex release activity (25), might impede immune complex deposition in tissues. Therefore the indiscriminate blockade of the complement system may not always be of therapeutic value.

The antiinflammatory action of corticosteroids in the treatment of human GN is generally accepted, while the importance of their immunosuppressive action is still discussed. The corticosteroids modify all lymphocytes and macrophage functions, from antigen recognition to antibody synthesis. Modified antibody synthesis (34) can influence the physico-chemical and biological characteristics of the immune complexes. Their pathogenetic role in human GN results from in situ formation or from renal deposition of circulating immune complexes.

When immune complexes are formed in situ, the pathogenetical mechanism could be evidently modified (24). Circulating immune complexes alter their solubility properties after corticosteroid treatment: the shift from moderate to high antigen excess renders the immune complexes more soluble and less pathogenic; a shift from equivalence to moderate antigen excess produces nephrotoxic immune complexes (37).

Considering as positive the effect of corticosteroids, besides their immunosuppressive action, one can suppose a blockade of the anatomical damage from these antiinflammatory effects.

As the basis of corticotherapy in GN is largely empirical, so are the current therapeutic schedules, often derived from uncontrolled trials, balancing: the need for high doses of corticosteroids against the possible side effects. Apart from prednisone, various corticosteroids with different characteristics of duration of action, sodium retention and antiinflammatory effect, have been used. Considering their antiinflammatory dose equivalent, nonsignificant difference has been found among the various corticosteroids. Therefore prednisone remains the drug most widely employed (table 2).

MODALITIES AND RESULTS OF TREATMENT WITH CORTICOSTEROIDS

Some of the antinflammatory and immunosuppressive actions of the corticosteroids are transient and dose-time dependent (34). To obtain the highest effect it is necessary to give frequent high doses of corticosteroids for a long time.

The daily dose of prednisone varies from 10 to 30 mg/day in hypersensitivity or autoimmune diseases, to 1 g or more of methyl-

Table 2. Side effects of various modalities of corticosteroid therapy.

	Daily prednisone	Alternate day prednisone	Pulses of methyl-prednis.	Corticosteroids with 9 F1 configuration
Reduced resistance to a number of bacterial, viral, fungal and parasitic diseases or activation of latent diseases	++	+	±	
Osteoporosis	++	+		
Aseptic necrosis of bone	++	++	+++	
Muscle wasting, myopathy	++	++		+++
Growth failure	+++	±		
Negative effect on wound healing and scar fittue formation	++		+	
Diabetes	+++	+		
Sodium retention and hypertension	++	+		±
Adrenal insufficiency	++	+	±	+++
Cushingoid apparance moon face and centripetal redistribution of fat	++	±		
Gastritis, peptic ulcer	++	+		
Posterior subcapsular cataracts	++	+		
Glaucoma	++	++		
Depression, psychoses, mania	+	+		
"Pseudotumor cerebri"	±	±		+++

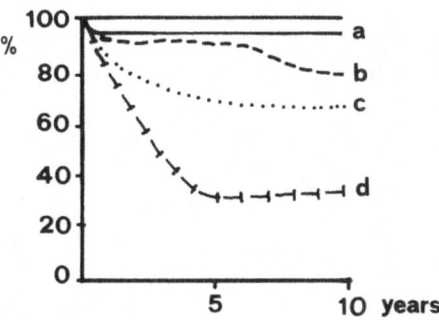

Fig. 1. Survival of children with Minimal Change G.N.
a) as '55-'70 + cyclophosphamide in relapsing
   patients (after '70);
b) antibiotics, diuretics, corticosteroids and
   treatment of hypovolaemia ('55-'70);
c) antibiotics and diuretics ('40-'55)
d) before antibiotics and corticosteroids;
                    (modified from Cameron '78)

prednisolone in hyperacute rejections, or in acute autoimmune diseases.

The vasting effect of the high dose daily therapies (1, 75) confined such schedules only to exceptional situations (21, 59, 78). The immunologic and inflammatory reactions at the beginning of therapy often require high doses of corticosteroids (pulsed methyl-prednisolone therapy), while, when the disease is well controlled, it is possible to reduce the dosage to levels which suppress the disease, with minimal side effects.

A single daily dose of corticosteroids, given in the morning between 6 and 8 A.M., is better tolerated and influences the normal ACTH-cortisol levels less markedly than divided doses (49). This concept can be extended to the alternate day therapy. In nephrology it is recognized that this kind of treatment can give good results with minimal side effects (26).

As some controlled trials have demonstrated, the nephrotic syndrome of minimal change GN responds very well to corticosteroids (13, 2, 76, 6, 57, 60, 43, 63, 8).

Nevertheless it is still controversial as to whether the

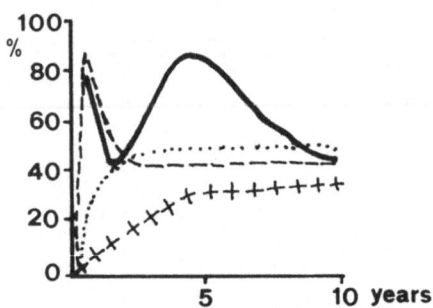

Fig. 2.   Percentage of children with minimal change G.N. free
           from proteinuria.
           +++ before antibiotics and corticosteroids;
           ... antibiotics and diuretics ('40-'55);
           --- antibiotics, diuretics, corticosteroids and
               treatment of hypovolaemia ('55-'70)
           ─── as '55-'70 + cyclophosphamide in relapsing
               patients (after '70).
                               (modified from Cameron '78)

corticosteroids control the origin of the disease or whether they
accelerate spontaneous evolution (8).

     From the 10 year survivals it is clear that the main improvement
was obtained after the introduction of antibiotic and diuretic
therapy, while the effect of corticosteroids and cyclophosphamide
has been less dramatic (16) (Fig. 1).

     The percentage of patients free of proteinuria after 10 years
from the beginning of this nephropathy is not so different from
that observed when corticosteroids were not still in use (Fig. 2)
(16).

     Nevertheless the positive effects of the corticosteroids can
reduce the risks of the major nephrotic syndromes. The most commonly
employed schedules of corticotherapy in this disease start with
dose of 1-2 mg/Kg/day of prednisone for 4-8 weeks, followed by a
progressive reduction through alternate day therapy. Previously
lower doses were employed, but the results were often unsatisfactory.
When remission is obtained treatment with alternate day therapy is

generally useful. Employing these schedules (table 3), the proteinuria generally falls after this period of time: 93-98% of the
children (13, 2, 76, 6) and 61-93% of adults (57, 60, 43, 68, 8)
are proteinuria-free after eight weeks; 80% of our cases responded
to the treatment (table 4). This wide range probably depends on
the preselection of patients, or on different histological evaluation (table 5).

OTHER FORMS OF TREATMENT. PROBLEM OF THERAPY IN NON-RESPONDERS AND
IN RELAPSING PATIENTS

The major problem of the use of corticosteroids is that about
10% of patients are "non responders".

Cyclophosphamide is often useful in these cases (72, 73, 4, 5,
14, 17, 18, 54, 45, 46, 69). Another problem is the frequent relapses
which so often discourage both doctor and patient. Nevertheless
relapses do not mean a poor prognosis, in spite of the risk of the
reappearance of the nephrotic syndrome together with the consequent
resumption of high doses of corticosteroids (68).

Table 3. Corticosteroid treatment in MC nephrotic syndrome.

INITIAL DOSE : 4 - 8 weeks

      Adults: 1-1.5 mg/Kg/day
    Children: 60 mg/m^2/day or 2 mg/Kg/day

1-2 WEEKS AFTER REMISSION :

  Alternate day therapy

2   MONTHS LATER :

  Progressive reduction

4-6 MONTHS LATER :

  Stop

Table 4. Effects of corticosteroid treatment in 15 patients with MCGN in the Nephrology and Dialysis Units, San Giovanni Hospital, Torino (Italy).

TOTAL REMISSIONS	PARTIAL REMISSIONS	NO RESPONSE
12 (80%)	2 (13%)	1 (7%) (remission after cyclophosphamide)

RELAPSES: 1/12

	BEFORE corticosteroids (mean values)	AFTER corticosteroids (mean values)
PROTEINURIA (g/day)	9.28	0.24
Cr CLEARANCE (ml/min)	86	138

FOLLOW-UP (months):   13

BEFORE REMISSION :   1.5
        (months)

CONTINUOUS CORTICOSTEROID TREATMENT: Nil

OTHER DRUGS AFTER CORTICOSTEROID   : 2-Indomethacin

Some patients present with relapses over 10 years or more, without an unfavourable course. Cyclophosphamide and sometimes chlorambucil can induce prolonged remission (69, 72, 73, 4, 5, 14, 17, 18, 54, 45, 46, 39, 9) even in corticodependent patients (9). Not a few patients still have some proteinuria during follow-up, though the amount of proteinuria is generally very slight (Fig. 3). In our experience, such minimal proteinuria generally eventually disappears and therefore it is not usually treated.

The use of corticosteroids and of other treatment modalities in the nephrotic syndrome of focal sclerosing GN is very debatable,

Table 5. Clinical responses according to histological patterns (from Barnett 1976).

		Number of patients	Response at 4 weeks
NO CHANGE	:	187	94
+ focal glom. obsolescence	:	79	90
+ focal tubular athophy	:	18	89
M.C.			
+ focal hypercellularity	:	19	84
+ mesangial thickening	:	17	74

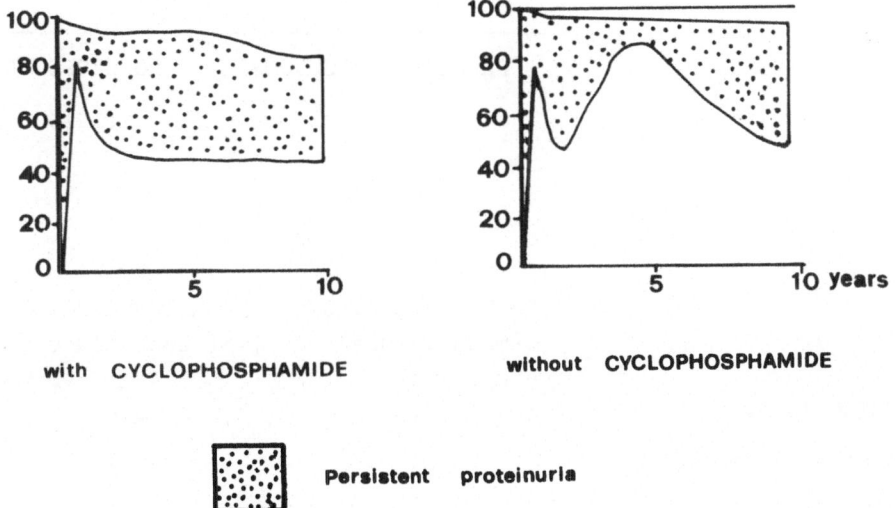

with CYCLOPHOSPHAMIDE   without CYCLOPHOSPHAMIDE

Persistent   proteinuria

Fig. 3. Persistent proteinuria in children with minimal change nephrotic syndrome, treated with antibiotics, with or without cyclophosphamide (modified from Cameron 1978).

especially since the anatomo-pathologic interpretation of these
lesions is still controversial (40, 53).

Some reports suggest a subpopulation (about 50%), characterized
by a severe nephrotic syndrome with no response to corticosteroids
and cyclophosphamide and with an unfavourable course (40, 69, 76),
but a 20% of these may still have an apparently lasting remission
(16). Some cases of severe nephrotic syndrome can respond to increased doses of corticosteroids (more than 1 mg/Kg/day) (7).

Other patients responded to corticosteroids, as probably they
represent a subgroup of focal sclerosing GN. In our series we
observed a remission induced by corticosteroids in 7 out of 16
cases: however four patients relapsed during the follow-up (table 6).
The cases responding to the treatment with a decrease or a disappearance of the proteinuria, generally present a favourable course,
even if frequently with various relapses.

Table 6.   Effects of corticosteroids treatment in 16 patients
           with focal sclerosing GN in Nephrology and Dialysis
           Units, S. Giovanni Hospital, Torino (Italy).

TOTAL REMISSIONS	PARTIAL REMISSIONS	NO RESPONSE
7(44%)	7(44%)	2.(12%)

RELAPSES: 4/7

	BEFORE CORTICOSTEROIDS (mean values)	AFTER CORTICOSTEROIDS (mean values)
PROTEINURIA (g/day)	10.1	2.5
CR CLEARANCE (ml/min)	90	122.4

FOLLOW-UP (months): 18

BEFORE REMISSION (months) 3

CONTINUOUS CORTICOSTEROID TREATMENT: 4 cases

Table 7. Short-term prednisone treatment in adult membranous
nephropathy (from Coggins et al. (20 bis)).

45-80 Kg: 125 mg prednisone every other morning

a) No response in 8 weeks: tapering in 4 weeks

b) Partial $^{(\circ)}$ or complete $^{(*)}$ response: 25 mg per dose
                                    each week   25 mg and then
                                    5 mg per dose

            (°) proteinuria 0.21-2   g/24 h
            (*) proteinuria 0.1-0.2 g/24h

Some Authors have tried cyclophosphamide treatment in these
patients. The worst results were observed in "non responder" patients. Also in membranous GN it is very difficult to evaluate
whether the corticosteroids are useful or not. A few studies, including controlled trials (64, 42, 13, 61, 63, 37, 43, 30, 66, 60,
8), could not demonstrate a real therapeutic effect of corticosteroids in this nephropathy; also a higher mortality rate was found
in treated patients. Therefore many people do not treat membranous
GN with corticosteroids, because of the possibility of spontaneous
remissions and of the frequency of prolonged periods of clinical
stability (57).

However membranous GN has a variable course. The Necker group
report a survival of 75% at 10 years (57), but data from other
centres suggest a lesser favourable course, of 30% survival at 10
years.

Therefore we prefer the use of antiproteinuric drugs, especially
Indomethacin, while other Authors still maintain corticotherapy (29).
A prospective investigation bu Coggin's group (20) has recently
presented the problem in a critical discussion. Employing the alternate day therapy schedule (table 7) a greater incidence of remission was observed although at follow-up this difference
was no longer significant. It was pointed out that a decrease in

Table 8.   Short-term prednisone treatment of membranous
nephropathy (from Coggins (20 bis)).

TREATMENT	REMISSION	TOTAL No.	FINAL
	Complete	8	4
Prednisone	Partial	14	8
(34 patients)	None	12	22
	Complete	5	4
Placebo	Partial	6	3
(38 patients)	None	27	31
	P value	< 0.02	n.s.

renal function was observed to be more rapid in cases receiving a
placebo than in cortisone-treated patients (table 8). The mean
glomerular filtration rate reduction/year, and the percentage of
cases redoubling their blood creatinine levels, were significantly
greater in non treated cases. One can argue that non treated patients presented with an unusually unfavourable course.

It is very difficult to explain the conflicting results obtained. Absence of true randomization, or differences in the systematic testing for the proteinuria and the employement of renal
biopsies, have probably led to these contrasting conclusions.

Our results show 57% total remissions and 14% partial remissions
(table 9). Our cases are selected, because we preferentially treated
with corticosteroids cases with initial histological lesions and
selective proteinuria.

It must be noted that membranous GN can be induced by various
antigens, and the disorder is consequently very heterogenous and
often not comparable between cases. Therefore the problem of how
to treat the nephrotic syndrome of membranous GN, is still open to
further evaluation. In clinical practice it seems justified to try
a treatment with steroids, at least in cases with an usually un-

Table 9. Effects of corticosteroid treatment in 14 patients
          with membranous GN in the Nephrology and Dialysis
          Units, S. Giovanni Hospital, Torino (Italy).

TOTAL REMISSIONS	PARTIAL REMISSIONS	NO RESPONSE
8 (57%)	2 (14%)	4 (29%)

RELAPSES: 1/8

	BEFORE corticosteroids (mean values)	BEFORE corticosteroids (mean values)
PROTEINURIA (g/day)	8.9	3.8
Cr CLEARANCE (ml/min)	106.1	106.5

FOLLOW UP (months)   : 29

BEFORE REMISSION   : 2.6
       (months)

CONTINUOUS CORTICOSTEROID TREATMENT: 7 cases

OTHER DRUGS AFTER CORTICOSTEROIDS: 8 cases

6 indomethacin
2 azathioprine
1 azathioprine +
  indomethacin

favourable course: -males, adults, and cases with heavy proteinuria.
In cases of acute glomerulonephritis the frequency of spontaneous
healing is such that corticotherapy is not indicated. Corticotherapy
may be tried in protracted forms and especially in rapidly progressive glomerulonephritis, where usually immunosuppressive drugs
and heparin are employed together (3, 11, 12, 15, 35, 47, 48, 67,).
We observed a stabilization or an improvement of the renal function
in 40% of rapidly progressive GN treated with corticosteroids in
association with immunosuppressive drugs and heparin (table 10).

Table 10.   Effects of corticosteroid treatment in 8 patients
with rapidly progressive GN in the Nephrology and
Dialysis Units, S. Giovanni Hospital, Torino (Italy).

UNCHANGED OR IMPROVED RENAL FUNCTION	NO RESPONSE
3	5

	BEFORE corticosteroids (mean values)	AFTER corticosteroids (mean values)
PROTEINURIA (g/day)	3	1.2
Cr CLEARANCE (ml/min)	25	.17.5

FOLLOW-UP (months) :   12.6

CONTINUOUS CORTICOSTEROID TREATMENT: NIL

It is a common belief that corticotherapy is not useful in
cases of mesangiocapillary G.N. (8, 41, 52). The reports of some
positive results obtained with corticosteroid and immunosuppressive
drugs (53) probably confirm the lack of a homogeneous pathogenesis
of this histological form.

In our experience (table 11) we felt that a good cortico-
sensitivity existed in those forms characterized by heavy subendothelial deposits of early complement fractions (C1q, C4) as
shown by immunofluorescence. We tried corticotherapy in 16 such
cases with a total remission of nephrotic syndrome in 19% and
partial remission in 62%. During a 38 months follow-up the mean
creatinine clearance values were substantially unchanged.

Table 11. Effects of corticosteroids in 16 patients (of whom
           13 received immunosuppressives and/or heparin) with
           mesangiocapillary GN in the Nephrology and Dialysis
           Units, S. Giovanni Hospital, Torino (Italy).
           14/16 cases presented Ig and C_4-C_{1q} deposits.

TOTAL REMISSIONS	PARTIAL REMISSIONS	NO RESPONSE
3 (19%)	10 (62%)	3 (19%)

RELAPSE : 1/3

	BEFORE corticosteroids (mean values)	AFTER corticosteroids (mean values)
PROTEINURIA (g/day) 9.4		2.7
Cr CLEARANCE (ml/min) 81.7		113.2

FOLLOW-UP (months) : 38

BEFORE REMISSION   : 3.3
        (months)

CONTINUOUS CORTICOSTEROID TREATMENT : 2 cases

TOTAL REMISSION TIME (months)   : 3.3

CONCLUSIONS

     In the treamtent of idiopathic GN, corticosteroids sometimes
produced excellent results. The choice of this treatment still
depends on the histological classification of GN, even if differing
results obtained in groups of patients with similar histological
lesions are not infrequent.

     In order to correctly institute therapy with corticosteroids,
a knowledge of the immunological abnormalities of each GN, and a
careful individualisation of each schedule and dosage is necessary.

Although the impact of corticosteroids on the interrelated functions
of the immune system is currently being investigated, the need
for corticotherapy and its real therapeutic usefulness in human
idiopathic nephritis is still largely unknown.

REFERENCES

1. Abeles, M., Urmann, Y. and Rothfield, N.F.: Aseptic necrosis
    of bone in systemic Lupus Erythematosus. Arch. Intern. Med.
    138: 750, (1978).
2. Abramowicz, M., Arneil, G.C., Barnett, H.L., Barron, B.A.,
    Edelmann, C.M., Gordillo, P.G., Greifer, I., Hallmann, N.,
    Kobayashi, K.O. and Tiddens, H.A.: Controlled trial of aza-
    thioprine in children with the nephrotic syndrome. Lancet 1:
    959, (1970).
3. Arieff, A.I. and Pinggera, W.F.: Rapidly progressive glome-
    rulonephritis treated with anticoagulants. Arch. Int. Med.
    129: 77, (1972).
4. Barratt, T.M. and Soothill, J.F.: Controlled trial of cyclo-
    phosphamide in steroid-sensitive relapsing nephrotic syndrome
    of childhood. Lancet 2: 479, (1970).
5. Barratt, P.M., Cameron, J.S., Chantler, C., Ogg, C.S. and
    Soothill, J.F.: Comparative trial of 2 weeks and 8 weeks
    cyclophosphamide in steroid-sensitive relapsing nephrotic
    syndrome of childhood. Arch. Dis. Child. 48: 287, (1973).
6. Barnett, H.L.: The natural and treatment and history of glo-
    merular diseases in children. In Proc. 6th Int. Cong. Nephrol.
    Firenze 1975, page 470.
7. Beaufils, H., Alphonse, J.C., Guedon, J. and Legrain, H.:
    Focal Glomerulosclerosis: natural history and treatment.
    Nephron 21: 75, (1978).
8. Black, D.A.K., Rose, G.A. and Brewer, D.B.: Controlled trial
    of prednisone in adult patient with the nephrotic syndrome.
    Brit. Med. J. 3: 421, (1970).
9. Bergstrand, A., Bollgren, I., Samuelsson, A., Tornroth, T.,
    Wasserman, J. and Winberg J.: Idiopathic nephrotic syndrome
    of childhood: cyclophosphamide-induced conversion from steroid-
    refractary to highly steroid-sensitive disease. Clin. Nephrol.
    1: 302, (1973).
10. Boggs, D.R., Athens, J.W., Cartwright, G.E. and Wintrobe, M.M.:
    The effect of adrenal glucocorticoids upon the cellular com-

position of inflammatory exudates. Am. J. Pathol. 44: 763, (1964).

11. Brown, C.B., Wilson, D., Turner, D.R., Cameron, J.S., Ogg, C. S., Chantler, G. and Gill, D.: Combined immunosuppression and anticoagulation in rapidly progressive glomerulonephritis. Lancet 2: 1166, (1974).

12. Cade, J.R., deQuesada, A.M., Shires, I.L., Levin, D.M., Hackett, R.L., Spooner, G.R., Schlein, E.M., Pickering, M.J. and Holcomb, A.: The effect of long-term high-dose heparin treatment on the course of chronic proliferative glomerulonephrits. Nephron 8: 67, (1971).

13. Cameron, J.S.: Histology, protein clearances, and response to treatment of the nephrotic syndrome. Br. Med. J. 4: 352, (1968).

14. Cameron, J.S., Turner, D.R., Ogg. C.S., Sharpstone, P. and Brown, C.B.: The nephrotic syndrome in adults with "minimal change" glomerular lesions. Quart. J. Med. 43: 471, (1974).

15. Cameron, J.S., Gill, D., Turner, D.R., Chantler, C., Ogg, C.S., Vosnides, G. and Williams, D.G.: Combined immunosuppression and anticoagulation in rapidly progressive glomerulonephritis. Lancet 2: 923, (1975).

16. Cameron, J.S.: The natural history of glomerulonephritis. In: Renal diseases. Eds: Black, D. and Jones, N.F., Blackwell, 1979.

17. Chiu, J.M.B., Mc Laine, P.N. and Drummond, K.N.: A controlled prospective study of cyclophosphamide in relapsing, cortico-steroid-responsive minimal lesion nephrotic syndrome in child-hood. J. Pediat. 82: 607, (1973).

18. Chiu, J.M.B. and Drummond, K.N.: Long-term follow-up of cyclo-phosphamide therapy in frequent relapsing minimal lesion nephrotic syndrome. J. Pediat. 84: 825, (1974).

19. Cochrane, C.G., Unanne, E.R. and Dixon, F.J.: A role of poly-morphonuclear leukocytes and complement nephrotoxic nephritis. J. Exp. Med. 122: 99, (1965).

20. Coggins, C.M.: The effects of short-term prednisone treatment in adult nephrotic with minimal change (M.C.) and membranous (M) histology. In: Abstract of the VIIth International Congress of Nephrology. Montreal, June 1978, p. 136.

20. bis. Coggins, C.M.; Principal investigator of collaborative study of the adult idiopathic nephrotic syndrome: A controlled study of short-term prednisone treatment in adults with membranous nephropathy. N. Eng. J. Med. 301: 1301, (1979).

21. Cole, B.R., TrevorBrocklebank, J., Kienstra, R.A., Kissane, J.M. and Robson, A.M.: "Pulse" methylprednisolone therapy in

the treatment of severe glomerulonephritis. J. Pediat. 88: 307,
(1976).

22. Cope, C.: Adrenal steroids and disease. Philadelphia, J.B.
     Lippincott Co, 1972.

23. Cotran, R.S.: Monocytes, proliferation, and glomerulonephritis.
     J. Lab. Clin. Med. 92: 837, (1978).

24. Couser, W.G., Stenmuller, D.R., Stilmann, M.M., Salant, D.J.
     and Lowenstein, L.M.: Experimental glomerulonephritis in the
     isolated perfused rat kidney. J. Clin. Invest. 62: 1275, (1978).

25. Czop, J. and Nussenzweig, V.: Studies on the mechanism of
     solubilization of immune aggregates by complement. J. Exp.
     Med. 143: 615, (1976).

26. Dale, D.C., Fauci, A.S. and Wolff, S.M.: Alternate-day pred-
     nisone: leukocyte kinetics and sysceptibility to infections.
     N. Engl. J. Med. 291: 1154, (1974).

27. Dale, D.C., Fauci, A.S., Guerry IV, D. and Wolff, S.M.: Com-
     parison of agents producing a neutrophilic leukocytosis in
     man. Hydrocortisone, prednisone, endotoxin, and etiocholanolone.
     J. Clin. Invest. 56: 808, (1975).

28. Dixon, F.J., Vazquez, J.J., Weigle, W.O. and Cochrane, C.G.:
     Pathogenesis of serum sickness. Arch. Pathol. 65: 18, (1958).

29. Ehrenreich, T., Porush, J.G., Churg, J., Garfinkel, L., Glabman,
     S., Goldstei, M.M., Grishman, E. and Yunis, S.L.: Treatment
     of idiopathic membranous nephropathy. New Engl. J. Med. 295:
     741, (1976).

30. Epstein, W.L.: Granulomatous hypersensitivity. Prog. Allergy
     11: 36, (1967).

31. Erwin, D.T., Donadio, J.V. Jr. and Holley, K.E.: The clinical
     course of idiopathic membranous nephropathy. Mayo Clin. Proc.
     48: 697, (1973).

32. Fauci, A.S. and Dale, D.C.: The effect of in vivo hydrocortisone
     on subpopulations of human lymphocytes. J. Clin. Invest. 53:
     240, (1974).

33. Fauci, A.S. and Dale, D.C.: Alternate day prednisone therapy
     and human lymphocyte subpopulations. J. Clin. Invest. 55: 22,
     (1975).

34. Fauci, A.S., Dale, D.C. and Balow, J.E.: Glucocorticosteroid
     therapy: mechanism of action and clinical considerations. Ann.
     Intern. Med. 84: 304, (1976).

35. Fye, K.H., Hancock, D., Moutsopoulos, H., Humes, H.D. and
     Arieff, A.I.: Low-dosage heparin in rapidly progressive glo-
     merulonephritis. Arch. Intern. Med. 136: 995, (1976).

36.   Germuth, F.G. and Rodriguez, E.: Immunopathology of the renal
      glomerulus. Little Brown and company. Boston, 1973.
37.   Germuth, F.G., Valdez, A.J., Jentrfit, L.B. and Pollack, A.D.:
      A unique influence of cortisone on the transit of specific
      macromolecules across vascular walls in immune complex disease.
      John Hopkins Med. J. 122: 137, (1968).
38.   Gluck, M.C., Gallo, G., Lowenstein J. et al: Membranous glo-
      merulonephritis. Ann. Int. Med. 78: 1, (1973).
39.   Grupe, W.F., Makker, S.P. and Ingelfinger, J.R.: Chlorambucil
      treatment of frequently relapsing nephrotic syndrome. New
      Engl. J. Med. 295: 746, (1976).
40.   Gubler, M.C., Broyer, M. and Habib, R.: Signification des
      lésions de sclerose hyalinose segmentaire et focale (S-HSF)
      dans la néphrose. In: Proceedings of the VIIth International
      Congress of Nephrology. Montreal, June 1978, p. 437.
41.   Habib, R., Kleinknecht, C., Gubler, M.C. and Levy, M.: Idio-
      pathic membrano-proliferative glomerulonephritis in children.
      Report of 105 cases. Clin. Nephrol. 1: 194, (1973).
42.   Habib, R.: Focal glomerular sclerosis. Kidney Int. 4: 355,
      (1973).
43.   Hardwicke, J., Blainey, J.D., Brewer, D.B. and Soothill, J.F.:
      The nephrotic syndrome. Proc. 3rd Int. Congr. Nephrol.,
      Washington 3: 69, (1966).
44.   Hayslett, J.P., Kashgarian, M., Bensch, K.G. et al.: Clinico-
      pathological correlations in the nephrotic syndrome due to
      primary renal disease. Medicine (Baltimore) 52: 93, (1973).
45.   International Study of Kidney Disease in Children: International
      Workshop on Risk/Benefit assessment of cyclophosphamide in
      renal disease. Kidney Int. 2: 352, (1972).
46.   International Study of Kidney Disease in Children: Prospective,
      controlled trial of cyclophosphamide therapy in children with
      nephrotic syndrome. Lancet 2: 432, (1974).
47.   Kincaid-Smith, P., Saker, B.M. and Fairley, K.F.: Anticoagulants
      in "irreversible" acute renal failure. Lancet 2: 1360, (1968).
48.   Kincaid-Smith, P., Laver, M.C. and Fairley, K.F.: Dipyridamole
      and anticoagulants in renal disease due to glomerular and
      vascular lesions: a new approach to therapy. Med. J. Aust. 1:
      145, (1970).
49.   Klinefelter, H.F., Winkenwerder, W.L. and Bledsoe, T.: Single
      daily dose prednisone therapy. JAMA 241: 177, (1979).
50.   Mallick, N.P.: The pathogenesis of minimal changes nephropathy.
      Clin. Nephrol. 7: 87, (1977).

51. Meadous, R.: Renal histopathology, London Oxford Univ., New York, 1973.

52. Medical Research Council Working Party: Controlled trial of azathioprine and prednisone in chronic renal disease. Brit. Med. J. 2: 239, (1971).

53. Mc Adams, A.J., Mc Enery, P.T. and West, C.D.: Mesangiocapillary glomerulonephritis: changes in glomerular morphology with long-term alternate day prednisone therapy. J. Ped. 86: 23, (1975).

54. Mc Donald, J., Murphy, A.V. and Arneil, G.C.: Long-term assessment of cyclophosphamide therapy for nephrosis in children. Lancet 2: 980, (1974).

55. Monga, G., Mazzucco, G., Coppo, R., Piccoli, G. and Coda, R.: Glomerular findings in mixed IgG-IgM cryoglobulinemia. Virchof Arch. of B cell Pathology, 1976.

56. Neifeld, J.P., Lippman, M.E. and Tormey, D.C.: Steroid hormone receptors in normal human lymphocytes. Induction of gluco-corticoid receptor activity by phytohemagglutin stimulation. J. Biol. Chem. 252: 2972, (1977).

57. Nesson, H.P., Sproul, L.E., Relman, A.S. and Schwartz, W.B.: Adrenal steroids in the treatment of idiopathic nephrotic syndrome in adults. Ann. Intern. Med. 58: 268, (1963).

58. Noel, L.H., Zanetti, M. and Droz, D.: Long-term prognosis of idiopathic membranous glomerulonephritis. Study of 116 untreated patients. Am. J. Med. 66: 82, (1979).

59. O' Neil, W.N., Whitson, B.E. and Bloomer, A.: High-dose corticosteroids. Their use in treating idiopathic rapidly progressive glomerulonephritis. Arch. Inter. Med. 139: 514, (1979).

60. Pearl, M.A., Burch, R.R., Carvajal, E., Mc Cracken, B.H., Woody, H.B. and Sternberg, W.H.: Nephrotic syndrome: a clinical and pathological study. Arch. Int. Med. 112: 716, (1963).

61. Pierides, A.M., Malasit, P., Morley, A.R. et al.: Idiopathic membranous nephropathy. Q. J. Med. 46: 163, (1977).

62. Pollack, V.E., Rosen, S., Pirani, C.L. et al.: Natural history of lipoid nephrosis and membranous glomerulonephritis. Ann. Int. Med. 69: 1171, (1968).

63. Robson, J.S., Mc Donald, M.K., Ruckley, V.A., Lambie, A.T., Petrie, J.J.B. and Mc Lean, P.R.: Unpublished results: cited in Robson J.S.. Renal disease; Blackwell, 1979, p. 275.

64. Rosen, S.: Membranous glomerulonephritis. Hum. Pathol. 2: 209, (1971).

65. Ross, E.J.: Effect of long-term steroid therapy in adults.

Presented at the 3rd International Congress of Nephrology. Washington. September 23-30, 1966.

66. Row, P.G., Cameron, J.S., Turner, D.R. et al.: Membranous nephropathy. Q.J. Med. 44: 207, (1975).

67. Shires, D., Holcomb, A., Cade, R. and Levin, D.: Treatment of chronic glomerulonephritis with heparin. Clin. Res. 14: 387, (1966).

68. Siegel, N.J., Goldberg, B., Krassner, L.S. and Hayslett, J.P.: Long-term follow-up of children with steroid-responsive nephrotic syndrome. J. Pediat. 81: 251, (1972).

69. Spitzer, A.: Cyclophosphamide in the treatment of the nephrotic syndrome in childhood. J. Pediat. 50: 358, (1972).

70. St. Hillier, Y., Morel-Maroger, L., Woodrow, D. and Richet, G.: Focal and segmental hyalinosis. Adv. Nephrol. 5: 67, (1975).

71. Thorn, G.W.: The adrenal cortex: reflections, progress and speculations. Trans. Assoc. Amer. Phys. 86: 65, (1973).

72. Trainin, E.B., Bochis, H., Spitzer, A., Eldelman, C.M. and Greifer, I.: Late nonresponsiveness to steroids in children with the nephrotic syndrome. J. Pediat. 87: 519, (1975).

73. Uldall, P.R., Frest, T.G., Morley, A.R., Tomlison, B.E. and Kerr, D.N.S.: Cyclophosphamide therapy in adults with minimal change nephrotic syndrome. Lancet 1: 1250, (1972).

74. Waldman, T.A., Border, S., Krakauer, R., Mac Dermott, R.P., Durm, M., Goldman, C. and Meade, B.: The role of suppressor cells in the pathogenesis of common variable hypogamma-globulinemia and the immunodeficiency associated with myeloma. Fed. Proc. 35: 2067, (1976).

75. Webel, M., Donadio, J., Woods, J. and Maher, F.: Effects of large dose of methylprednisolone on renal function. J. Lab. Clin. Med. 86, (1972).

76. White, R.H.R.: Glomerulonephritis in childhood. Brit. J. Hosp. Med. 3: 746, (1970).

77. White, R.H.R., Glasgow, E.F. and Mills, R.J.: Focal glomerulosclerosis in childhood. In Kincaid-Smith P. and Mathew T.H., New York, Wiley 1973, p. 231.

78. Woods, J.E., Anderson, C.F., DeWeerd, J.H., Johnson, W.J., Donadio, J.V., Leary, F.J. and Frohnert, P.P.: High-dose intravenously administered methylprednisolone in renal transplantation. JAMA 223: 896, (1973).

PLASMAPHERESIS AND CRYOPHERESIS

Q. Maggiore, A. L'Abbate, F. Bartolomeo, V. Misefari,
A. Caccamo and C. Martorano

Centro Fisiologia Clinica, C.N.R.
Via Sbarre Inferiori, 39
89100 Reggio Calabria (Italy)

ABSTRACT

Our studies on patients with essential mixed cryoglobulinemia
(EMC) and glomerulonephritis strongly support the hypothesis that
the circulating cryoglobulins play an important role in the pathogenesis of renal damage in EMC. In fact: a) deposits of IgM with
anti-IgG factor activity were found both in the glomeruli and in
the serum cryoprecipitate of the patients; b) IgM(k) anti-IgG/IgG
was eluted from a kidney biopsy specimen of one such patient; c)
after selective removal of circulating cryoglobulins by cryopheresis
and immuno-suppressive therapy, marked improvement in the renal
function and immunological activity of the disease occurred in all
5 patients so treated.

INTRODUCTION

Plasmapheresis is gaining increasing acceptance as a therapeutic
tool for an ever widening variety of renal diseases (table 1).

The rationale underlying its clinical use lies in the fact
that it effects the removal of circulating macromolecules some of
which are known to incite renal injury, such as the antibasement
membrane antibodies in Goodpasture's syndrome (10, 11, 16), the

183

Table 1. Indications for plasmapheresis.

- Goodpasture's syndrome

- Lupus erythematosus nephritis

- Rapidly progressive glomerulonephritis

- Wegener's granulomatosis

- Vasculitis

- Paraproteinemia

- Hemolytic uremic syndrome

- Thrombotic thrombocytopenic purpura

---

circulating immune complexes in rapidly progressive glomerulonephritis (12) and LE nephritis (17, 18), the light chains in the
acute renal failure which sometimes complicates myeloma (7, 15),
and the circulating cryoglobulins in glomerulonephritis associated
with essential mixed cryoglobulinemia (EMC). However, with the
possible ecception of Goodpasture's syndrome, where suggestive
evidence has been obtained that the improvement is associated with
the reduction of the serum levels of the offending antibody, the
relationship between the serological changes and the clinical results of treatment is far from being clear cut. There have even
been some suggestions that the benefit is related, at least in some
diseases, more to the supply of fresh donor plasma than to the
extraction of autologous plasma (4, 5). These uncertainties only
highlight the fact that with plasmapheresis one is effecting a
number of different manoeuvres, each of which is potentially capable
of exerting a therapeutic activity.

To gain an insight into the mechanism of therapeutic activity
of plasmapheresis, one should attempt to assess separately the
effect of the removal of the putative offending macromolecule from
that of plasma replacement.

One such approach is being carried out by our group in Reggio

Calabria. It is based on the specific removal of circulating cryoglobulins from the blood of patients with EMC. EMC appears uniquely
suited for the study of the effect of the removal of specific
immunoreactant from the circulation in that this condition fulfills
three important requisites:

1 - the immune complex mechanism is reasonably well established;

2 - the level and composition of the circulating immune com-
    plexes can be easily assessed so that the effect of treat-
    ment can be evaluated;

3 - the circulating immune complexes can be specifically re-
    moved.

Indirect evidence suggests that the circulating cryoglobulins
are involved in the renal damage seen in this condition. It has
been shown that the types of immunoglobulin found in the cryoprecipitate are nearly always represented within the renal immuno-
deposits (6, 19). Furthermore, the cryoprecipitate shows, at electron
microscopy, a fibrillar appearance similar to that of the subendothelial deposits in the glomeruli (3, 8).

In our laboratory we obtained more direct evidence in favour
of this hypothesis by showing that the glomeruli contain, like the
serum cryoglobulins, deposits of monoclonal rheumatoid factor (RF)
(14). Deposits of RF were demonstrated in the kidney biopsy frozen
sections by using a preparation of fluoresceinated aggregated human
IgG (Fig. 1B), The glomeruli fixed this reagent in a specific way
since their staining was blocked by pretreatment with an unfluoresceinated aggregated IgG. The tissue RF is likely due to IgM
deposits (Fig. 1A) since the pretreatment of the tissue sections
with the unfluoresceinated anti-IgM specific antiserum blocked their
staining with the fluoresceinated aggregated IgG (Fig. 1C). Pretreatment with unfluoresceinated sera against either IgG, or IgA,
or C1q had no blocking effect.

The proof that the glomerular RF activity was sustained by a
monoclonal IgM with anti-IgG activity was provided by elution studies
in one patient. With acid buffer treatment of the kidney biopsy an
immune complex was eluted which was composed of a polyclonal and
a monoclonal IgM (k) type having anti-IgG activity (Fig. 2).

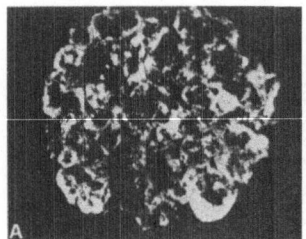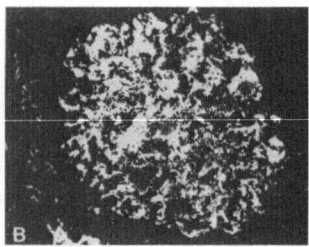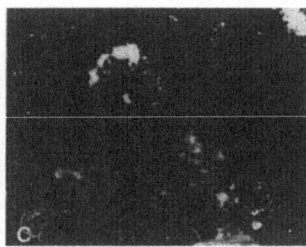

Fig. 1.   Glomerular staining with fluoresceinated anti-IgM
           serum (A); fluoresceinated aggregated IgG (B) and its
           blockade with the pretreatment with unfluoresceinated
           anti-IgM serum (C).

This complex closely resembled, even in the anti-IgG titer of
each mg of monoclonal IgM, that of the serum cryoglobulins (Table 2).

As to the second point, simple methods are available to assess
the serum level of cryoglobulins, their composition and the specific
activity of their components (9).

As to the third point, we showed some years ago that the
circulating cryoglobulins can be easily removed taking advantage
of their cold sensitivity (13). The procedure allows the selective
removal of the cryoglobulin without significant loss of other plasma
fractions, save some fibrinogen. We could not find any complement
activation from the application of such a procedure since serum C3
split products did not increase and total hemolytic complement did
not decrease.

After each treatment the cryoglobulin serum levels decreased
by 50% or more. But in the following days their levels rose again
reaching or overshooting the control values. The cryoglobulin components, the monoclonal IgM anti-IgG and the polyclonal IgG, followed
a similar trend (Fig. 3 left side). This pattern closely resembles
that which has been observed in animals after the selective removal
of circulating antibodies (2). Therefore, the removal seems useless
unless we succeed in depressing the cryoglobulin rebound so as to

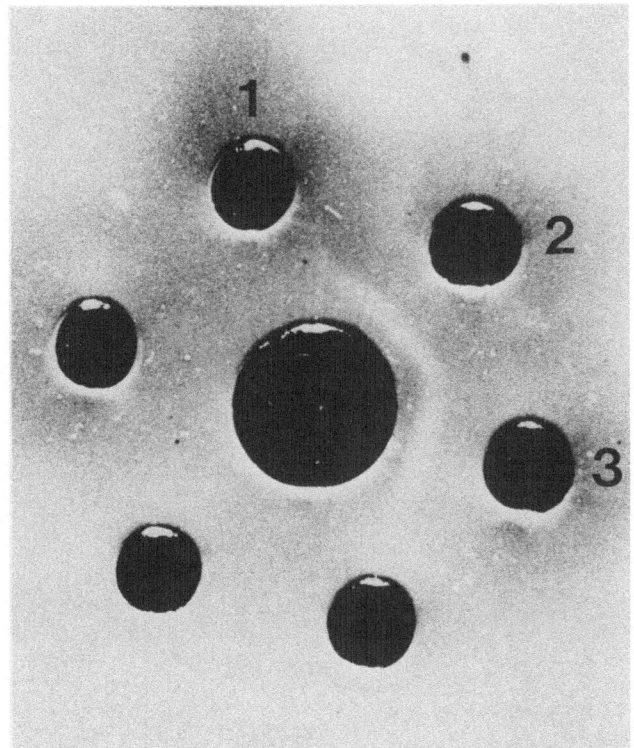

Fig. 2.   Double diffusion plate: the IgM gel–chromatography
          fraction of kidney eluate (central well) reacts with
          anti–IgM serum (well 2) and anti–kappa serum (well 3)
          with complete fusion of precipitation lines. No
          reaction with anti–lambda serum (well 1).

achieve a sustained reduction in their plasma concentration. To
this end we administered Cyclophosphamide to the patients, and in
4 of 5, Prednisolone. With this treatment, the rebound was dampened.
The attenuation in the rebound of serum cryoglobulin was shown in
all 3 patients in whom cryopheresis alone (Fig. 3; left side) was
compared with cryopheresis plus immunosuppressive therapy (Fig. 3;
right side).

Table 3 depicts the overall effect of the combined treatment
in the 5 patients treated so far. Concomitant with the marked drop
in serum cryoglobulin level there was a definite improvement in
creatinine clearance and an increase in C3 serum levels. Both changes

Table 2. Rheumatoid Factor activity (RF) in kidney eluate and serum cryoglobulins.

	Serum supernatant	Serum cryoglobulin	Kidney eluate
IgG mg/100 ml	350	250	3.3
IgM ml/100 ml	155	120	2.4
RF titer (reciprocal)	40	1400	24.0
RF titer : IgM	0.2	11	10.0

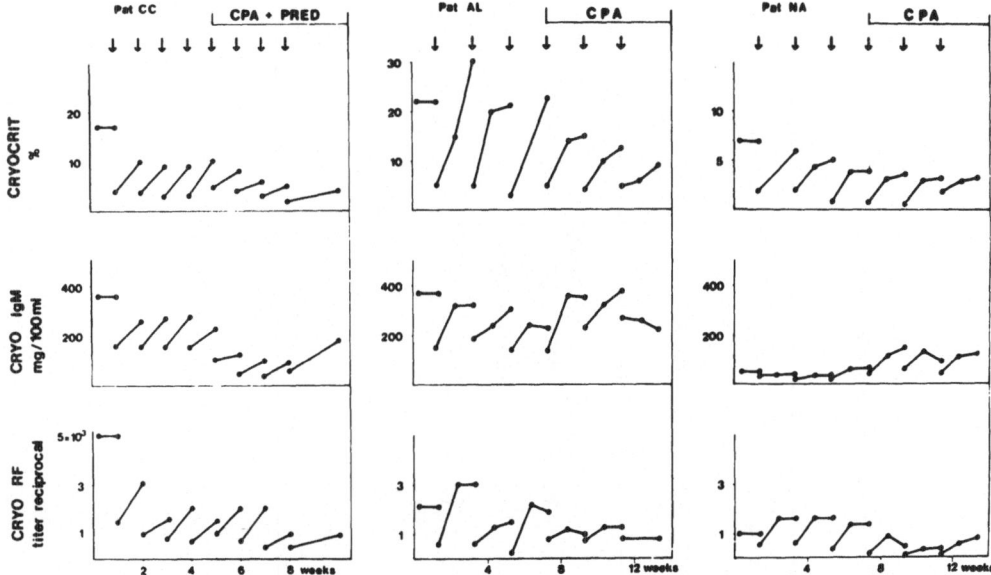

Fig. 3. Effect of Cyclophosphamide (CPA) and Prednisone
(PRED) on the cryoglobulin rebound following cryopheresis (1). Pretreatment values are indicated by
the horizontal bar in the left of each

suggest a decline of immunological activity of the disease. The
improvement outlasted the treatment duration (Fig. 4).

This study lacks a control period with immunosuppressive
drugs, so it cannot be ruled out that similar therapeutic effects
would have been produced by drug treatment alone. Evidence from
animal studies suggests that when an antibody is removed from the
circulation, the antibody-producing cells are specifically recruited
to proliferate and these rapidly dividing cells show a greater than
usual sensitivity to cytotoxic drugs (1). Therefore, the combined

Table 3. Effects of Cryopheresis and Immunosuppressive therapy.

Pts No.	Total No. of cryophereses	Duration of treatment (months)	Serum Cryocrit %		Serum RF titer (reciprocal)		Serum C3 mg/100 ml		Cr. Clearance ml/min	
			B	A	B	A	B	A	B	A
1	8	4	19	9	2000	1000	34	50	34	71
2	3	2	7	4	800	400	60	70	40	69
3	11	5	17	5	5000	1200	50	95	32	58
4	11	4	6	4	1200	800	50	70	125	125
5	8	2	6	3	800	160	30	90	30	70

B= Before; A= After.

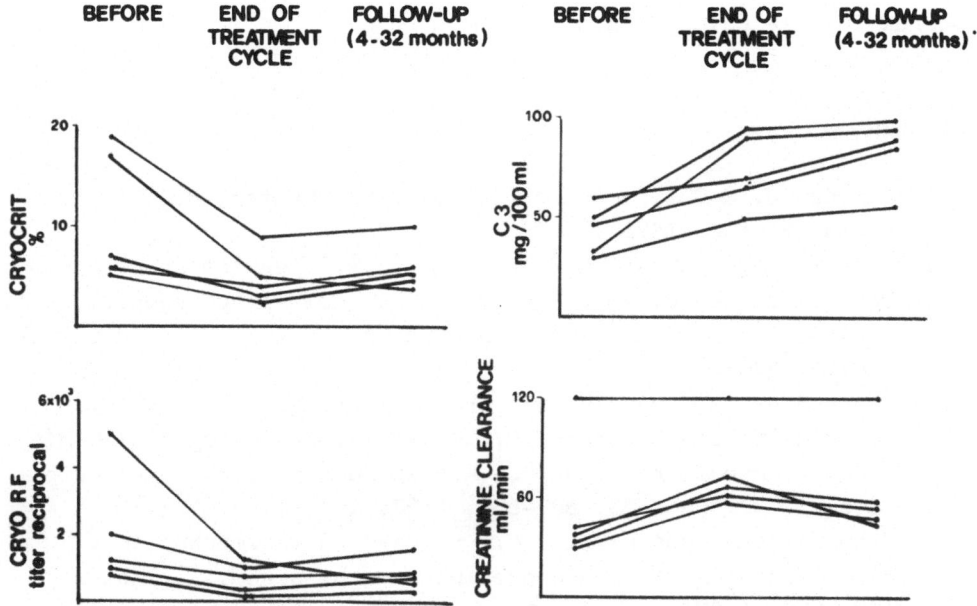

Fig. 4.   Follow-up of the effect of cryopheresis and immuno-
          suppressive therapy.

antibody removal plus cytotoxic drugs should be more effective in
suppressing the antibody synthesis than cytotoxic drugs alone.
Further studies are needed to prove this point. Even with the
present limitations, we have gained valuable information from this
kind of approach: the selective removal of cryoglobulins is feasible using a relatively simple technique; their removal is followed by a rebound which needs to be suppressed; in this setting
the rate of synthesis of an autoantibody, such as the monoclonal
rheumatoid factor, can be depressed.

Thus far, we have focused our efforts on the removal of cryoglobulin from the circulation and have not tried very hard to develop a method for the specific suppression of cryoglobulin synthesis. This will be a goal for future investigations.

AKNOWLEDGMENTS

We gratefully aknowledge the technical assistance of Mr. Nun-

ziato Marino for the immunofluorescence studies and Mr. Tiziano
Cerrai for the cryopheresis.

REFERENCES

1. Barenbaum, M.C.: Immunological disease (Samter, M., Little
    Brown, Boston, 1971).
2. Bystrin, J.C., Graf, M.W. and Uhr, J.W.: Regulation of anti-
    body formation by serum antibody. II. Removal of specific anti-
    body by means of exchange transfusion. J. Exp. Med. 132: 1279-
    1287, (1970).
3. Cordonnier, D., Martin, H., Groslambert, P., Micouin, C.,
    Chenais, F. and Stoebner, P.: Mixed IgG-IgM cryoglobulinemia
    and ultrastructural study of kidney and in vitro cryoprecipi-
    tate. Am. J. Med. 59: 867-872, (1975).
4. Czop, J. and Nussenzweig, V.: Studies on the mechanism of
    solubilization of immune precipitates by serum. J. Exp. Med.
    143: 615, (1976).
5. Editorial: Plasma exchange in thrombotic thrombocytopenic
    purpura. Lancet 1: 1065-1066, (1979).
6. Druet, P., Letonturier, P., Contet, A. and Mandet, C.: Cryo-
    globulinemia in human renal disease: a study of 76 cases.
    Clin. Exp. Immunol. 15: 483-496, (1973).
7. Feest, T.G., Burge, P.S. and Cohen, S.L.: Successful treatment
    of myeloma kidney by diuresis and plasmapheresis. British Med.
    J. 1: 503-504, (1976).
8. Feiner, H. and Gallo, G.: Ultrastructure in glomerulonephritis
    associated with cryoglobulinemia. Am. J. Pathol. 88: 145-162,
    (1977).
9. Gray, H.M. and Köhler, P.F.: Cryoimmunoglobuline. Sem. Hemat.
    10: 87-112, (1973).
10. Kincaid-Smith, P. and d'Apice, A.J.F.: Plasmapheresis in
    rapidly progressive glomerulonephritis. Am. J. Med. 65: 564-
    566, (1978).
11. Lockwood, C.M., Pearson, T.A., Rees, A.J., Evans, D.J., Peters,
    D.K. and Wilson, C.B.: Immunosuppression and plasma-exchange
    in the treatment of Goodpasture's syndrome. Lancet 1: 711-715,
    (1976).
12. Lockwood, C.M., Rees, A.J., Piching, A.J., Russell, B., Sweny,
    P., Uff, J. and Peters, D.K.: Plasma-exchange and immunosup-
    pression in the treatment of fulminating immune complex

crescentic nephritis. Lancet 1: 63–67, (1977).

13. L'Abbate, A., Paciucci, A., Bartolomeo, F., Misefari, V.,
     Nobile, F., Cerrai, T. and Maggiore, Q.: Selective removal of
     plasma cryoglobulinemia; Proceedings of E.D.T.A. 14: 486–494,
     (1977).

14. Maggiore, Q., L'Abbate, A., Bartolomeo, F., Misefari, V.,
     Caccamo, A., Barbiano di Belgiojoso, G., Tarantino, A. and
     Colasanti, G.: Cryopheresis in cryoglobulinemia. A model for
     the in vivo removal of immune complexes. La Ricerca Clin. Lab.
     10: 67–73, (1980).

15. Misiani, R., Remuzzi, G., Bertani, T., Licini, R., Levoni, P.,
     Crippa, A. and Mecca, G.: Plasmapheresis in the treatment of
     acute renal failure in multiple myeloma. Am. J. Med. 66: 684–
     688, (1979).

16. Rosenblatt, S.G., Knight, W., Bannayan, G.A., Wilson, C.B. and
     Stein, J.K.: Treatment of Goodpasture's syndrome with plasma-
     pheresis. A case report and review of the literature. Am. J.
     Med. 66: 689–696, (1979).

17. Verrier Jones, J., Bucknall, R.C., Cumming, R.H., Asplin, C.M.,
     Fraser, I.D., Bothamley, J., Davis, P. and Hamblin, T.J.:
     Plasmapheresis in the management of acute systemic lupus
     erythematosus? Lancet 1: 709–711, (1976).

18. Verrier Jones, J., Cumming, R.H., Bacon, P.A., Evers, J.,
     Fraser, I.D., Bothamley, J., Tribe, C.R., Davis, P. and Hughes,
     G.R.V.: Evidence for a therapeutic effect of plasmapheresis
     in patients with Systemic Lupus Erithematosus. Quart. J. Med.
     N.S. 48: 555–576, (1979).

19. Zimmerman, S.W., Dreher, W.H., Buckholder, P.M., Goldfarbs,
     S. and Weinstein, A.B.: Nephrophathy and mixed cryoglobulinemia:
     evidence for an immune complex pathogenesis. Nephron 16: 103–
     115, (1976).

NEW THERAPEUTIC APPROACH TO IgA MESANGIAL GLOMERULONEPHRITIS

(BERGER'S DISEASE)

J. Egido, J. Sancho, M. Lopez Trascasa, M. Sanchez Crespo,
F. Rivera, V. Alvarez, A. Barat and L. Hernando

Servicio de Nefrologia
Fundación Jiménez Díaz
Avda.Reyes Catolicos, 2
Madrid - 3 (Spain)

ABSTRACT

In mesangial IgA glomerulonephritis deterioration of renal
function is not uncommon, occurring in about 20% of patients after
several years. In this paper we present preliminary results of a
therapeutic trial with phenytoin in thirty-three patients affected
by this nephropathy. Up to 1 year of follow-up no significant changes
in renal function, proteinuria or haematuria were seen. A significant decrease in serum IgA and a normalization of the high polymeric
IgA levels frequently found in these patients, were observed at this
time. Of seven patients rebiopsied, markedly reduced IgA mesangial
deposits were seen in one and disappearance in the other, by immunofluorescence and by electron microscopy.

INTRODUCTION

Mesangial IgA glomerulonephritis is the most common primary
glomerular disease found in our country. Initially it was considered
a benign entity but further reports (3, 4, 13) have shown that
deterioration of renal function is not uncommon, occurring in about
20% of patients after several years. Actually this nephritis accounts for approximately 10% of all patients in chronic dialysis
programmes in large renal units (3). Until now no available treatment has been considered to be useful (2, 10). The lack of knowl-

edge about the exact immunopathologic mechanism involved in this
form of nephritis made the design of new therapeutic approaches
rather difficult.

We have recently provided some data showing the presence of a
large amount of serum IgA with a high molecular weight, partially
in the form of immune complexes, in a certain number of patients
with Berger's disease. Some biochemical characteristics of this IgA,
such as the presence of a J-chain and its affinity for the free
secretory component, confirmed the presence of high levels of true
polymeric IgA (8, 9). Since a reduction of serum IgA is a common
finding in epileptic patients receiving phenytoin therapy (1, 11),
this being a drug-induced effect, we decided to use this drug in a
group of patients with Berger's disease.

Material and Methods

The diagnosis of IgA glomerulonephritis was based on clinical
and laboratory data and on histological and immunofluorescent studies
of percutaneous renal biopsy specimens. Patients with clinical or
biochemical evidence of liver disease, systemic lupus erythematosus,
Henoch-Schönlein syndrome and other systemic diseases were excluded.
Informed consent was obtained from the patients. The dose of
phenytoin was 300 mg/day. This was the only drug taken by the majority of the patients during the study. Some patients also received a
thiazide diuretic, and the doses of hydralazine and propranolol which
were needed to control their blood pressure. Clinical, laboratory
and immunological assessment was made at each hospital visit. Serum
IgG, IgA, IgM as well as C3 and C4 were serially determined by
radial immunodiffusion before treatment and after 6, 12, 18, 24
months on phenytoin. Antinuclear antibodies were evaluated by direct
immunofluorescence.

Special studies of IgA purification by starch electrophoresis,
J-chain examination and affinity of the secretory component for
polymeric IgA have been published previously (8, 9). The different
percentages of serum IgA according to their molecular weights were
established in 5-40% sucrose density gradient ultracentrifugation.
Fractions between 5-9 S, 9-13 S, 13-17 S, 17-21 S were assayed for
IgA by radioimmuno-analysis using a competitive double antibody
method (9).

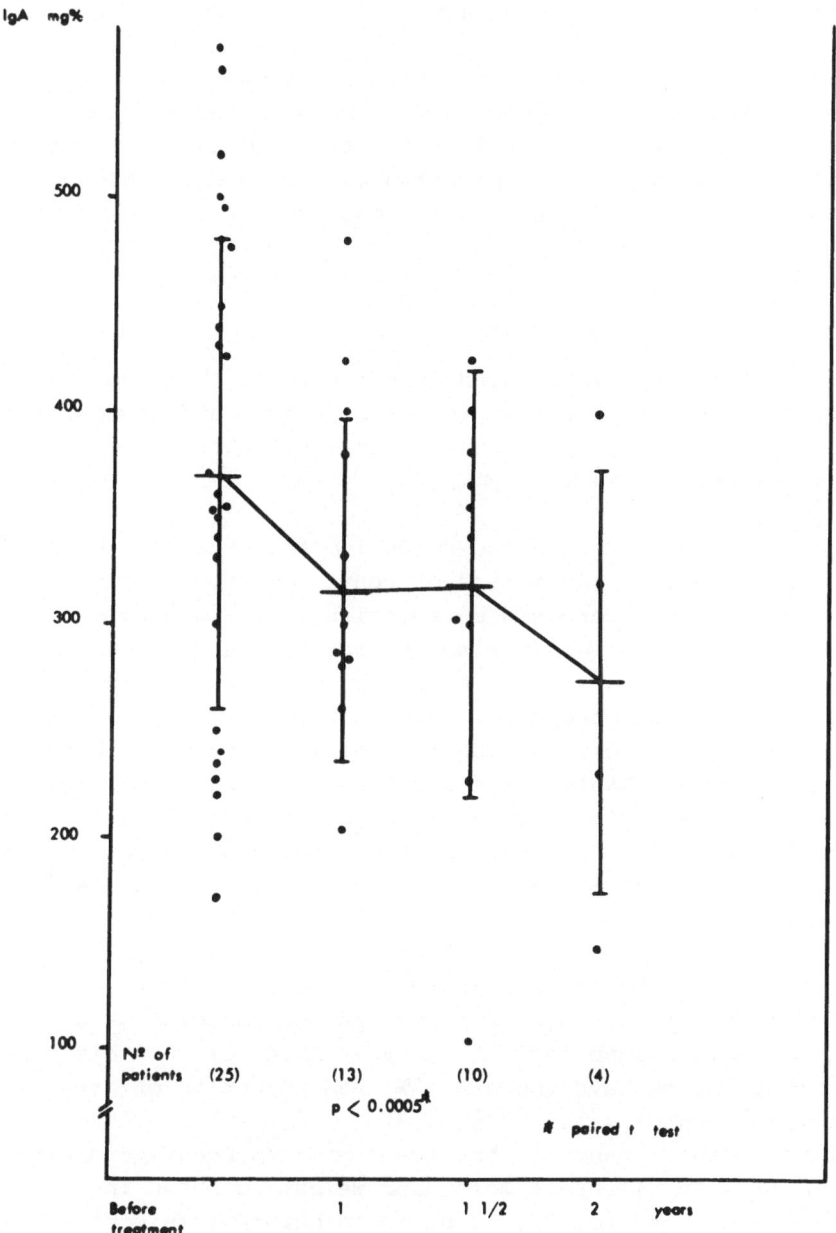

Fig. 1: Serum IgA levels before and after phenytoin treatment at
        several periods of time.

The serum concentration of phenytoin was measured by a com-
mercially available radioimmunoassay (Amersham, England).

Renal material obtained by percutaneous puncture was processed
for examination by light microscopy, immunofluorescence and in some
cases by electron microscopy. Specificity of the antisera was checked
by double diffusion gel precipitation and by immunoelectrophoresis.
The amount and density of granular deposits were semiquantitatively
scored on a scale from 0 to 3.

Results

Thirty-three patients, aged 28.9 $\pm$ 9 years, have now entered
the phenytoin trial. Follow-up is now between 3 and 28 months in 29
patients. Twenty-two patients have been treated for 6 months, and
13 for one year.

Up to 1 year of the follow-up period there has not been any
significant change in parameters of renal function, proteinuria or
haematuria. However there was a reduction in the number of episodes
of gross haematuria compared with the preceeding two years in the
same patients and with another group of untreated patients.
Four episodes of macroscopic haematuria related to infection occurred
in two patients, and one episode in another in relation with the
local reaction to a Mantoux test, while they were receiving
phenytoin. In one patient, this phenomenon occurred shortly after
starting therapy. In other patients, intercurrent infection during
therapy caused no macroscopic haematuria or only a few days of
microhaematuria.

No important changes were noted in the serum levels of IgG and
IgM; a significant decrease in serum IgA was observed as early as 6
months, being more important at one year (Fig. 1). At this moment
58% of the patients have shown a 13% mean reduction of serum IgA
below pretreatment levels.
The effect of phenytoin treatment on the percentage distribution
of serum IgA with different molecular weight is shown in Table 1. As
previously described (8, 9), in patients before treatment there was
a significant increase in the percentage of IgA in the fractions
with sedimentation constants between 9–13 S, 13–17 S and 17–21 S and a
significant decrease between 5–9 S in relation to the controls. In a
group of patients with high serum polymeric IgA levels, the percentage

Table 1. Effect of phenytoin treatment on serum IgA percentage distribution after ultracentrifugation in sucrose density gradients (pH 7.4) in patients with IgA mesangial glomerulonephritis.

	N°	5-9 S	9 - 13 S	13-17 S	17-21 S
Before treatment	6	59.06 ± 2.14* $p < 0.0025$**	32.55 ± 2.04 $p < 0.025$**	8.43 ± 2.74 $p < 0.025$**	0.74 ± 0.17 NS
After 1 year treatment	6	77.32 ± 5.18	20.33 ± 5.33	1.94 ± 0.87	0.4 ± 0.31
Controls	9	75.08 ± 2.98	21.12 ± 3.25	3.41 ± 0.89	0.37 ± 0.15

* Mean ± Standard deviation

** Paired t-test

distribution with different molecular weights was established before
and after one year of phenytoin treatment. There was a decrease in
polymeric IgA and an increase in monomeric IgA adopting a pattern similar to the controls. As previously described (8, 9) even in patients
with a normal IgA percentage distribution there was also a decrease in
polymeric forms and an increase in monomeric forms, suggesting that the
action of phenytoin is more marked on the IgA of high molecular weight.

After one year of phenytoin treatment in 7 patients with stable
renal function and microhaematuria, and minimal or no proteinuria,
a kidney biopsy was repeated. In five patients there was no striking change in the amount of the IgA deposited at the mesangial
level. In one patient IgA deposits were markedly reduced. In the
other no deposits could be found either by immunofluorescence or by
electron microscopy (Fig. 2).

Discussion

In mesangial IgA glomerulonephritis a progressive decline of
renal function is not uncommon, towards end-stage renal failure, in
about 20% of the patients after several years. No treatment has as
yet been considered to be useful. Treatment with steroids and/or
immunosuppressive drugs does not seem to have any effect on the
course of this nephritis (2, 10). Knowing the secondary effects of
these drugs, their use seems rather contraindicated in the majority
of these patients due to the slow and indolent course of this
nephropathy. Unfortunately we do not know which patients are initially at risk. Furthermore these drugs could not avoid the recurrence of the nephritis on kidney allografts. Tonsillectomy was
followed by an immediate increase of haematuria in 74% of the 19
patients in whom it was performed (7). Its effects on long term
follow-up were difficult to evaluate on clinical grounds.

The diminution of serum IgA, and what is more important, of the
high polymeric IgA levels observed after 1 year of phenytoin treatment, is of particular interest. It is possible that the normalization of polymeric IgA levels could decrease the load of abnormal
IgA presented to the mesangium, thereby allowing it to eliminate
the polymeric IgA, which has also been demonstrated locally (5).

The overall experience obtained from serial renal biopsies
performed in IgA mesangium nephritis is not very large, but several

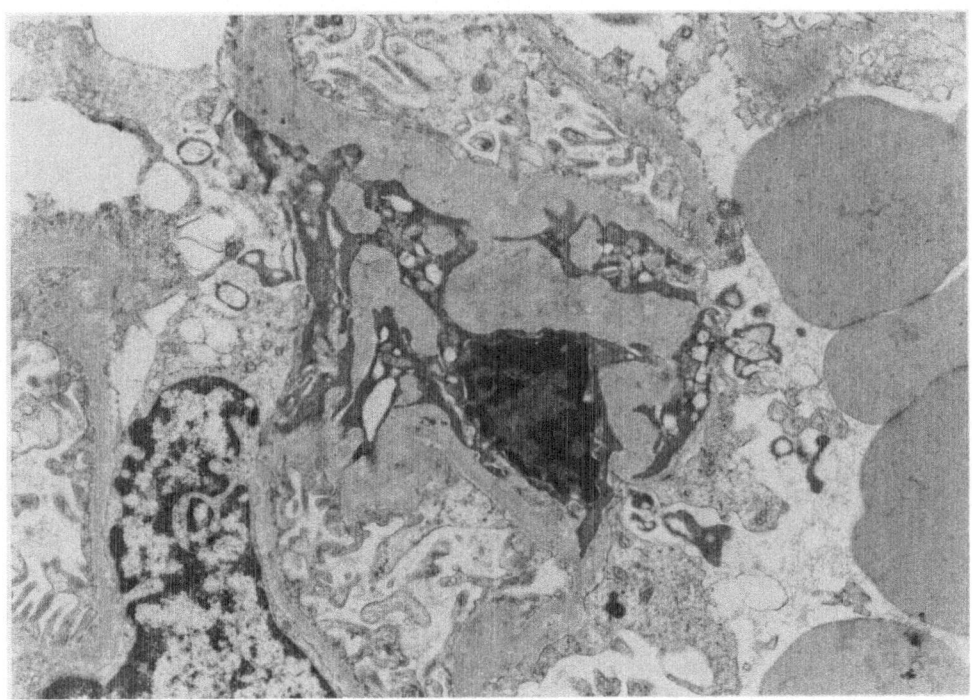

Fig. 2. Electron micrograph showing the absence of the electron dense
         deposits at the mesangial level. Glomerular basement mem-
         branes are also normal (x 15.000).

authors have repeatedly claimed that the IgA remains unchangeable at
the renal mesangium along the years. Although our data on kidney
biopsies after phenytoin treatment are still limited, the total
disappearance of the IgA in one patient and the important diminution
in another suggest that this therapy may be of value.

    How phenytoin reduces the serum IgA levels is not known (12),
but a certain individual predisposition to phenytoin-induced low
levels could exist in these patients, similar to the one described in
epileptic patients (6).

    Taking into account the lack of serious effect seen in the pa-
tients on phenytoin therapy, at least after two years of follow-up, we
think that the preliminary findings described in this paper offer a
new pathogenic and therapeutic approach to this form of glomerulonephritis.

ACKNOWLEDGEMENTS

This work was supported in part by a grant from the Instituto
Nacional de Salud (INSALUD) 163-79. We thank Miss Isabel Navajos
for secretarial assistance.

REFERENCES

1. Aarli, J.A. and Tonder, O.; Effect of antiepilectic drugs on
    serum and salivary IgA. Scand. J. Immunol. 4:391-396, (1975).
2. Berger, J., Yaneva, H., Nabarra, B. and Barbanel, C.: Recurrence
    of mesangial deposition of IgA after renal transplantation.
    Kidney Int. 7: 232-241, (1975).
3. Berger, J., Yaneva, H. and Crosnier, J.: La glomerulonéphrite a
    dépôts mesangiaux d'IgA: une cause fréquente d'insuffisance
    rénale terminale. Nouv. Presse Méd. 9: 219-221, (1980).
4. Droz, D.: Natural history of primary glomerulonephritis with
    mesangial deposits of IgA. Contr. Nephrol. 2: 150-157, (1976).
5. Egido, J., Sancho, J., Mampaso, F., Lopez Trascasa, M., Sanchez
    Crespo, M. and Hernando, L.: A possible common pathogenesis of
    the mesangial IgA glomerulonephritis in patients with Berger's
    disease and Schönlein Henoch. Proc. EDTA, vol 17: 660-666,
    (1980).
6. Haldorsen, T. and Aarli, J.A.: Immunoglobulin concentrations
    in first-degree relatives of epileptic patients with drug-
    induced IgA deficiency. Acta Neurol. Scandinav. 56: 608-612,
    (1977).
7. Jarlot-Déchelette, E.: Contribution a l'étude de la glomerulo-
    néphrite à dépôts mesangiaux d'IgA. Thèse de Médicine. Grenoble
    1975.
8. Lopez Trascasa, M., Egido, J., Sancho, J. and Hernando L.:
    Evidence of high polymeric IgA levels in serum of patients with
    Berger's disease and its modification with phenytoin treatment.
    Proc. EDTA vol. 16: 513-519, (1979).
9. Lopez Trascasa, M., Egido, J., Sancho, J. and Hernando L.: IgA
    glomerulonephritis (Berger's disease): Evidence of high levels
    of polymeric IgA. Clin. Exp. Immunol. 42: 247-254, (1980).
10. Levy, M., Beaufils, H., Gubler, M.C. and Habib,R.: Idiopathic
    recurrent macroscopic haematuria and mesangial IgA-IgG deposits
    in children (Berger's disease). Clin. Nephrol. 2: 63, (1973).
11. Seager, J., Jamison, D.L., Wilson, J., Hayward, A.R. and

Soothill,J.F.: IgA deficiency, epilepsy and phenytoin treatment. Lancet., 2: 632-635, (1975).

12. Sorrell, T.C. and Forbes, I.J.: Depression of immune competence by phenytoin and carbamazepine. Clin. Exp. Immunol. 20: 273-285, (1975).

13. Van der Peet, J., Arisz, L., Bretjens, J.R.H., Marrink, J. and Hoedemacker, P.J.: The clinical course of IgA nephropathy in adults. Clin. Nephrol. 8: 335-340, (1977).

PART 2: MEDICAL AND METABOLIC ASPECTS OF UROLITHIASIS

METABOLIC AND MEDICAL ASPECTS OF UROLITHIASIS - INTRODUCTION

R. Maiorca

Department of Nephrology
Civic Hospital
Brescia (Italy)

The second part of the course will be devoted to urolithiasis, an old issue for the urologist, but a rather new field of research for the nephrologist.

The recent information on the basic role of the kidney in the metabolism of calcium, phosphate and vitamin D opened new therapeutic approaches for the treatment of uraemic osteodystrophy.

In this sense the new knowledge of the pathogenesis of absorptive hypercalciuria represents an important advance in our understanding of urolithiasis.High levels of 1,25 cholecalciferol (1,25 CC) as well as an intestinal hypersensitivity to this hormone might explain the high urinary calcium excretion. Further investigations on absorptive hypercalciuria with high serum levels of vitamin D suggest that the hypophosphataemia, as a consequence of a primary renal leak of phosphate (3, 6), is able to stimulate renal synthesis of 1,25 CC (2).

Despite Pak's criticism of this hypothesis (11), high levels of 1,25 CC have been demonstrated in a relatively high percentage of patients (5, 7, 12) with absorptive hypercalciuria which ranges between 30 and 50%. Such a finding requires an explanation and it is unlikely that an increase in 1,25 CC remains without any effect upon intestinal calcium absorption.

Furthermore, it has been suggested that hypercalciuria (13) may depend on the impairment of tubular reabsorption of calcium, with secondary hyperparathyroidism and hyperphosphaturia (so-called "renal hypercalciuria"). According to others, hypercalciuria in such cases is due to hypercalcemia secondary to intestinal hyperabsorption of calcium with secondary decrease in serum parathyroid hormone (PTH) and PTH-dependent tubular reabsorption of calcium (2, 11, 12).

Finally, the investigations concerning the stimulating action of PTH on renal 1-alpha-hydroxylase (4) and the synthesis of 1,25 CC might explain the finding of vitamin D-dependent intestinal hyperabsorption of calcium both in renal hypercalciuria and in primitive hyperparathyroidism.

These recent acquisitions point to a renewed, interdisciplinary interest, both for the nephrologists and the urologists, in the investigational aspects of the pathophysiology of stone disease.

Nowadays we have a better knowledge of the broad group of hypercalciurias than we had ten years ago. What was formerly defined as "idiophatic" is now often given a more precise identification. This notwithstanding, our research must be directed to the minority (about 25%) of stone patients for whom no metabolic abnormalities can be clearly identified. Hopefully, in the next few years important contributions on physico-chemical aspects of stone disease will come from mineralogists, biochemists and urological investigators.

Moreover it would be very useful to have a better knowledge of the physico-chemical processes that can induce crystallization, crystal aggregation, and epitaxial induction of crystallization, as well as of the factors that inhibit crystal formation and aggregation. Furthermore our efforts must tend to clarify the role that tests measuring all factors favouring the crystallization or inhibition phenomena actually play in daily clinical work, in the hope of creating the basis for a rational therapeutic approach to renal stone disease.

These pathogenetic problems, together with advances in the fields of epidemiology and of genetics will be discussed during this course. I would also like to call your attention to recent research showing that absorptive hypercalciuria presents a male/female ratio of 3:1 whereas in renal hypercalciuria such a ratio is only 2:1. A high familial incidence has also been observed in all types of hypercalciuria, but particularly in the absorptive form.

A correct approach to the diagnosis of nephrolithiasis still
presents many problems. Multidisciplinary laboratories are essential,
if competent investigation is to be performed into the various
aspects of crystallography, physico-chemical factors, biochemistry
and endocrinology. These considerations led to the concept of
creating large regional diagnostic centres, some of which already
exist in some parts of Italy.

Undoubtedly no one can ignore the great difficulties inherent
in the measurement of serum PTH, of vitamin D metabolites and of
serum oxalate, just to mention some of the most critical points.
In the field of radioimmune PTH assay it is well known that many
different types of non-human PTH standards are employed and that
commercial immune sera are often devoid of sufficient specificity and
constant sensitivity. In my opinion, so far, PTH assays do not enable
one to differentiate between renal and absorptive hypercalciuria, no
matter which serum is used, either active against COOH or NH2 terminal
fragment. Perhaps it is still more useful to employ urinary cyclic AMP
assay as a proof of parathyroid activity on renal tubular cells. Nevertheless the measure of daily urinary excretion of cAMP is not fully reliable, as it is influenced by glomerular filtration rate, and by various, not easily evaluable hormonal effects. Thus, only the cyclic AMP
fraction, PTH-induced, secreted by tubular renal cells should be taken
into consideration. However, nephrogenic cAMP can diffuse (10) not only
into the tubular lumina but also into the capillaries and the interstitial spaces, in variable relative amounts. It is now well known
that tubular cyclic AMP synthesis is widely dependent on renal function,
rising in renal insufficiency in inverse relationship to the decrease
of glomerular filtration (8). Accordingly the urinary levels of nephrogenic cyclic AMP do not appear to be a fully reliable index in
the diagnostic approach to the urinary stone disease.

The same difficulties arise with regard to the evaluation of
urinary calcium, since significant differences have been found
between continents, countries and even regions in the same country.
In my opinion it is essential to refer to statistically significant
normal parameters on which one can rely before concluding that the
diagnosis is that of hypercalciuria. The diet of every single patient
should also be considered. Martinez-Maldonado has recently reported
his personal experience from San Juan in Puerto Rico (9). He showed

that in two hundred male urinary stones formers the mean daily
urinary calcium excretion was 220 mg, less than the comparable values
found in other countries (often > 300 mg/day). Nevertheless, this
"normal" urinary calcium excretion was twice as high as that of one
hundred healthy men of the same race, country and with a calcium
intake of 800 mg/day. It is evident that the normal urinary calcium
excretion in San Juan in Puerto Rico is quite different from that
reported by Coe in Chicago, or by Nordin, Robertson and Hodgkinson
in Leeds.

One can ask how much of the urinary calcium excretion comes
from the calcium content of the ingested water and what is the real
value of drinking the so called oligo-mineral waters, of low calcium
content, so widely accepted and used in some western European countries.

According to most reports in the literature the use of low calcium water appears useful as a preventive measure in urinary stones
disease. However the results reported by Balliker and Mallinson (1)
in 1979 were not very satisfactory. These Authors treated a group
of hypercalciuric patients, who usually drank water with high
mineral content, with soft water and other dietetic restrictions. In
a few patients, with marginal hypercalciuria, there was an effective
decrease in urinary calcium excretion; in the others the hypercalciuria was not controlled.

The interest in this course will not be focused only on calcium
nephrolithiasis but mention will also be made of struvite lithiasis,
and uric lithiasis, not forgetting the close relationship between
calcium lithiasis and hyperuricuria and the rare "renal" hyperuricuria. Soft renal calculi, cystine stones, nephrocalcinosis and
primary hyperparathyroidism will also be extensively treated.

Hyperoxaluric lithiasis, both in the primary and particularly
in the secondary form, is still in need of a definitive pathogenic
and clinical assessment. The relatively high incidence of bowel
disease as compared with the low occurrence of calcium lithiasis
secondary to intestinal hyperabsorption of oxalate is still unexplained. Again our problems stem from the lack of simple and reliable methods for the assay of oxalate.

The last problem to debate is the rational therapeutic ap-

proach to urinary stone disease. We have previously mentioned the
dietetic and the other prophylactic measures which should be followed. In the therapeutic field we have nowadays many drugs and new
methods of treatment that have shown some effect. The thiazides
are already known to be effective in hypercalciuria of renal origin
and also in the absorptive form, notwithstanding the many unanswered
questions about their mechanism of action. Their use as a diagnostic
tool, though not completely devoid of risks, is very helpful in
detecting cases of latent primary hyperparathyroidism, diagnosed by
hypercalcemia after the thiazide test.

As in other therapeutic situations, the question is not to
determine the effectiveness of single drugs, such as thiazides,
orthophosphates, or cellulose phosphate, but to assess the need for
the use of such drugs in each single patient. Then I would recall
to your mind the complete lack of any controlled trials which con-
clusively demonstrate the need for these drugs to be given to every
patient for his or her lifetime.

It is hoped that our ideas will be clarified, at least in part,
as a result of fruitful discussion during this course in Erice.

REFERENCES

1. Baker, L.R.I. and Mallinson, W.J.W.: Dietary treatment of
    idiopathic hypercalciuria. Brit. J. Urology 51: 181–183, (1979).
2. Bordier, P., Ryckewart, A., Gueris, J. and Rasmussen, H.: On
    the pathogenesis of so-called idiopathic hypercalciuria. Amer.
    J. Med. 63: 398–409, (1977).
3. Dominguez, J.H., Gray, R.W. and Lemann, J.: Dietary phosphate
    deprivation in women and men: effects on mineral and acid
    balances, parathyroid hormone and the metabolism of 25-OH –
    vitamin D. J. Endocrinol. Metab. 43: 1056–1068, (1976).
4. Garabedian, M., Holick, M.F., De Luca, H.F. and Boyle, L.T.:
    Control of 25-hydroxycholecalciferol metabolism by parathyroid
    glands. Proc. Natl. Acad. Sci. U.S.A. 69: 1673–1676, (1972).
5. Gray, R.W., Wilz, D.R., Caldas, A.E. and Lemann, J. Jr.: The
    importance of phosphate in regulating plasma 1,25 – (OH) –
    vitamin D levels in humans: studies in healthy subjects, in
    calcium stone formers and in patients with primary hyperpara-
    thyroidism. J. Clin. Endocrinol. Metab. 45: 299–306, (1977).

6. Hughes, M.R., Brumbaugh, P.F., Haussler, M.R., Vergedal, J.E. and Baylink, D.J.: Regulation of serum 1α, 25-dihydroxyvitamin D by calcium and phosphate in the rat. Science 190: 578-580, (1975).

7. Kaplan, R.A., Haussler, M.R., Deftos L.J., Bone, H. and Pak, C.Y.C.: The role of 1α, 25-dihydroxyvitamin D in the mediation of intestinal hyperabsorption of calcium in primary hyperparathyroidism and absorptive hypercalciuria. J. Clin. Invest. 59: 756-760, (1977).

8. Maiorca, R., Cristinelli, L., Cancarini, G.C., Mioni, G., Gregorini, G., David, S., Cecchettin, M., Pizzocolo, G. and Albertini, A.: Parathyroid hormone, nephrogenous cAMP, and tubular handling of phosphate and calcium in renal patients with reduced glomerular filtration rates. In: A. Leaf, G. Giebisch, L. Bolis and S. Gorini (Eds) - "Renal Pathophysiology - Recent Advances. pp. 261-268, Raven Press, New York, 1980.

9. Martinez-Maldonado, M.: Continuing challenges to the understanding of the definition and pathophysiology of hypercalciuria. Nephron, 24: 209-211, (1977).

10. Massry, S.G.: Open discussion. Proc. E.D.T.A. Vol. 14, p. 460, (1977).

11. Pak, C.Y.C.: Physiological basis for absorptive and renal hypercalciurias. Amer. J. Physiol. 237:

12. Shen, F.H., Baylink, D.J., Nielsen, R.L., Ivey, L. and Haussler, M.R.: Increased serum 1,25-dihydroxyvitamin D in idiopathic hypercalciuria. J. Lab. Clin. Med. 95: 80, (1977).

13. Smith, D.A. and Mackenzie, J.C.: In: Calcified tissues, edited by H. Fleisch, H.J.J. Blackwood and M. Owen, Springer, p.211, (1965).

EPIDEMIOLOGY OF UROLITHIASIS

M. Pavone-Macaluso, U. Rotolo, G. Caramia and D. Melloni

Institute of Urology
University Polyclinic Hospital
90127 Palermo (Italy)

INTRODUCTION

The prevention of urolithiasis might perhaps be achieved if its aetiological factors were known. In particular, the environmental factors, if clearly detected, might be avoided or eliminated.

It is hoped that at least some factors can be identified through epidemiological studies.

As discussed in our previous review (1) any epidemiological study of urolithiasis must consist of two phases:

1. The description of any difference observed, with regard to the incidence of stones in various areas and to any variability in stone formation in relation to space or time.

2. The analysis of various factors which may be responsible for the observed differences.

Such a study is far from easy in either phase.

In the first place, it is difficult to collect reliable statistical information. Data should refer only to those long resident in a given area.

Also statistics from hospitals or research groups may not reflect the real situation, as the admission of patients is usually biased by some form of selection. Necropsy reports or death certificates are also of doubtful reliability.

When epidemiological studies are aimed at the detection of a relationship between stone composition and aetiological factors, it is assumed that the chemical composition of the stones has been established without any shadow of doubt. This is rarely so, especially if we try to compare data reported by different authors.

Not only are the various analytical methods likely to give different results, but also the way of reporting data varies considerably. In our hands agreement between x-ray diffractometry and chemical analysis was found in less than 50% of cases (7).

The second phase, i.e. interpretation of data, is even more difficult than the first.

Though one or more significant epidemiological differences may be detected, the analysis of the responsible factors may be unrewarding as the causes that are responsible for such differences often remain obscure. A thorough analysis of the composition of the drinking water, nutrition, race and various genetic and environmental factors is necessary, but it seldom explains the observed epidemiological variations.

A comprehensive review of the epidemiology of urolithiasis was presented in 1978 by Blacklock (1).

In 1979 this topic was selected for a report to the International Society of Urology. We presented some data concerning epidemiology in Italy, whilst other reports and a large series of communications gave an extensive review of the present status of the investigations being undertaken in various parts of the world.

In this article we shall try to summarize some of most interesting data that emerged from these recent reports and to make a short reference to our personal experience.

PERSONAL CONTRIBUTIONS

We shall discuss, in particular, the following points:

1. A description of our personal research about the historical differences in composition of stones in one region of Italy (Sicily).

2. A review of published reports from various Italian centres.

3. A statistical evaluation of the distribution and composition of urolithiasis in various regions of Italy. The latter is mainly based upon data obtained through the Italian National Social Security (I.N.A.M.) and the National Centre for Research on Urolithiasis sponsored by the "Ente Fiuggi". These data were collected and analysed by Dr. Lucio Miano, from the Urological Clinic, University of Rome.

1. Historical Studies in Sicily

A family of urologists, to which one of the authors belongs, has been working in Palermo, Sicily, since the last two decades of 19th century. The first urologist of the family published a report in 1892, describing the composition of stones, their incidence and sites. This and his further report were based on a large series of bladder stones treated with litholapaxy (2, 3). His collection, with accurate recording of names of patients, their ages and sites of origin, permitted a recent investigation of these old calculi with x-ray diffraction studies. This demonstrated that 82% of bladder stones operated upon between 1892 and 1894 were composed of uric acid, thereby confirming, with a sufficient approximation, the estimates made by the senior author. A representative sample of bladder stones of this age was also submitted to complementary analytical methods: thermogravimetry, differential thermic analysis and infrared spectrometry, in addition to qualitative and quantitative chemical analysis. Preliminary results have shown that substantial amounts of ammonium acid urate were present in some of these ancient stones. This compound is no longer found in urinary stones in Italy, although it is still present in areas with endemic vesical lithiasis. It should be mentioned however that as late as 1966 traces of ammonium acid urate in association with calcium oxalate monohydrate and uric acid were discovered in 2 out of 3

primary bladder stones in boys or young adults operated upon in
Sicily and analysed in London by Dr. Kathleen Lonsdale.

Further statistical studies of urolithiasis in Sicily were
published in 1925, 1937 and 1952 (4, 5, 6) by Michele Pavone Junior,
the nephew and pupil of the former. In 1964 we published (7) the
results of 100 consecutive x-ray diffractometric studies performed
on all urinary stones that were spontaneously eliminated by or
surgically removed from patients followed in our ward. Our data
were compared with those of the previous authors (8, 9). Our results
were reviewed by Dr. Andersen who was leading similar and more
extensive studies in India, Norway and England (10, 11). In his
masterly review of environmental factors in the aetiology of urolithiasis, Andersen (11) extended his work to a detailed retrospective study of primary bladder stone disease in the Children's
Hospital in Palermo, Sicily. His data showed a remarkably consistent
inverse relationship with animal protein intake and the disappearance
of primary bladder stones in 1963.

He described a rise in the incidence of bladder stones co-
inciding with protein shortage towards the end of the war. This was
followed by a fall within a relatively short time, following improvement in the dietary protein content from 1944. As other possible
aetiological factors (climate, water supply, etc.) were little
changed during the period studied, Andersen suggested that dietary
changes were the major factor in bringing about this reversal in
the incidence of bladder stones in Western Sicily.

We repeated a similar study in 1978 (unpublished observations),
in order to detect any major difference that may have appeared after
an interval of over a decade. The results are shown in table 1.

It can be concluded that the composition of calculi has remained
unchanged in the last 15 years. The apparent rise in the percentage
of uric acid-containing stones in the upper urinary tract is probably
of little significance. It should be noted that over one half of all
bladder stones in Sicily still contain uric acid, pure or in association with oxalate. These are secondary bladder stones in elderly
patients with prostatic obstruction but without major disturbances
of uric acid metabolism and excretion.

Despite the obvious causes of error in comparing data obtained

Table 1. Percentage composition of urolithiasis in Palermo,
          Sicily, in 1964 and 1978.

	Kidney and Ureters		Bladder	
	1964	1978	1964	1978
a) Uric acid, alone or mixed	11.5	21	58.9	55.5
b) Phosphate, oxalate or both	88.5	79	41.1	44.5

with different criteria and techniques, we felt that some conclusions
at least could be reached and that the following trends were apparent
(as shown in figures 1 and 2):

a) a reduction in the incidence in bladder stones as compared
to that of renal and ureteric calculi;

b) a decrease in the percentage of uric acid calculi, with a
parallel increase of calcium-containg stones (especially oxalate).
The percentage of calculi containing uric acid still remains higher
that that reported in other westerh countries (Sweden, U.S.A., etc).
This is especially obvious in patients with secondary vesical
stones;

c) a reduction in primary bladder stones found in young adults
with almost complete disappearance of such disease in children;

d) a stabilization of the examined parameters has apparently
been reached in the last decade.

It should be noted that although most patients were from Palermo and its province, a significantly high number of urolithiasis
patients seen in Palermo came from the neighbouring provinces of
Trapani, Agrigento and Caltanissetta. No significant difference
was found between these various areas, despite observed differences
in the composition of the drinking water.

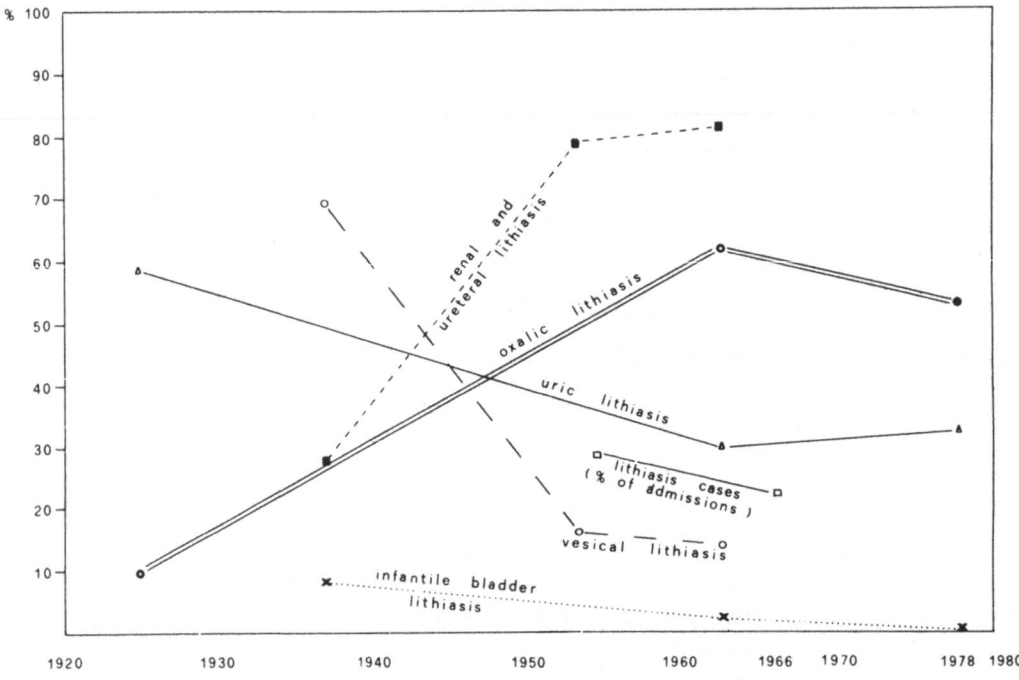

Fig. 1. Trends in urinary stone disease in Palermo, Sicily,
         from the beginning of this century to 1978. Redesigned
         from Pavone—Macaluso et al. (9) with the inclusion of
         more recent data.

2. Review of Published Reports from Italian Centres

Several reports have recently been published from different
Italian centres. Reprints and unpublished observations were kindly
submitted following our request. A detailed analysis of these
reports would be limited by space. The interested reader is therefore referred to the original papers (12 - 18). We only mention the
stimulating observation by Corrado and co-workers (19), who described
a correlation between renal oxalic lithiasis and cholesterol gall-
bladder stones found in Bologna in 18% of patients with urolithiasis.
As similar observations have not been described elsewhere, it is
as yet uncertain whether such an association has passed unnoticed
in other parts of the world or whether it represents an interesting
epidemiological factor.

The composition of urinary stones, as given in the various reports, is summarized in table 2.

The series are not uniform and the analytical methods were different. It is apparent, however, that the rural, non industrialized areas on the Adriatic coasts, uch as Abbruzzi and Marche, present the highest percentage of uric acid-containing stones, along with the highest incidence of urolithiasis in Italy. Torino, one of the richest and most industrialized area of northern Italy, shows the converse.

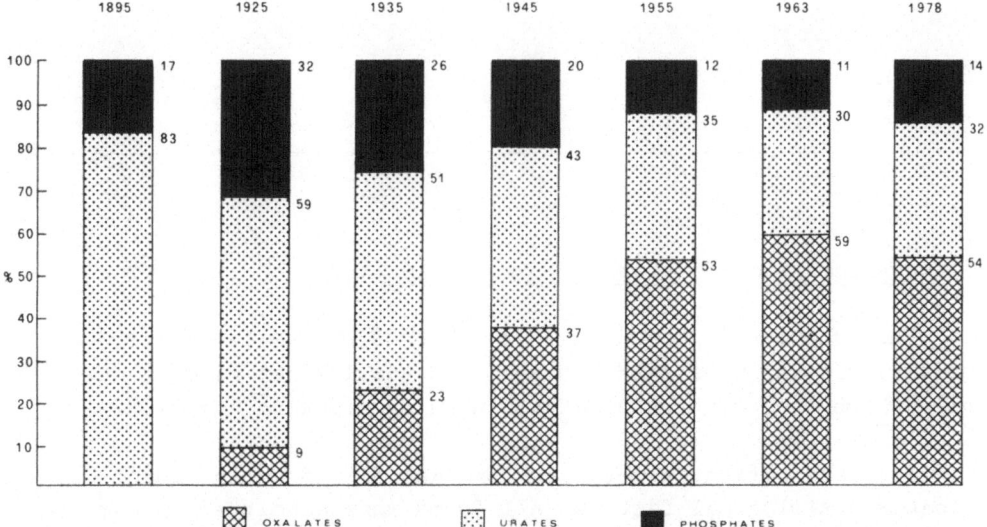

Fig. 2.   Changes in composition of stones at Palermo, Sicily. Redesigned from Andersen (11), whose figures were adapted from our data of 1963 (10). Two new columns have been added. The first (1895) refers to X-ray crystallographic analysis performed in 1978 on 12 ancient bladder stones from our collections. Following the addition of the 1978 column, it is apparent that little change has occurred since 1955.

Table 2. Prevalence of uric acid lithiasis (per cent of total
          of analyzed stones) in some Italian cities (from
          published data; for reference: see 1).

C I T Y	Uric acid, with or without oxalate
Torino	2.0
Casale Monferrato	53.1
Varese	13.1
Bologna males	19.0
females	11.7
Ancona	24.4
Pescara	22.6
Roma	9.0
Bari	15.2
Palermo	30.0

3. Geographic and Stastistical Studies in Italy.

This chapter is based on a study of 60,000 patients with
urolithiasis who were seen between 1975 and 1977 at Fiuggi, near
Rome, where they were referred for a water-drinking treatment.

For each patient a detailed form was filled in, including
possible aetiological factors. All forms were submitted to computerized statistical analysis. In about 6000 cases chemical and/or
crystallographic analysis of stones was also obtained.

Fiuggi, a well known spa, is centrally located in Italy. As
the expenses are partly covered by the State Social Security system,
large numbers of patients from all parts of Italy visit this attractive thermal resort yearly.

Although the regional distribution is not uniform, the large number of patients lends itself to a thorough statistical analysis.

Further data from a population of 8 million workers were
supplied from the National Institute of Insurance against Diseases
(I.N.A.M.), responsible for the social security of two thirds
of the Italian population (20). We thought it might be useful to
compare data obtained from both sources, in view of the errors
inherent in the uneven distribution of cases from Fiuggi (many patients from Rome and surroundings, relatively few from the islands)
and in the great prevalence of male workers in the cases from
I.N.A.M.. The data from Fiuggi consist exclusively of renal and
ureteric stones and it is likely that there are other differences in
patient recruitment and selection.

The results can be summarized as follows:

The frequency of urolithiasis is increasing in Italy, especially
in agricultural areas, namely southern Italy, Sicily and the central
and southern regions along the Adriatic coast. Lithiasis appears
to be more frequent in immigrants than in the native population.
The male to female ratio is higher in immigrants. It also higher
in the highly industrialized northwestern Italian regions and in
the largest towns.

The prevalence of urolithiasis is greater in the urban than in
the rural population that includes the inhabitants of the smallest
towns. Such a difference is higher in the southern than in the
northern regions and is mainly due to an increased incidence of
oxalate stones. With regard to the chemical composition of urinary
calculi in the various parts of Italy, the difference is not striking
as far as renal and ureteric stones are concerned. The prevalence
of stones containing uric acid in the upper urinary tract remains
relatively high in virtually all Italian regions, with a frequency
that averages 26%. It appears, however, that bladder stones containing uric acid are more frequent in the southern than in the
northern regions. A family history could be demonstrated in 39% of
renal and ureteric stones. Such a factor appears to be more
preminent in southern than in northern regions and affects oxalate
and phosphate to a slightly greater extent than uric acid stones.

So far, differences in climate, nutrition and water composition
have not given a definite clue to the cause of the variables observed.

Table 3.   Main trends in urolithiasis in Italy, in relation
           to time.

---

1.   Increase of renal and ureteric lithiasis.

2.   Increase of calcium lithiasis.

3.   Decrease of uric acid lithiasis.

4.   Slight decrease of infectious lithiasis.

5.   Disappearance of primary bladder stones in children.

6.   Almost complete disappearance of ammonium acid urate stones.

Table 3 summarizes some of the most striking findings derived
from our epidemiological survey in Italy. Table 4 gives some details
concerning the increase in the prevalence of urolithiasis in Italy
during the last two decades. The present series refers almost exclusively to renal and ureteric calculi.

Table 4. Prevalence of urolithiasis in Italy in 1954 and 1974.

	1954	1974	% Variation 1954 - 1974
Total prevalence (per 10,000 persons under insurance by INAM)	16.68	30.72	+ 84.3
Frequency of hospital admissions (per 10,000 persons under insurance by INAM)	4.96	9.73	+ 96.1
Risk of stone formation in this population	7%	13%	

Table 5. Individual and genetic factors.

- Age

- Sex

- Race

- Familial incidence

- Biochemical and metabolic factors

- Diseases

A REVIEW OF THE WORLD LITERATURE

A very detailed review would be too time-consuming. We have tried to summarize some recent data, mainly in the form of tables.

The various data can be analyzed with special reference to individual and genetic factors on the one hand and to environmental factors on the other.

The main individual and genetic factors are shown in table 5.

Table 6. Influence of age on prevalence of stones, with regard to their chemical composition (modified from Schneider and Hientsch) (21).

Composition	Most affected decade	Percentage below 20 years
Uric acid and urates	60–70	1.5
Cystine	40–50	14.8
Oxalate	40–50	6.5
Struvite	60–70	10.2

Table 7. Male to female ratio with regard to chemical
          composition and location of stones (modified from
          Schneider and Hientsch) (21).

Composition	Ratio	Location	Ratio
Uric acid	2.35	Kidney and ureter	0.54
Oxalate	2.04	Bladder	4.4
Calcium phosphate	0.86		
Struvite	0.98		

Some data about the role of age and sex are given in tables 6 and 7 respectively.

The influence of race is of some epidemiological interest. As shown in table 8, the black race appears to suffer from urolithiasis less than the white in both America and South Africa.

Table 8. Influence of race in the prevalence of urolithiasis.

a) In U.S.A.: North Carolina   (Boyce, 22)

Prevalence per 10,000 inhabitants:

	Males	Females
Whites	36.4	14.4
Blacks	9.7	3.4

b) In South Africa: (Modlin, 23)

Lower prevalence and incidence in the Bantu population:

possible role of sodium intake in addition to genetic factors.

Table 9. Influence of family history of urolithiasis in
          Italy with regard to chemical composition. Percentage
          of patient with at least one relative having suffered
          from lithiasis (from Pavone-Macaluso and Miano,
          modified) (24).

COMPOSITION

Cystine	67
Phosphate	39.7
Oxalate	39.6
Uric acid	34.4

Male to female ratio = 0.80

The incidence of a family history of urolithiasis is relatively
high, especially in Italy, as shown in table 9.

Contrary to a common belief, both oxalate and phosphate stones

Table 10. Diseases of possible epidemiological importance in
           urolithiasis.

Bilharzia:	Bladder stones in Egypt.
Malaria:	Real relationship or coincidence?
Enteritis:	Diarrhea, increase of uric acid lithiasis
Neurosis:	Described in Germany, especially in females.
Intestinal resection for obesity:	
	Oxalate lithiasis, especially in U.S.A.
Obesity per se:	Described in several countries.
Hydatid cysts:	Greece.
Diabetes:	Described in France.

Table 11. Environmental factors as causative agents for uro-
         lithiasis.

Seasonal variations - Climate.

Residence - Other geographic factors.

Activity - Type of work - Sport - Social class.

Diet: qualitative and quantitative.

Water intake: total intake, composition of water.

Other drinks.

---

showed a positive familial incidence in almost 40% of cases, which
is slightly higher than the values observed for uric acid. The
high familiarity rate for cystine stones is well known.

As shown in table 10, a few diseases have been associated
with an increased incidence of urolithiasis. Such relationship
usually been described in limited areas and their epidemiological

Table 12. Variability in incidence of stone formation with
          regard to time. Modifications in different epochs.

RAPID MODIFICATIONS
related to historical events

        War
        Famine
        Epidemics

SLOW MODIFICATIONS
related to:

        Changes in diet
        Migrations, etc.

Table 13. Role of migration upon composition of stones.

In Belgium   (Fuss and Simon, 25)

Calcium oxalate is predominant (80% of stones) both in natives and in immigrants from Mediterranean lands (France, Spain, Italy) or other countries (Czechoslovakia) where uric acid lithiasis is relatively more common. This suggests that environmental factors predominate over the genetic predisposition.

In Tyrol, Austria   (Joost et al, 26)

High incidence of apatite stones in immigrants from Yugoslavia and Turkey as compared to the local population.

---

relevance is doubtful.

Several environmental factors have been identified or suspected, as listed in table 11, With regard to time, apart from the phenomena alluded to when describing our own personal observations, the so-called war waves should be recalled (table 12).

Table 14. Hospital admissions for idiopathic renal and ureteric calcium oxalate lithiasis in Forsyth County, North Carolina, U.S.A., by sex and working activity. (From Boyce, 22).

Class of workers	A. "White collar"	B. "Blue collar"	C. Labourers
Type of activity	Sedentary, indoors	Manual, indoors	Manual, outdoors
Males	58 (64%)	16 (18%)	16 (18%)
Females	26 (93%)	2 (7%)	0
T O T A L	84 (71%)	18 (15%)	16 (14%)

Table 15.   Prevalence of renal and ureteric lithiasis among
            lifeguards in Israel (Kedar et al) (27)

- 24% of lifeguards are affected.

- Prevalence is 20 times higher than in the rest of the population.

- Calcium lithiasis predominates over uric acid lithiasis.

   The role of migration is of some interest (table 13). It is
of some importance in Italy and in Austria, but no difference
between the local and the immigrant population was observed in
Belgium.

   The type of occupation, intellectual or manual, seems to affect
the incidence of nephrolithiasis. Sedentary work was associated
more often with lithiasis than manual work, be it indoors or in the
open air (table 14).

   A strikingly high incidence of nephrolithiasis was detected
in Israel in a particular category of workers, the lifeguards, as
shown in table 15. This seems to be due not only to dehydration
from sweating and reduced water intake, but also to excessive
exposure to the sun's rays, with formation of active vitamin D me-

Table 16.   Role of some dietary factors (per cent). Modified
            from Steg et al. (28).

Predominant food	Subjects with lithiasis	Subjects without lithiasis
Meat	14.1	22.0
Cheese	33.7	29.0
Chocolate	44.1	7.0
Mixed	38.1	38.1

Table 17.   Primary bladder lithiasis - Epidemiological factors
in Thailand.

More frequent in towns than in villages.

- It depends on: a) Diet of the newborn:

- glutinous rice within the first week
- fermented fish and bananas
- oxalate-rich vegetable (fed also to nursing mothers).

b) frequent dehydration (diarrhea).

c) increased oxaluria, oxalic crystalluria and, in Laos, hyperuricaemia.

d) decrease of urinary Na: Ca ratio and of total phosphate and pyrophosphate in the urine.

---

tabolites in the skin.

Other categories of workers have also been found to be at risk. Medical practitioners, cooks, sailors, pilots and other members of flying crew, cadmium workers and electricians apparently belong to these categories. However, people working in the beer factories in Czechoslovakia were found to show a very low incidence of uro-lithiasis. Considering that some of these observations were made in limited geographic areas, their epidemiological significance is doubtful.

The diet probably has a higher epidemiological value, especially with regard to primary bladder stones. The relatively high intake in France of a calcium-rich food, such as cheese, does not seem to greatly influence stone formation. Also in France, it appears instead that a diet rich in chocolate markedly predisposes to oxalate lithiasis (table 16). It is fairly clear that reno-ureteral lithiasis, especially calcium lithiasis, is steadily increasing in the western industrialized countries. It has been noted that the diet in af-

Table 18.   Features of primary vesical lithiasis.

- Frequent in Thailand, but also in Laos, Indonesia, Iran, India, Pakistan, Egypt.

- Rare in South America and in Central and Southern Africa.

- Greatest incidence in children below 5 years of age.

- Frequent in males, rare in females (10:1 ratio).

- Is not associated with renal lithiasis.

- No tendency to recur.

- Composition: calcium oxalate, ammonium acid urate.

---

fluent countries contains larger amounts of animal proteins, lipids and sugar than in poorer countries. According to Blacklock (1), sucrose enhances intestinal absorption of calcium if the diet is poor in vegetable fibres. On the contrary, bran ingestion reduces calciuria.

The role of the diet also appears evident as far as primary bladder lithiasis is concerned, as shown in table 17. The data presented in table 18 seem however to suggest that qualitative rather than quantitative food intake is responsible for the greater incidence in Asian as opposed to African countries, although genetic factors cannot be ruled out.

It is doubtful whether dietary differences play an important role in the observed variability in stone composition in diverse countries. Uric acid stones are much more frequent in some European countries, including Italy, than in others.

This is also true for bladder stones. In our experience, about one half of bladder stones secondary to prostatic obstruction are made of uric acid in Sicily, whereas struvite stones are found almost exclusively in other countries under identical circumstances. It is interesting to note that a very high percentage of bladder stones were also predominantly composed of uric acid in the 19th century even in some of those European countries where uric acid bladder stones are nowadays almost never seen (table 19).

Dietary factors such as increased animal protein intake are very likely to play a role in this regard.

Nephrolithiasis is more frequent in hot than in cold climates. Quantitative water intake appears to be the critical factor. However, the role of the quality of drinking water and, in particular, of its calcium content, has not yet been clearly identified. Paradoxically an inverse relationship between tne calcium content of the water and the prevalence of reno-ureteral lithiasis has been observed along the Adriatic coast of Slovenia.

CONCLUSION

We did not attempt to prepare a complete review of published work related to epidemiology of urolithiasis, but only to pinpoint some topics· of a certain interest, especially in the light of recent reports.

Although some useful data have already emerged, much remains to be studied in this fascinating but difficult field.

Table 19.   Main chemical component of bladder stones in various European countries during the past century (per cent of all vesical calculi).

COMPOSITION	England	Denmark	Germany	Sicily
Uric acid and urates	65.3	89.4	70.3	83.0
Cystine	0.4	0	0.3	0
Oxalate	23.1	1.6	18.8	0
Phosphate	11.1	9.0	10.7	17.0

ACKNOWLEDGMENTS

Our thanks are due to Prof. M. Leone, University of Palermo, for performing the x-ray diffractometric studies; to Prof. G. D'Ascenzo, University of Rome, for performing the infrared spectrometry, thermogravimetry and differential thermic analysis; to colleagues from various part of Italy (Bianchi, Ancona; Bongi and Pirani, Pescara; Bono and Roggia, Varese; Corrado and Fini, Bologna; Piccinno, Bari; Miano, Rome), who have kindly supplied their data and publications.

REFERENCES

1. Blacklock, N.J.: Epidemiology of Urolithiasis. In: The scientific basis of Urology. Edited by D.I. Williams and G.D. Chisholm. Heinemann, London, 1976, vol. 1, 235-243.
2. Pavone, M. Senior: Trattato sulle malattie delle vie urinarie. Spinnato, Palermo, 1892.
3. Pavone, M. Senior: La litotribolapassi, Idelson, Naples, 1922.
4. Pavone, M. Junior: Sui costituenti dei calcoli urinari e sulla natura dei calcoli vescicali in Sicilia. Cult. Med. Mod. 4: 250, (1925).
5. Pavone, M. Junior: Cura medica e dietetica della calcolosi urinaria. L. Pozzi, Roma, 1937.
6. Pavone, M. Junior: La mia statistica sulla calcolosi urinaria in Sicilia. Atti Soc. Ital. Urol. 25: 267, (1952).
7. Pavone-Macaluso, M., Leone, M. and Piazza, B.: Determinazione della composizione dei calcoli urinari mediante roentgen-diffrattometria. Urologia (Treviso), 31: 270, (1964).
8. Pavone-Macaluso, M.: Trends in urinary stone disease in Sicily from the beginning of this century to the present day. Panminerva Medica, 9: 78, (1967).
9. Pavone-Macaluso, M., Piazza, B. and Madonia, S.: Modificazioni di alcuni indici statistici nella urolitiasi in Sicilia. Rass. Urol. Nefrol. 3: 17, (1965).
10. Andersen, D.A.: Historical and geographical differences in the pattern of incidence of urinary stones considered in ralation to possible aetiological factors. In: Renal stone research symposium. Edited by A. Hodgkinson and B.E.C. Nordin. Churchill, London, 1969, p. 7-31.
11. Andersen, D.A.: Environmental factors in the aetiology of

urolithiasis. In: Urinary calculi. Recent advances in aetiology, stone, structure and treatment. Edited by L. Cifuentes Delatte, A. Rapado and A. Hodgkinson. S. Karger, Basel 1973, p. 130-144.

12. Borgno, M. and Marten-Perolino, R.: Aspetti chimici e cristallografici delle calcolosi di ossalato di calcio. Minerva Urol. 14: 1, (1962).

13. Bono, A.V., Roggia, A., Comeri, G.C., Gianneo, E., Benvenuti, C. and Santoro, C.: Indagine statistica su un'ampia popolazione di nefrolitiasi. Presented at 27th meeting of Urological Society of Northern Italy, Modena 20th June 1978.

14. Corrado, F. and Fini, M.: Personal communication.

15. Bianchi, F.: Dati statistici in tema di urolitiasi. Urologia (Treviso), 41: 480, (1974).

16. Bongi, G., Di Tizio, A. and Pirani, A.: Incidenza e natura dell'urolitiasi osservata nella divisione di Urologia dello Ospedale di Pescara. In press.

17. Bracci, U. and Miano, L.: Quoted by: Il Centro Studi e ricerche sulla calcolosi urinaria e malattie metaboliche correlate. Edited by: Ente Fiuggi, Roma 1977.

18. Piccinno, A., Di Pierro, M., Pagliarulo, A. and Bruno, L.: Aspetti metabolici in corso di calcolosi renale. Puglia Chirurgica, 17: 3, (1974).

19. Corrado, F., Fini, M., Severini, G. and Roda, E.: Correlations between renal oxalic lithiasis and cholesterol gall bladder lithiasis. In: Urolithiasis research. Edited by H. Fleisch, W.G. Robertson, L.H. Smith and W. Vahlensieck. Plenum Publ. Corp. New York, 1976, p. 417.

20. I.N.A.M.: Annuario statistico 1974-75. Arti grafiche Panetto Petrelli, Spoleto, 1978.

21. Schneider, H.J. and Hienzsch, E.: Epidemiology of urolithiasis. In: XVIII Congrès de la Societé Internationale d'Urologie. Abstracts, 1979, p. 73.

22. Boyce, W.H.: Epidemiology of lithiasis in the United States. In: XVIII Congrès de la Societé Internationale d'Urologie. Tome 1 - Rapports, 1979, p. 79-86.

23. Modlin, M.: Renal calculus in the republic of South Africa. In: A. Hodgkinson and B.E.C. Nordin: Renal stone research Symposium. Churchill, London, 1969, p. 49-58.

24. Pavone-Macaluso, M. and Miano, L.: Epidemiology of urolithiasis in Italy. In: XVIII Congrès de la Societé Internationale d'Urologie. Tome 1 - Rapports, 1979, p. 113-137.

25. Fuss, M. and Simon, J. Renal stone composition in 768 Belgian patients. XVIII Congrès de la Societé Internationale d'Urologie.

Abstracts, 1979, p. 55.

26.   Joost, J. Marberger, H., Egger, G. and Hohlbrugger, G.: The epidemiologic aspects of upper urinary bladder calculi in Tyrol. XVIII Congrès de la Societé Internationale d'Urologie. Abstracts, 1979, p. 66.

27.   Kedar, S., Shabtai, M., Haimowitz, C. and Better, O.S.: Increased incidence of urolithiasis in lifeguards in Israel. XVIII Congrès de la Societé Internationale d'Urologie. Abstracts, 1979, p. 65.

28.   Steg, A., Landier, J.F., Teyssier, P., Champagnac, A. and Thomas, J.: Epidemiologie et étiologie de la lithiase oxalique en France. XVIII Congrès de la Societé Internationale d'Urologie. Abstracts, 1979, p. 76.

CRYSTALLOGRAPHY: AN INTRODUCTION

L. Riva di Sanseverino

Istituto di Mineralogia
Università di Bologna (Italy)

ABSTRACT

Renal calculi are predominantly formed from crystals of one or
more substances. Crystals can be defined as groups of particles
accurately and orderly arranged in 3 dimensions.

A crystal is different from glass, in that particles are not
orderly arranged in the latter. It is sufficient to examine the
fracture planes of both materials to have a hint of the regularity
of the first and of the disorder of the second, at the level of the
disposition of the respective constituent units.

The particles forming the crystal, i.e. atoms, ions or molecules, are set at a reciprocate distance of about 10^{-8} cm. It is
fortunate for the investigators that such a value coincides with
the wavelength of x rays. Crystal-x ray interaction, called "diffraction", allow us to "see" these distances, which are as charac-
tersitic, as an identity card for each crystalline "phase."

This is why crystallographic methods enable us to recognize
the crystals of a given substance. The observed distances are different in the case of a salt (e.g. calcium oxalate) and its hydrate
derivative (calcium oxalate bihydrate). They are different in the
case of tricalcium phosphate, $Ca_3 PO_4$, and of acid calcium phosphate,
$Ca HPO_4$. They are different even in the case of the two classical

natural modifications (or "crystalline phases") of Ca Co$_3$, namely calcite and aragonite.

Along with progress in knowledge, crystallography has developed new methods, always based on x ray-crystal interaction, which allows us to obtain a quantitative detection of a substance present in a renal stone, even in very small amounts.

More recently, with the aid of electron microscopy, interesting speculations have been put forward about the modalities of crystal growth, based on morphological observation of crystals. The investigations were carried on together with the mathematical analysis of energetic phenomena on the specific deposition surfaces and with the study of dimensions of elementary granules. It is assumed that elementary particles per se do not give rise to growth by regular arrangement, but only in the presence of small pre-aggregated nuclei, forming the growth units.

Together with qualitative and quantitative diagnosis, the problem of the growth of a renal calculus is one of the most fascinating aspects for a crystallographer. It is to be hoped that close co-operation with urologists will help in solving such a basic problem.

———

A crystal consists of units (atoms, molecules or ions) accumulated in a three-dimensional regular arrangement. In two dimensions a familiar example may be given by a wall-paper pattern. But the units in the crystal are arranged in such a way as to minimize their energy interactions, and thus may be described as spacefilling bodies; therefore, again in two dimensions, they are represented graphically in the best way by the world famous Escher's drawings (1).

The repeat distances in three non coplanar directions are typical of each crystalline material (Fig. 1): a significant response to the question "Which substance......" or better "Which crystalline phase are we analyzing?" is given to us when appropriate methods succeed in unravelling them. Three distances therefore identify unequivocally the chemical crystalline phase.

The method most commonly used is x-ray diffraction by crystals: the wave nature of this penetrant and energetic radiation is posi-

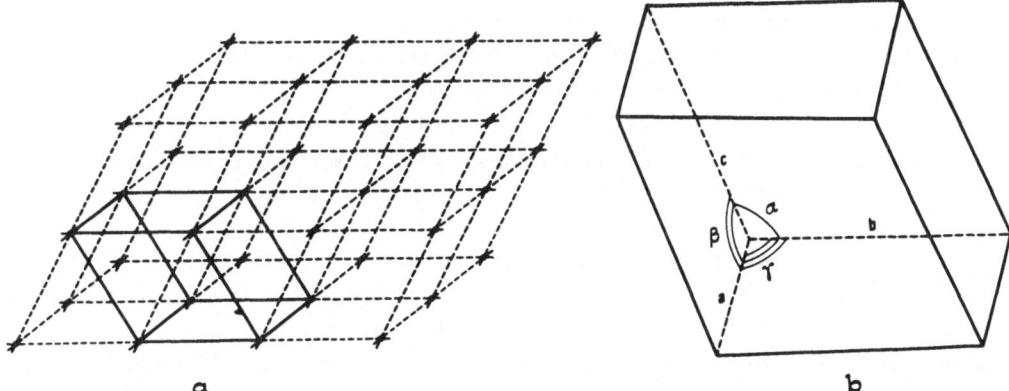

Fig. 1. (a) The crystal lattice schematically reduced to
         point-like lattice: an alternative unit cell is in-
         dicated, (b) detail of an elementary (or unit) cell,
         a, b, c are the periodic distances, α, β and ɣ the
         angles between their directions.

tively "impressed" by the regular network (lattice) of equivalent
"points" distributed in three dimensions, the x-ray wavelength being
of the same order of magnitude as the periodic distances in the
lattice.

One should remember now that the "points" are actually atoms,
molecules or ions; the term and the representation are used only to
simplify drawings and for mathematical treatment.

The resulting effect on the impact of x-rays with crystals is
an interference, a phenomenon of selective nature: coherent scat-
tering of x-rays occurs only in a few directions, depending on the
geometry of the experiment: x-ray beams are then trasmitted from
the crystals.

Easy measurement of the topology of this phenomenon, called
diffraction, gives quantitative information on the repeat distances:
quite quickly therefore we obtain a series of figures from which we
can define with certainty the crystalline phase. The method, currently adopted by mineralogists, chemists, metallurgists and many others,
is called x-ray diffractometry.

Morever, evaluation of the intensity of the x-ray diffracted
beam would give us information on the chemical and also stereochemical nature of the diffracting matter and, after some time-consuming
mathematical approaches, on the distribution of atoms or ions in
the space delimited by the elementary cell, i.e. the repeat unit
of the lattice (see Fig. 1, b).

This last result of the diffraction of x-rays by crystals
represents a contribution of increasing importance in current studies
in chemistry, pharmacology and molecular biology, because crystal-
lographers are finally able to draw molecular shapes and build
models of the most important compounds in these fields. The first
results have consequences of relevance in the study of urinary
stones. A small fragment or a few milligrams of matter are sufficient to determine which compound is present in a stone in a matter
of minutes.

The most recent advances in the study of crystal growth lead
us to expect further contributions from crystallography to urology.

Until now we have considered ideal lattices, where the elementary unit is accumulated regularly in a three dimensional array.
Recently it has been generally recognized that crystals do not grow
as   perfectly as described theoretically, and both the stages of
nucleation and of aggregation are being investigated to discover
the mechanism of growth.

In examining the growth of stones from solution, the only
aspect of interest to the urologist, the phenomenon of supersaturation, has been given the major weight. Because of an excessive
presence of the solute, a process of accumulation on incidental
surfaces or "seeds" starts.

Recently, theoretical studies and experimental results have
shown that the growth of crystals would be quite difficult from an
energetic point of view, if left to supersaturation effects only.

The most accepted theory of successive accumulation of matter
today relies on crystal defects, that is on slight irregularities
occurring during the regular deposition of single or poly-units; it
is easily shown that such defects allow for an increased velocity
of deposition.

Today there is great interest in these defects and in trying
to cause those most productive of crystal growth. It will not be
forgotten here that the most modern technological improvements are
based on faults in crystal lattices, which award to the solid phase
such peculiar properties as semiconduction, elasticity, and fast ionic
transport.

REFERENCES

1. Escher, M.C.: Graphiek en Tekeningen (Erven J.J. Tijl N.V.,
    Zwolle 1959).

CRYSTALLOGRAPHIC CONTRIBUTIONS TO THE PATHOGENESIS OF RENAL

LITHIASIS

R. Mongiorgi

Istituto di Mineralogia e Petrografia
University of Bologna
Bologna (Italy)

ABSTRACT

The present work aims to explain some aspects of renal
lithiasis, especially taking into account the growth phenomena of
crystals.

The integration of x-ray diffractometric analysis with scanning
electron microscopy (S.E.M.), allows the use of new tools for the
interpretation of growth processes involved in urinary stone formation. As a first result, the growth of an alternative crystal
component upon a basic lattice with clear similarity in one or more
cell dimensions, defined more generally as epitaxy, was detected and
confirmed by an accurate analysis of cell parameters of both compounds. Secondly, S.E.M. analysis ascertained the prevailing presence of a dendritic morphology, through which the primary origin
of stone growth may be involved. It is suggested that the interstitial spaces of a dendritic formation assume the role of "active
sites" for further growth: this hypothesis is based on the observation of partial dissolution within cavities, a condition showing
the metastability of this crystalline morphology.

INTRODUCTION

More than 99% of the human calculi are composed of crystalline

substances. Most of these are present in nature and known as minerals
and consequently the urologists very often indicate them with their
mineralogic names.

However they may be divided into two fundamental classes of
stones: the first includes organic substances, and the second a
large percentage of inorganic substances. This simple classification,
sufficient to comprehend the basic differences between the two large
families, allows us to detect that the succession of elementary
processes involved in the formation of a crystal must be the same,
as the nucleation and the growth environment are the same. In fact —
apart from the metabolic processes involved, which characterize the
urinary solution (multicomponent solution) step by step — the crystalline phase or phases which will constitute the calculus are formed
by a process common to both the above mentioned large families of
compounds. The morphological aspect is of great significance in the
interpretation of this phenomenon, as shown later in this report.

EXPERIMENTAL METHODS AND RESULTS

A. X-ray diffractometric analysis

X-ray diffractometry, apart from allowing an accurate qualitative, and in some cases quantitative determination of the crystalline phases present in the urinary stone (1), has been employed
to collect the necessary data in order to calculate the crystallographic constants (a, b, c, α, β, γ,) of the single crystalline
components of the urinary stones. The values obtained, refined by
the least squares method, did not show any substantial variations
from those reported in the literature. A direct comparison between
the cell constants obtained in this way showed structural analogies
representing a favourable factor for epitaxial growth (2). Further
elements supporting the hypothesis are presented by the remarkable
number of interplanar distances (d_{hkl}) with equal or similar values
found in the different crystalline phases. This increases the probability that this growth will occur in a predictable way, as suggested by the theory.

B. Scanning electron microscopic analysis.

Previous studies on the external and fractured surfaces of the urinary stones carried out by means of the scanning electron microscope (S.E.M.), allowed the identification of zones with crystalline morphologies which may be useful in the interpretation of the chemical and physical conditions of the urinary solution facilitating the development of these processes.

Experimental investigation by SEM on a large number of samples allows the observation of:

1) stratified growth with microlayer formation (Photo 1) and regular associations between crystals of the same species (Photo 2), both connected with epitaxial growth;

2) formation of geminates, constituted by the union of two or more crystals of the same species following well defined and characteristic laws (Photos 3, 4);

3) rose-shaped structures (Photo 5);

4) many fibrous and radial-fibrous aggregates, placed above all in the numerous urinary stone cavities (Photo 6, 7);

5) spherulic structures, typical of hydroxyapatite crystallization (Photo 8);

6) numerous dendrite formations, present in all crystalline species, constituting the urinary stones. These formations presumably constitute the so-called "crystalline skeleton", the basic element for subsequent   growth (Photos 9, 10, 11, 12).

DISCUSSION

The careful experimental investigations carried out with both x-ray diffractometry and SEM, allowed the presence of the epitaxial phenomenon to be established. It should be stressed that this phenomenon is not the most important factor in the crystalline growth of urinary stone, its contribution being limited. We do not yet have

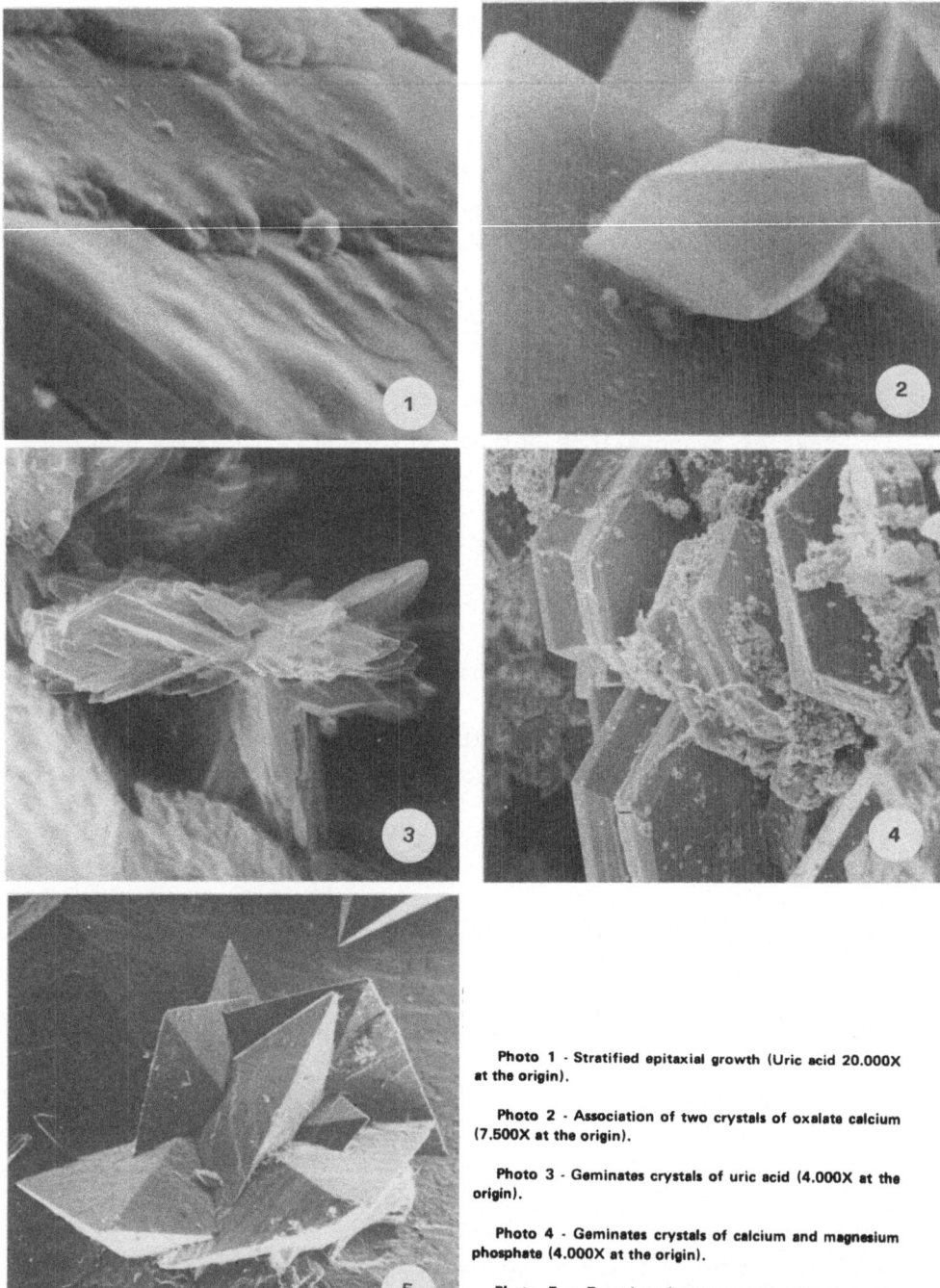

Photo 1 - Stratified epitaxial growth (Uric acid 20.000X at the origin).

Photo 2 - Association of two crystals of oxalate calcium (7.500X at the origin).

Photo 3 - Geminates crystals of uric acid (4.000X at the origin).

Photo 4 - Geminates crystals of calcium and magnesium phosphate (4.000X at the origin).

Photo 5 - Rose-shaped structure of calcium oxalate (2.500X at the origin).

Photo 6 · Fibrous aggregates of uric acid (5.500X at the origin).

Photo 7 · Radial-fibrous aggregates of uric acid (5.500X at the origin).

Photo 8 · Partially crystallized spherulitic structure of hydroxiapatite (8.000X at the origin).

Photo 9 · Microdendrites of Whewellite (6.500X at the origin).

Photo 10 · Crystals of /-Cystine, (example of foliated dendritism, 4.000X at the origin).

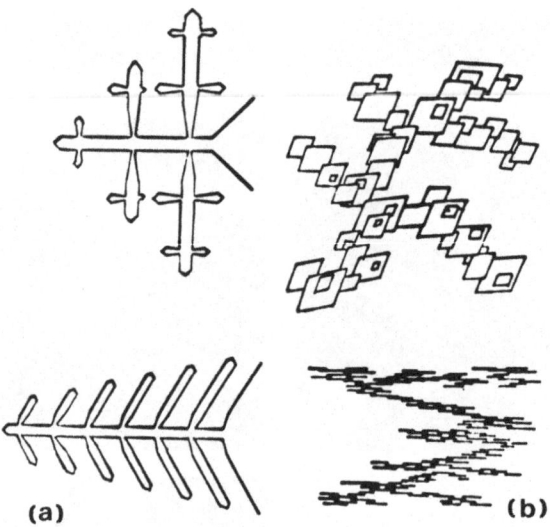

Fig. 1. Various forms of dendritism. (a) ramified; (b) foliated.

a complete picture of the causes of the formation of geminates.
They are however rarely found in human renal stones. It is considered that a geminate comes into being when there is an accidental
equilibrium deviation, in the formation of the first crystalline
nucleus. It is therefore clear that crystal gemination, whether due to
energetic causes or explicable by a kinetic process, must be influenced by the conditions in which the crystal is formed. Over-
saturation, the extremely small dimensions of the crystalline nuclei
and rapidity of growth are favourable conditions to the development
of the geminate. Such conditions, with a certain variability, are
also common to the rose-shaped structure formation, to radial-
fibrous aggregates and also to dendrites.

    Dendritism, both in its ramified and foliated forms (Figs 1a
and 1b) deserve particular attention. Dendritism is easily formed
in a multicomponent solution, like urine containing impurities, at
almost constant temperature, about 37° - 38° C (3). Important local
thermal variations or heat waves are caused by the crystallization.
In this way, there is a rapid growth of dendrites from oversaturated
solutions. Under these conditions, in fact, the molecules of solvent
have no time to leave the crystal surface, because of lack of con-

Photo 11 - Foliated dendritism in uric acid (4.000X at the origin).

Photo 12 - Microdendrites of Weddellite (5.500X at the origin).

Photo 13 - Formation of new dendrites branches in "active site" (4.500X at the origin).

Photo 14 - Partial dissolution in uric acid (5.500X at the origin).

vection or diffusion. The dendrites rapidly increase in size because of the high thermal conductivity of the solution (anisotropical characteristic). If both crystal and solution conducted heat equally, there would not be any reason for the form to deviate from the normal growth model. Furthermore, it was demonstrated (4) that growth appears on planes with lower surface tension, that is in the most populated ones: by consequence the result is not a crystal in equilibrium, but a dendrite.

By assuming that the numerous dendritic formations constitute the basic crystalline skeleton for further growth, it is possible to infer that the interstices formed between the crystalline dendrite aggregates, act as "active sites" for growth.

According to the laws regulating dendrite formation and following changes in temperature, numerous microcrystals (crystalluria) will be subjected to dissolution and to re-crystallization, enlarging the surrounding crystals and producing new ramified dendrites, radial-fibrous aggregates and geminates, which represent a metastable condition (Photo 13).

Furthermore, the urinary solution at an approximate temperature equal to that of the human body surely favours the orientation of free crystalloids and cells, making them crystallographically capable of a rapid increase. Crystal growth from water solutions presents an absolute maximum at about 37° C, equal to the body temperature (5).

In accordance with the hypothesis concerning the fundamental role of the cavities as "active sites" in the growth of further generations of microcrystals, dendritism is a type of growth especially favourable to the supposed mechanism.

The presence of extraneous substances (e.g. inorganic and organic impurities, muco-proteins) favours this rapid increase, and explains the dendritic morphologies.

Hence dendritism represents unstable crystallization; there is a natural approach to equilibrium, as has already been demonstrated for some other substances "in vitro".

This assumption is confirmed by the partial dissolution observed in crystals not treated with drugs (Photo 14).

Dendritism has been considered as the primary cause of volumetric growth of the urinary stones (6).

These assumption indicate that the presence of a muco-protein matrix is fundamental. It is more likely that it plays a complementary role in nucleation processes.

For further information the reader is referred to reference 6.

REFERENCES

1. Mongiorgi, R.: Sulla composizione chimico-mineralogica dei calcoli renali umani. Analisi quali-quantitativa per via diffrattometrica e fluorescenza X. Gior. Clin. Med. 52: 681-699, (1972).
2. Castellano, A., Krajewski, A. and Mongiorgi, R.: Sulla crescita epitassiale dei calcoli renali. Atti Acc. Sc. Lett. Arti di Palermo, Serie IV. 24: 77-89, (1975-1976).
3. Saratovkin, D.D.: Dendritic crystallization. Consultants Bureau, Inc. New York, 1959.
4. Kern, R.: Croissance épitaxique (aspects topologiques et structuraux). Bull. Minéral. 101: 202-233, (1978).
5. Bedarida, F.: La crescita dei cristalli. Rend. Soc. Ital. Min. e Petr. 29: 135-152, (1973).
6. Krajewski, A., Mongiorgi, R., Sabatino, P. and Castellano, A.: Correlation between crystalline micromorphology and type of growth of human renal calculi by scanning electron microscopy. Miner. Petrogr. Acta 21: 101-107, (1976-1977).

CRYSTALLURIA: A VALUABLE PARAMETER IN THE STUDY OF UROLITHIASIS

A. Martelli, V. Pulini and P. Buli

Department of Urology
University of Bologna
Via Massarenti, 9
40138 Bologna (Italy)

ABSTRACT

The finding of crystals in the urinary sediment is of no clinical importance when their number is low and no aggregates are formed. Crystals can occasionally be formed in urines in normal individuals in the absence of any significant and persistent physico-chemical abnormality.

Precipitation and crystallization are indeed constantly secondary to urinary changes (pH variations, decrease of inhibitory power, hyperconcentration) leading to supersaturation. Crystalluria is strictly dependent upon urinary supersaturation. Crystal formation is an obligatory intermediate step between physico-chemical alterations and calculus formation.

From a study of crystalluria that we have conducted in a group of 50 patients suffering from various types of urolithiasis, as well as in a group of normal subjects as controls, interesting results have emerged with regard to volume and aggregation of crystals.

In recurrent stone-formers crystals are usually greater than 150 - 200 μ, whereas in patients having had a single episode of lithiasis, crystals only rarely exceed 100 μ.

Morphologic and ultrastructural analysis of crystals and of crystal aggregates has often revealed epitaxial and dendritic growth patterns.

In addition, crystalluria has been investigated with regard to diurnal variations. A circadian rhythm in crystal excretion has been frequently but not constantly found, involving variations in the number, volume and morphologic features of crystal phases.

Crystalluria is therefore a reliable index of urinary supersaturation (or of lack of inhibitory factors), thereby representing a valuable prognostic marker.

———

The finding of urinary crystals of different composition is a fairly common occurrence in all subjects (3). Variations in number, size and their state of aggregation are representative, however, of those patients suffering from renal stones (5, 6, 7, 8) and may represent a useful parameter in the evaluation of the therapeutic response (9, 10).

METHODS AND MATERIAL

The technique used to evaluate crystalluria may be summarized as follows:

- rapid filtration on S&S 589 filters;

- washing of the filters with distilled water in order to separate the crystals from the other urinary components;

- drying of filters at 37°C;

- transfer of crystals over slides previously treated with a gluewater solution (4);

- addition of fluids with known refraction index (tetrahydronaphthalene; m-xylene; etc.).

For qualitative recognition the following parameters were considered:

- morphology, colour and trasparency of the crystals;

- interference colours by crossed nicols;

- determination of refraction index by means of Becke line
  method;

- state of aggregation and cleavage lines;

- observation of type of growth (epitaxial or dendritic).

The preliminary phase of our study included a control group
of 60 patients without any urologic pathology and 50 patients with
renal stones, 25 of whom had only had a single episode (12 stone
formers at time of first examination, 13 after surgical removal or
spontaneous espulsion), and 25 with plurirecurrent renal stones (at
least 4 episodes in the last two years). In the selection of patients, no importance was given to the type of renal stone, but
only to the pattern of crystalluria.

RESULTS

Our results can be summarized as follows:

a) Control group

- absence of crystals;

- microcrystalluria (crystals smaller than 10 μ);

- presence of very few crystals of size between 10 and 50 μ .

b) Non recurrent stone-formers

- absence of aggregates or, only occasionally, aggregates
  smaller than 50 μ (Fig. 1);

- presence of a few crystals smaller than 100 μ , usually in
  greater number than in control group;

- size of crystals and of aggregates rarely greater than 100 μ.

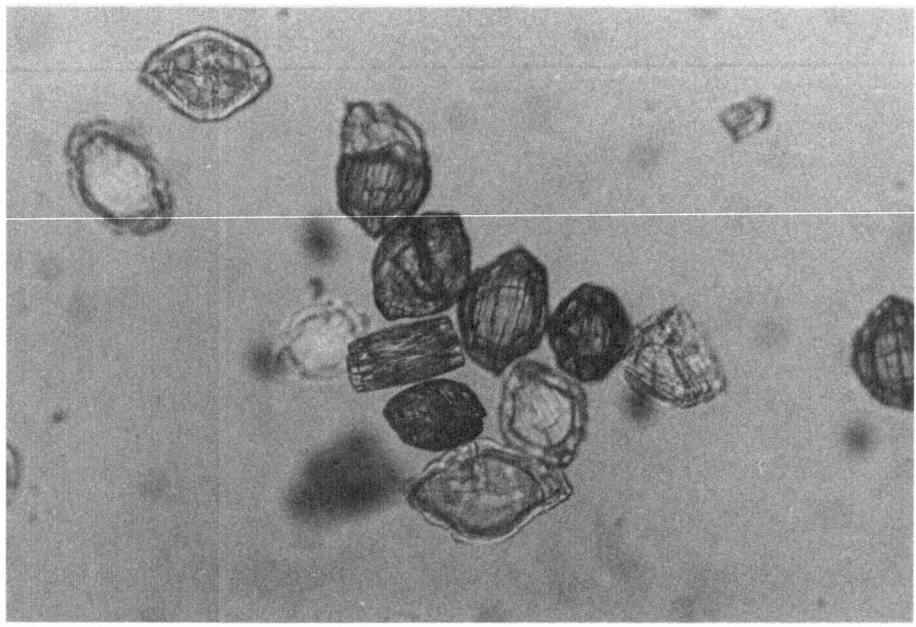

Fig. 1.   Uric acid crystalluria in a non recurrent stone former.
          The absence of aggregates should be noted.

c) <u>Recurrent stone formers</u>

- presence of aggregates, greater than 150 μ (Fig. 2);

- number of crystals greater than 100 per litre of urine.

- presence of crystals and of aggregates greater than 100μ.
  In struvite stones, however, smaller crystals (50-60μ) may
  be present.

These data are presented in figures 3 and 4.

CONCLUSIONS

    The data obtained confirm the importance of crystalluria as a

Fig. 2. Uric acid crystalluria in a recurrent stone former.

diagnostic test in patients with renal stone disease. This affirmation is strengthened by the following _in vivo_ observations:

1) pathological crystalluria is the result of an imbalance in the relationship between saturation and inhibition in the urine.

2) the degree of crystalluria is proportional to urinary saturation.

From a qualitative point of view some observations need to be stressed:

- in struvite crystalluria the volume of crystals is much more variable than in the other crystalline phases;

- the monohydrated form of calcium oxalate (whewellite) is much more frequent than the dihydrated one (weddelite). This result, which emerged from our work, is partially in contradiction with the data from other Authors (1, 2, 5).

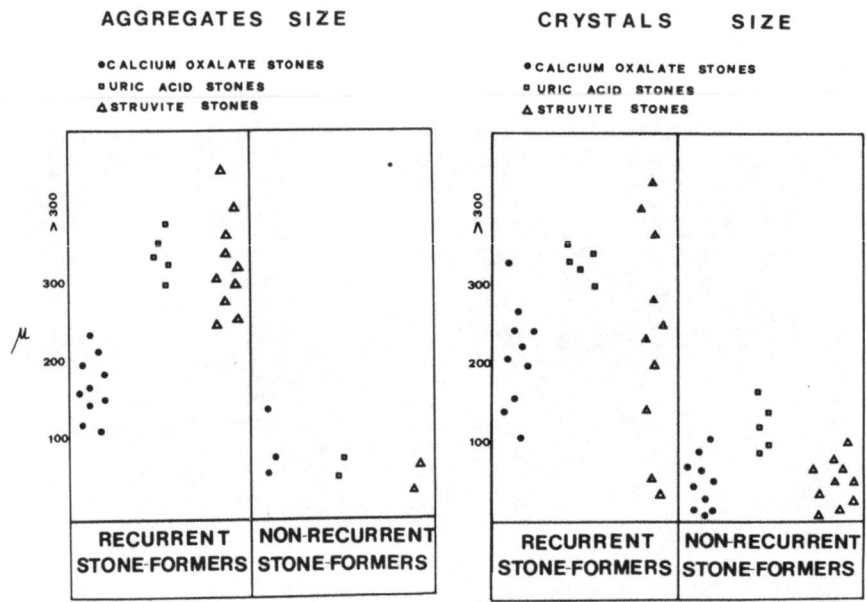

Fig. 3.   Number of crystals and of aggregates in recurrent
and non recurrent stone formers.

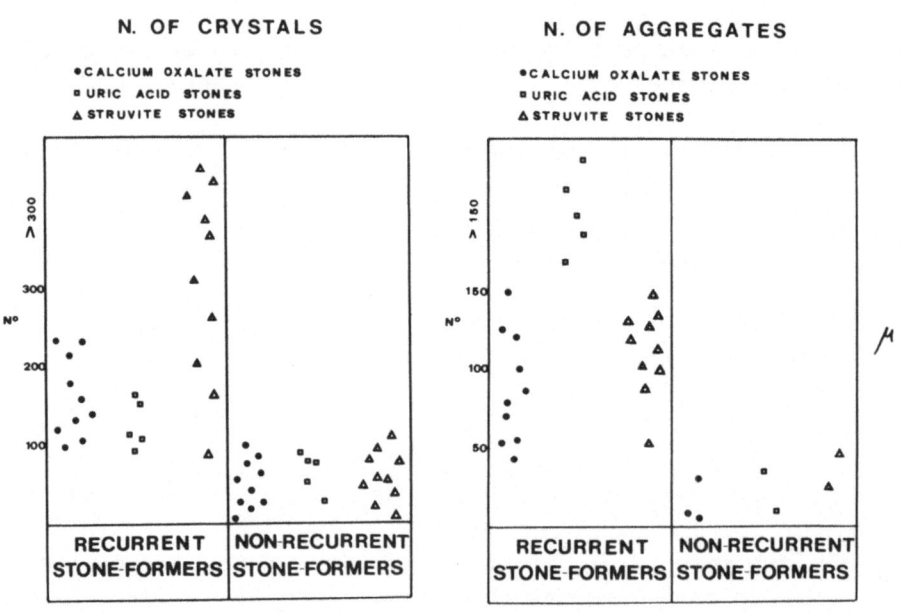

Fig. 4.   Size of crystals and aggregates in recurrent and non
recurrent stone formers.

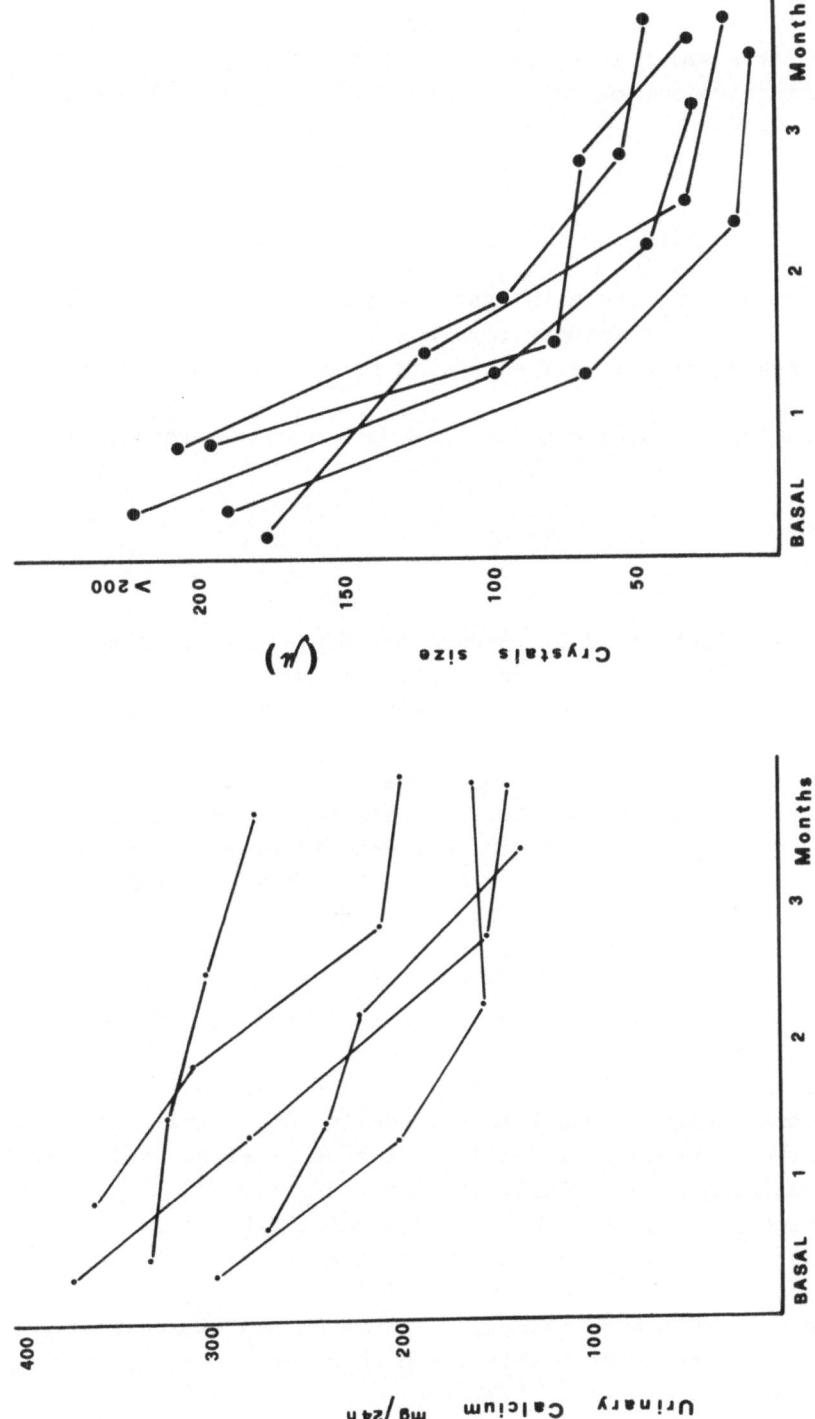

Fig. 5. Variations of urinary calcium and crystal size during therapy with hydrochlorothiazide (100 mg/day).

Finally, an investigation of crystalluria can supply a useful parameter for a valid interpretation of the therapeutic response during administration of the drugs used in preventive medical therapy (Fig. 5).

SUMMARY

If crystalluria is investigated using appropriate techniques, it will be found that urinary crystals vary in number, size and degree of aggregation in patients with recurrent renal stones.

Patient with a single stone episode may have crystalluria within normal limits.

REFERENCES

1. Berg, W., Schnapp, J.D., Schneider, H.J. and Hienzsch, E.: Crystal-optical and spectroscopical findings with calcium oxalate crystals in the urine sediment. Eur. Urol. 2: 92-97, (1976).
2. Berg, W., Schneider, H.J. and Hesse, A.: Crystal-optical findings on calcium oxalate of uric concretion. In Urolithiasis Research, edited by Fleisch, H., Robertson, W.G., Smith, L.H. and Vahlensieck, W., Plenum Press, New York, 241-247, (1976).
3. Elliot, J.S. and Rabinowitz, I.N.: Crystal habit, structure and incidence in the urine of a hospital population. In Urolithiasis Research, edited by Fleisch, H., Robertson, W.G., Smith, L.H. and Vahlensieck, W., Plenum Press, New York, 257-260, (1976).
4. Gazzi, P., Zuffa, G.G., Gandolfi, G. and Paganelli, L.: Provenienza e dispersione litoranea delle sabbie delle spiaggie adriatiche fra le foci dell'Isonzo e del Foglia: inquadramento regionale. Mem. Soc. Geol. It. 12: 1-37, (1973).
5. Robertson, W.G., Peacock, M. and Nordin, B.E.C.: Calcium oxalate crystalluria and urine saturation in recurrent renal stone-formers. Clin. Sci. 40: 365-374, (1971).
6. Robertson, W.G. and Peacock, M.: Calcium oxalate crystalluria and inhibitors of crystallization in recurrent stone formers. Clin. Sci. 43: 499-506, (1972).
7. Robertson, W.G., Marshall, R.W., Peacock, M and Knowles, M.R.C.:

The saturation of urine in recurrent, idiopathic calcium stone-
formers. In Urolithiasis Research, edited by Fleisch, H.,
Robertson, W.G., Smith, L.H. and Vahlensieck, W., Plenum Press,
New York, 335-338, (1976).

8. Robertson, W.G.: Physical chemical aspects of calcium stone-
    formation in the urinary tract. In Urolithiasis Research,
    edited by Fleisch, H., Smith, L.H. and Vahlensieck, Plenum
    Press, New York, 25-29, (1976).

9. Smith, L.H.: Application of physical, chemical, and metabolic
    factors to the management of urolithiasis. In Urolithiasis
    Research, edited by Fleisch, H., Robertson, W.G., Smith, L.H.
    and Vahlensieck, W., Plenum Press, New York, 199-211, (1976).

10. Van Den Berg, C.J., Cahill, T.M. and Smith, L.H.: Crystalluria.
    In Urolithiasis Research, edited by Fleisch, H., Robertson,
    W.G., Smith, L.H. and Vahlensieck, W., Plenum Press, New York,
    365-366, (1976).

URODYNAMIC INVESTIGATIONS IN NEPHROLITHIASIS

J. Hannappel

Department of Urology
Medical Faculty
Rheinisch Westfälische Technische Hochschule
Aachen (West Germany)

ABSTRACT

Urodynamics and urinary lithiasis are two subjects in urology which are normally thought to be well separated. Nevertheless, sometimes it is useful to have a new approach to well established questions. Three points will be discussed in this paper:

1. Urodynamic aspects of stone genesis.

2. Current urodynamic methods for functional studies of the upper urinary tract.

3. Physiology and pharmacology of the pyeloureteral system.

1. URODYNAMIC ASPECTS OF STONE GENESIS

Primarily, renal lithiasis is the consequence of metabolic disturbances, which can create intrarenal crystallizations or microliths in the urinary collecting system. The first beginnings of these processes are still under discussion. Mainly, there are two theories (table 1):

Table 1. Primary nucleation in urinary lithiasis.

THEORY OF FIXED PARTICLE NUCLEATION:

Nucleation in the parenchyma of the papillae
(Randall's Plaques).

THEORY OF FREE PARTICLE NUCLEATION:

1) Nucleation in the renal tubules.
2) Nucleation in the renal calyces.

---

1. Theory of fixed particle nucleation, i.e. nucleation in the
parenchyma of the papillae;

2. Theory of free particle nucleation, i.e. crystallization
starts in the urinary collecting system.

In 1936 Randall (8) was the first to demonstrate that little
crystalline plaques can sometimes be found in the papillary region
of the kidney. They are located subepithelially and are called
Randall's plaques. Recently Hautmann and Lutzeyer (4) have reported
evidence which supports the theory that renal lithiasis starts in
the parenchyma itself, by demonstrating an excessive supersaturation
for calcium and oxalate in the tissue of the papilla. Moreover, the
theory of renal nucleation has the advantage that microliths in the
parenchyma have plenty of time to grow, whereas calculi forming in
the collecting system should always be washed out by the urinary
stream. Hautmann and Lutzeyer have presented the calculations shown
in table 2.

It ensues that crystallization in the renal collecting system
in the absence of stasis create very small calculi.

On the other hand, Schultz and Schneider (9) demonstrated by
calculation that laminar flow should be present in all parts of
the upper urinary tract including the renal tubules. The mean urine
flow rates in the different regions are given in table 3. These
values are obviously largely depending on anatomical and functional
variations. Moreover, laminar flow, the suggested pattern in the

Table 2. Renal transit times.

Parenchymal transit time:	3 min
Pelvic transit time:	12 min
Growth rate of calcium oxalate stones:	2 μm/min
Potential increase in stone radius	0.03 mm

upper urinary tract, is characterized by highly different flow velocities (Fig. 1): Central flow is highest and it slows to zero at the inner wall of the tube. This means that there is no flow near the urothelium and consequently no force which could interfere with adhesive forces between urothelium and small calculi. This is probably the reason why even small calculi can be fixed in the upper urinary tract, are enabled to enlarge and finally cause obstruction.

All these facts demonstrate that over and above the metabolic disturbances which are obviously the most important reasons for urinary lithiasis, urodynamic aspects must also be taken into consideration to get a better understanding of this condition.

Table 3. Flow rates in the urinary tract.

Distal tubules:	0.3 - 3.8 mm/min
Papillary ducts:	80 - 980 mm/min
Renal pelvis:	1 - 10 mm/min
Ureter (systole):	600 - 4000 mm/min
Ureter (diastole):	0 mm/min
Urethra:	0 - 50000 mm/min

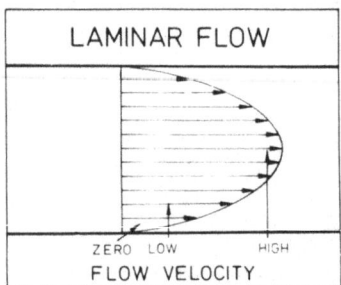

Fig. 1. Laminar flow, as throught to occur in the upper
         urinary tract is characterized by highly different
         flow velocities.

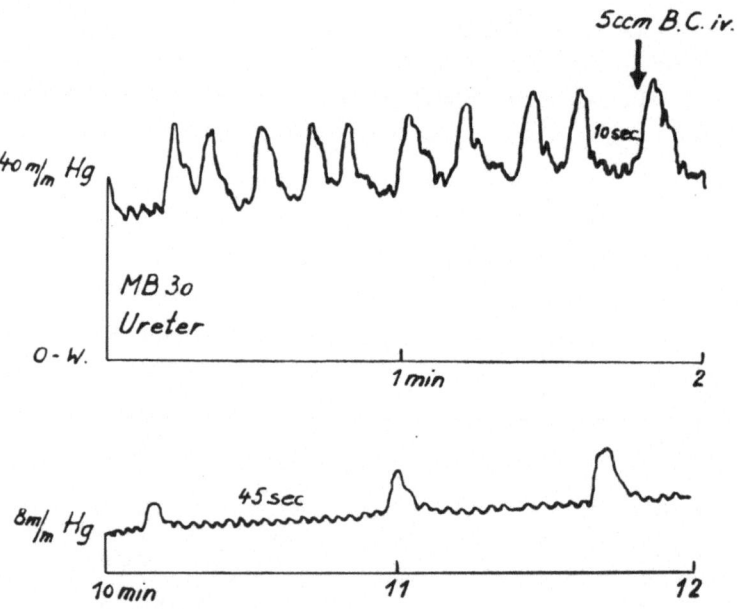

Fig. 2. Pressure waves in human ureter before and after 20 mg
         hyoscine-n-butylbromide intravenously (5 cc B.C. iv.)
         (from Lutzeyer, 6).

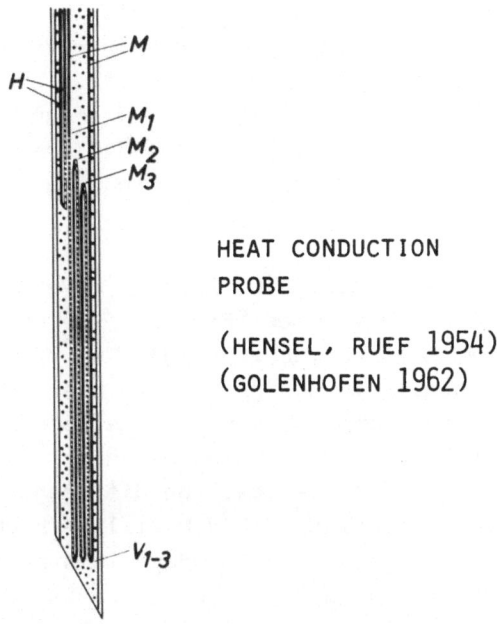

HEAT CONDUCTION
PROBE

(HENSEL, RUEF 1954)
(GOLENHOFEN 1962)

Fig. 3.   The heat conduction probe (from Hensel, Ruef (5) and
          Golenhofen (2).

2. CURRENT URODYNAMIC METHODS FOR FUNCTIONAL STUDIES OF THE UPPER URINARY TRACT

Intrapelvic and intraureteral pressure measurements have been
facilitated by the invention of electronic measuring instruments.
One of the first to do these investigations in the urinary tract
was Lutzeyer (6), who publishied his results in 1957. Fluid-filled
catheters were introduced transurethrally into the renal pelvis and
the ureter and connected to pressure transducers. The effect of
drugs on the peristalsis of the ureter could thus be investigated
(Fig. 2). In the following years, numerous results obtained by
pressure measurements were published and there is no doubt that
much of our knowledge of upper urinary tract function is due to this
method. Nevertheless, this technique has the disadvantage of being
quite invasive, which can be hazardous to the patient and which
can possibly cause artifacts. Probably, this is the reason why this
method is not used routinely in our hospitals.

Whitaker (11) has described a method to measure obstruction
in the upper urinary tract, in which a needle is introduced into
the renal pelvis transcutaneously, guided by x-rays or ultrasounds.
The pyeloureteral system is then perfused with a sterile solution
at a rate of 10 ml/min and the intrapelvic pressure is recorded
simultaneously. Pressure rises below 10 cm H_2O are regarded as
normal. Rises above 20 cm H_2O indicate obstruction in the pyelo-
ureteral system.

Hensel, Ruef and Golenhofen (6, 2) have invented a probe to
measure heat conduction in tissues (Fig. 3). For this purpose, a
needle is equipped with a heating zone (H). Two temperature measuring
points are mounted equally in the probe: one near by (M) and one
far from (V) the heating zone. This probe has been developed origi-
nally to measure the blood flow in tissues and it is easy to understand that, if the blood flow is low, the difference in temperature
between the two measuring points V and M will be high and vice versa.
Melchior and Simhan (7) have tried to adapt this probe in order to
measure intraureteral flows, and have combined it with an electronic
pressure device. Here too, a thermosensitive region of the probe is
heated by a constant current and cooled down by the passing urine
boli. Thus, the probe gives information about urine flow and urine
pressure in the ureter. But because of the very difficult calibration,
this method cannot be used routinely in clinical practice.

Another important method for investigating ureteral peristalsis
is electromyography. It gives important information concerning
physiology, pathophysiology and pharmacology of the pyeloureteral
system. For that purpose we have implanted circular electrodes
around the ureters of laboratory animals (Hannappel and Golenhofen,
3). The connecting leeds were conducted subcutaneously and left
the body of the animal in the region of the neck. By this method
it was possible to record ureteral peristalsis for several days or
even weeks without further disturbing the animal. This method, which
has great benefit in scientific research, is obviously not applicable
in humans, and as far as we know, electromyographic studies are not
yet used routinely to make functional studies on ureteral peristalsis in patients.

Tscholl et al. (9) invented videodensitometry. This method is
based on modern videotechnology including cineradiography and a
television-amplifier (Fig. 4). After having given a contrast medium

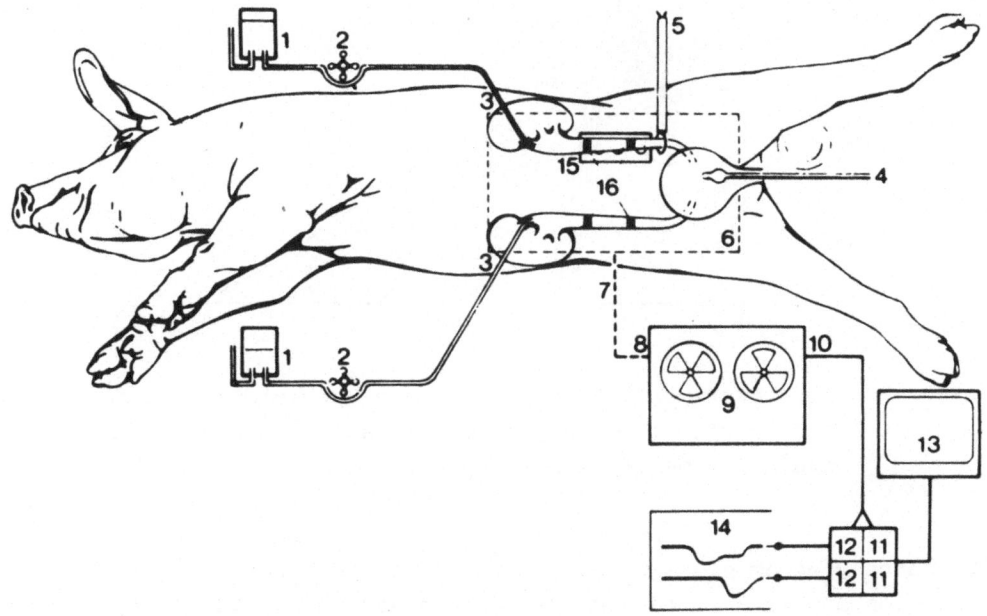

Fig. 4. Videodensitometry is based on modern videotechnology
including cineradiography and a television amplification (from Tscholl et al., 10).

intravenously, which is excreted by the kidney, two regions of
interest are chosen electronically. These small electronic windows
are localized exactly to a proximal and a distal part of the same
ureter. Whenever urine with dye is transported by the ureters and
is thus passing one of the two windows, the video signals change.
These changes are displayed by a direct recorder. Thus it is possible
to get records of the contraction frequency and conduction velocity
of the ureters. With more sophisticated electronics, this method
may become more widely employed. One of its greatest advantages is
that it is be non-invasive.

Durben and Gerlach (1) have recently presented their Time-
Distance Diagram (TDD). This method is also based on cineradiography:
ureteral peristalsis is filmed, 5 to 10 pictures being taken each
second; the length of the urine boli in these pictures is measured
and plotted in a Time-Distance Diagram. The abscissa gives the time,
and the ordinate the length of the pyeloureter (Fig. 5). This Time-

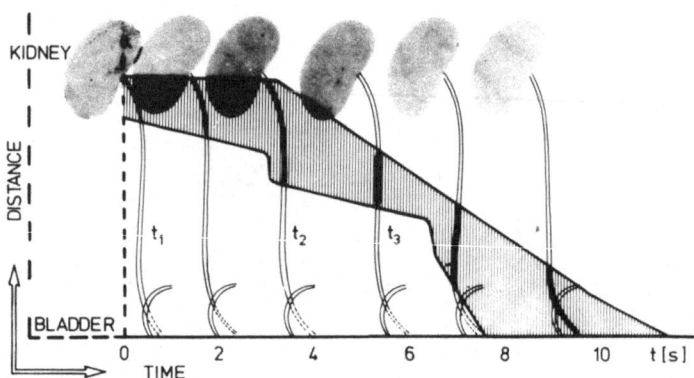

Fig. 5.   Time–distance diagram of ureteral peristalsis in dog.
          The length of the urine boli (X axis) is plotted
          against time (Y axis) (from Durben and Gerlach, 1).

Distance Diagram gives us good information concerning ureteral
function and is able to display early stages of disturbances.

     All methods described here are sometimes of great value if a
subtle investigation of ureteral function is requested. Nevertheless,
in most cases cineradiography with TV-amplification and tape recording is sufficient. This method is non-invasive, easy to do and
gives more information than simple urography (Fig. 6). The latter
has its advantages when high resolution is requested, but is largely
inferior in studies of functional disturbances of the ureter which
can well be responsible for recurrences of stone formation.

3. PHYSIOLOGY AND PHARMACOLOGY OF THE PYELOURETERAL SYSTEM

     The function of the pyeloureteral system can be compared to the
physiology of the heart: in both, generation and trasmission of
exitation are of myogenic nature. In the pyeloureter, the quickest
spontaneous activity is located in muscle strips from the calyces.
Much lower spontaneous activity is found in isolated muscle strips
from the renal pelvis and even lower contraction frequencies are
measured in isolated ureters (Fig. 7). As all these regions are
coupled myogenically via tight junctions, the region with the highest
spontaneous activity will act as a pacemaker for the whole system
So the ureter is mainly commanded by the calyceal musculature, as
the heart is controlled by the sinus node.

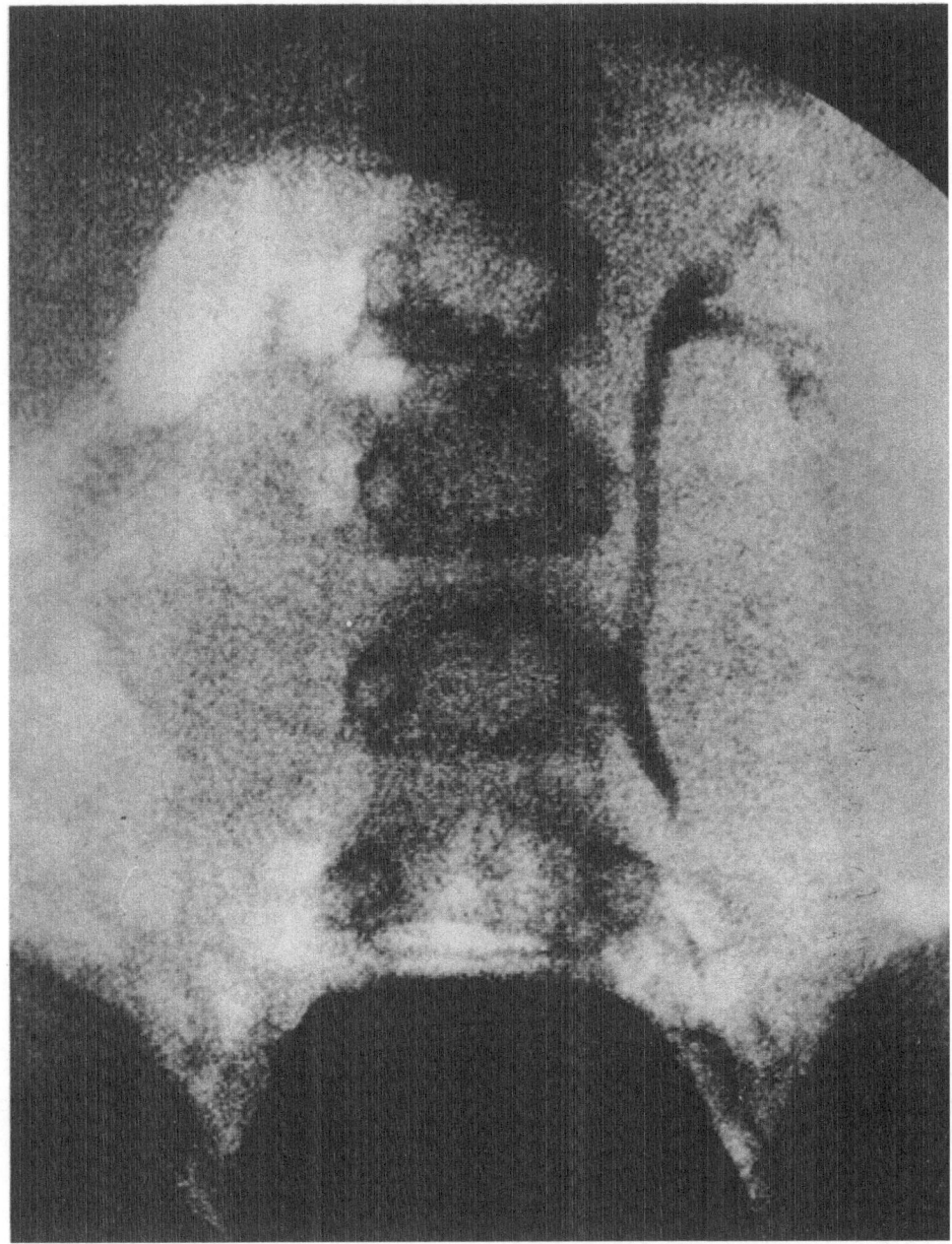

Fig. 6. Cineradiography with TV-amplification and tape re-
         cording gives the best information in functional
         disturbances of the ureter, but is inferior to simple
         urography if high resolution is required.

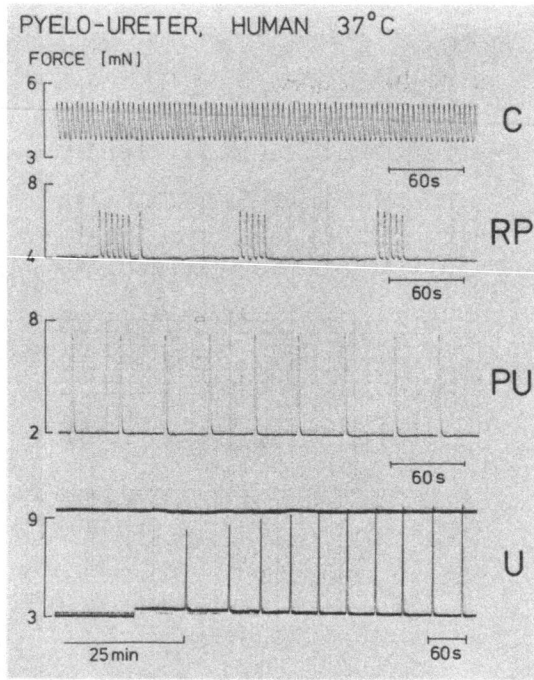

Fig. 7.   Spontaneous active muscle strips from different
          regions of the human pelvis and ureter. Decreasing
          contraction frequency from proximal to distal.
          (RP = Renal Pelvis; PU = Pyeloureteral Junction;
          U = Ureter).

         But in one point, there is an important difference between
the physiology of the heart and the pyeloureter: in the heart, the
contraction frequency is modified by sliding variations of the
pacemaker, i.e. sinus node frequency, whereas blocks in the myogenic
conduction are always pathologic signs.

         Contrarily, the calyceal pacemaker of the ureter seems to be
relatively stable and its frequency is nearly independent of diuresis.
The regulation of the transport capacity of the ureter takes place
at the level of the excitation conduction: if there is a high urine
output each excitation is conducted from the calyces and the pelvis
to the ureter, whereas at low urine flows a physiological conduction
block takes place. That means that many of the excitations are

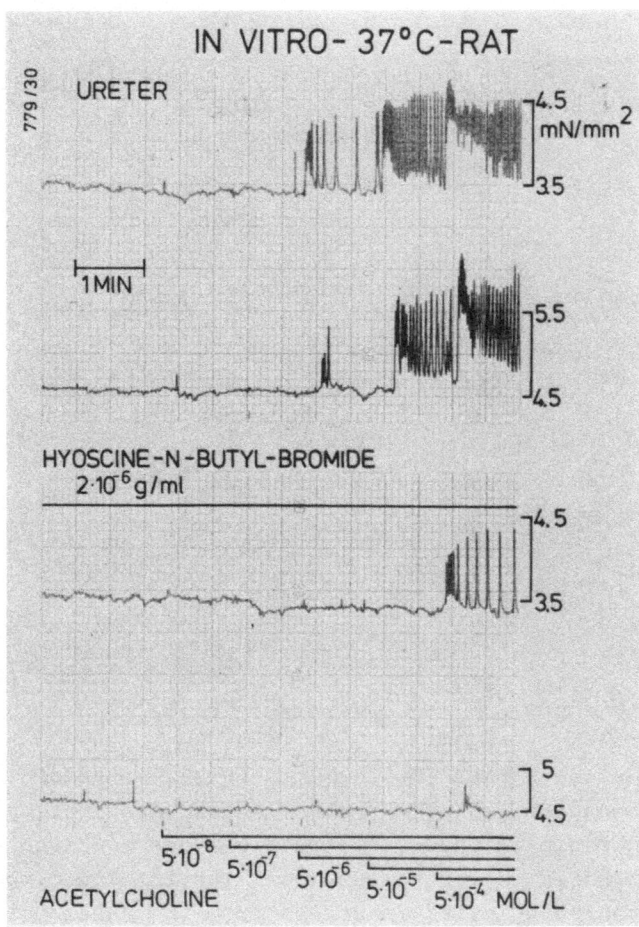

of acetylcholine can be inhibited by hyoscine-n-butylbromide.

stopped myogenically in the region of the pyeloureteral junction.

However the pyeloureteral system is not only regulated by myogenic activity and diuresis. The nervous system is able to modify its function. The parasympathetic system, which releases acetylcholine as transmitter substance in its postganglionic synapses augments the ureteral contractions.

This can be shown in an organ bath, where muscle strips from the ureter are exposed to increasing concentrations of acetylcholine. This effect can be inhibited by anticholinergic substances like

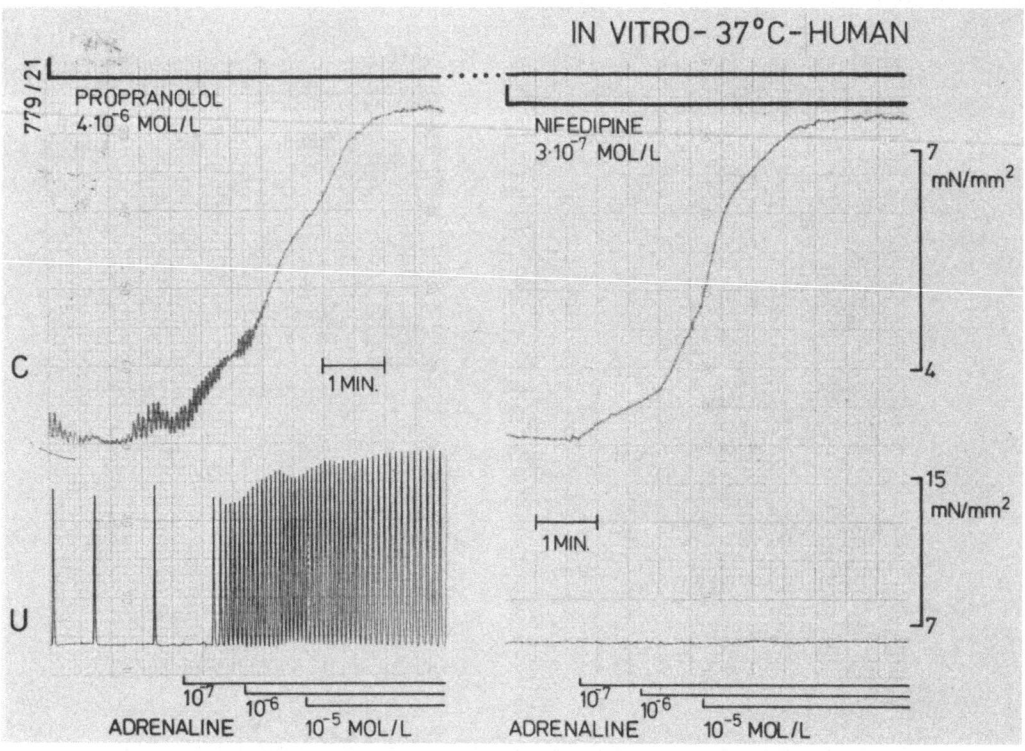

Fig. 9.   Muscle strips from human pelvis and ureter (RP =
          Renal Pelvis; U = Ureter). After application of
          nifedipine phasic contractions are completely sup-
          pressed. Nevertheless, long lasting tonic contractions
          can be elicited in strips from the renal pelvis (but
          not from the ureter).

hyoscine-n-butyl-bromide, which blocks the cholinergic receptors
without exiciting them (Fig. 8).

    Intraureteral pressure recordings revealed the same effects.
Here too, accelerated ureteral peristalsis is normalized after
application of hyoscine-n-butyl-bromide intravenously (see Fig. 2).

    The sympathetic system also influences ureteral ·peristalsis,
the latter in a double manner. There are alpha-adrenergic (ex-
citatory) and beta-adrenergic (inhibitory) influences on the ureter.
Via alpha-receptors the sympathetic system accelerates the pacemaker
processes, improves the conduction rate and augments the contraction

force of ureter. Via beta receptors contrary effects are provoked.
Beta-stimulators, like orciprenaline, are even capable of completely
inhibiting ureteral peristalsis. This effect was thought to be able
to block ureteral spams in patients with ureteric stones and thus
to accelerate their emission. So far however, this treatment has
failed to prove its effectiveness and it also has severe cardiac
side effects, e.g. tachycardia and is therefore applicable only in
younger patients.

Nifedipine, which belongs to the group of antagonists of calcium, is another pharmacologic substance with important influences
on ureteral peristalsis. With nifedine, phasic contractions of the
ureter are completely suppressed (Fig. 9).

Perhaps, some of this groups of drugs will enable us in future
to give more effective treatment of urinary lithiasis and to bring
more stones to spontaneous emission, quicker and with less pain.
Thus, it might be possible to interrupt those vicious circles,
which are shown in Fig. 10 and which favour the growth of renal
calculi at an earliest stage.

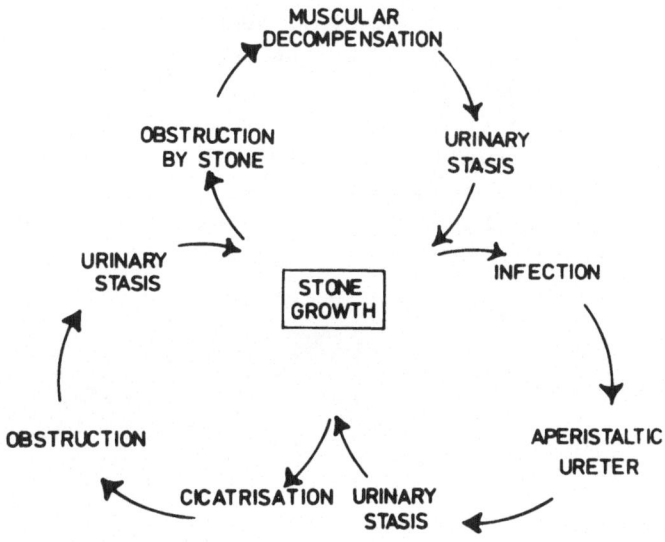

Fig. 10.   Three vicious circles, which may favour growth of
           renal calculi.

REFERENCES

1. Durben, G., Gerlach, R., Eichhorn, F., Friedrich, R., Schäfer, W. and Lutzeyer, W.: A new method to analyze ureteral peristalsis by cineradiography. Inv. Urol. 18: 207-208, (1980).
2. Golenhofen, K.: Zur Reaktionsdynamik der menschlichen Muskelstrombahn. Arch. Kreisl. 38: 202-223, (1962).
3. Hannappel, J. and Golenhofen, K.: Comparative studies on normal peristalsis in dogs, guinea-pigs and rats. Pflügers Arch. 348: 65-76, (1974).
4. Hautmann, R. and Lutzeyer, W.: Der Kalzium-Oxalat Stein. (Springer Verlag, Berlin, Heidelberg, New York, 1980).
5. Hensel, H. and Ruef, J.: Fortlaufende Registrierung der Muskeldurchblutung am Menschen mit einer Calorimetersonde. Pflügers Arch. ges. Physiol. 259: 267-280, (1954).
6. Lutzeyer, W.: Schmerzbekämpfung und Spasmolyse in der Urologie. Zeitschr. Urol. 50: 109-119, (1957).
7. Melchior, H. and Simhan, K.K.: Zur Problematik der Urorheomanometrie (Entwicklung eines neuen Urorheomanometers). Biomed. Tech. 16: 99-102, (1971).
8. Randall, A.: Hypothesis for origin of renal calculus. N. Engl. J. Med. 214: 234, (1936).
9. Schultz, E. and Schneider, H.J.: Der Einfluß der Strömungsverhältnisse im hydrodynamischen System der harnableitenden Wege auf die Harnsteingenese. Z. Urol. Nephrol. 72: 915-927, (1979).
10. Tscholl, R., Keller, U. and Spreng, P.: Investigation of urine transport in the ureter of the pig by x-ray videodensitometry. I. The effect of increasing diuresis on velocity and rhythm of contractions: relationship between pressure wave and urine bolus. Inv. Urol. 13: 404-410, (1976).
11. Whitaker, R.H.: Methods of assessing obstruction in dilated ureters. Brit. J. Urol. 45: 15, (1973).

INHIBITORS AND PROMOTERS OF STONE FORMATION*

H. Fleisch

Department of Pathophysiology
University of Berne
Berne (Switzerland)

Currently, three main mechanisms are thought to be important
in the formation of urinary stones: 1) the relationship between the
concentration of the precipitating substances in urine and the solubility of the mineral phase formed, 2) the role of promoters of
crystallization and aggregation, and 3) the part played by inhibitors of crystal formation and aggregation (Fig. 1).

SATURATION OF URINE

It is now widely accepted that even in normal people, urine
is ordinarily supersaturated with respect to calcium oxalate (1-6),
octocalcium phosphate (2, 3), hydroxyapatite (1, 2), and sometimes
with respect to brushite (1, 7, 8). The degree of supersaturation
is usually higher in patients with urinary stones (2-4, 7-10). This
is due mainly because these patients tend to excrete more calcium
(10-12), but also because urinary oxalate can be increased (10, 13).
Urine is also often supersaturated with respect to sodium urate and
ammonium urate (14, 15). Supersaturation with regard to magnesium
ammonium phosphate is restricted to the cases where the urine becomes alkaline because of ammonium production by bacteria.

---

* Reprinted from <u>Kidney International</u>, Vol. 13, 1978, pages 361-371
  (with permission).

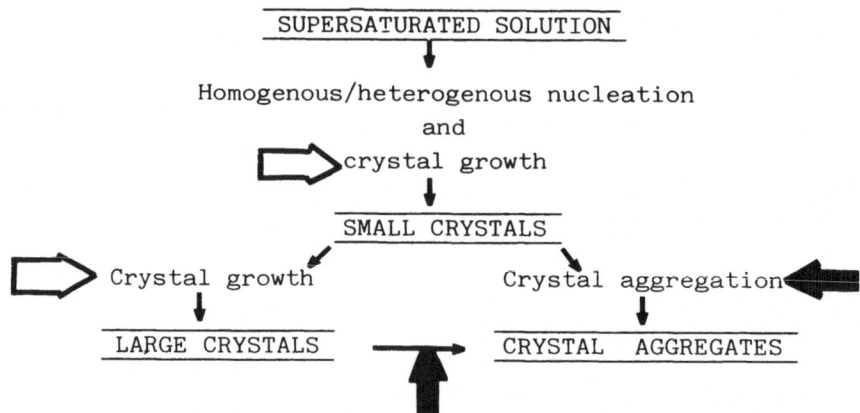

Fig. 1.   Mechanisms of mineral formation. Inhibition of crystal
          growth is denoted by open arrows; inhibition of crystal
          aggregation is denoted by the large closed arrows.

Supersaturation can vary in degree. It can be in the metastable range where precipitation may occur only when induced by
epitaxy or heterogenous nucleation, or it can be in the unstable region
where rapid spontaneous precipitation does occur. The limit between
the two ranges, which can be called the spontaneous formation product, is not a fixed number but will depend upon the duration of
incubation.

These theoretical considerations are relevant to what occurs
in vivo. It was found that when the urinary saturation measured
chemically is above the spontaneous formation product required for
a rapid induction of precipitation in inorganic solutions, crystals
can usually be detected in the voided urine specimen (3). On the
other hand, when the ion product is below this product, crystals are
absent.

CRYSTAL GROWTH AND CRYSTAL AGGREGATION

In the past, attention was devoted mostly to the formation and
growth of crystals. Recently, interest has been directed to an area
which, until now, had been neglected: the crystal aggregation. This
term describes the process of crystals binding one to another,
resulting in the formation of larger clusters. In vitro, aggregation
of both calcium oxalate (16, 17) and calcium phosphate crystals (18)

occurs readily when the solution is supersaturated. Aggregation could be the mechanism which distinguishes simple crystalluria, which occurs in most normal people, from stone formation. This line of thought is strengthened by the finding that while usually only individual calcium oxalate crystals are found in normal people, stone-formers often excrete large aggregates of this salt (16, 19).

PROMOTERS OF CRYSTALLIZATION

A few years ago, great emphasis was placed on the supposed activating role of organic compounds in urine (20). This was based on the finding that urinary proteins bind calcium (21) and can induce calcification in vitro (22) under certain conditions. Furthermore, proteins are increased in amount (22, 23) and are qualitatively different (24, 25) in urine samples from stone-formers than in that from normal people. Such an activating theory was invoked to account for the close morphological relation existing between the matrix and the mineral in the stones (26, 27). All these results, however, give no evidence that the change in the urinary proteins is the primary event in stone formation. Indeed, they just as well might be secondary to the irritation from the stone and to infection. The morphological relation between matrix and mineral could be due to a co-precipitation of the proteins with the mineral or their secondary binding into the crystals (28). Lately, the theory of the activating role of urinary proteins has been abandoned by most investigators. As such a role, however, could not be proven, it also has not been disproven.

In recent years, emphasis has been shifted towards the role of crystals of one salt inducing the crystallization of a salt of another nature. Such an epitactic induction between crystals having similarities in lattice dimensions is a well known phenomenon in crystallography. Relevant lattice similarities are present between uric acid , calcium oxalate, and calcium phosphate crystals (29), and epitactic induction does occur among them. Thus, the precipitation of sodium urate is favored by both hydroxyapatite and calcium oxalate (30). The precipitation of calcium oxalate can be induced from a metastable solution by hydroxyapatite (30, 31), brushite (30), and urate (30, 32, 33) but not consistently by uric acid (30, 33). Conversely, calcium phosphate precipitation is stimulated by both calcium oxalate and monosodium urate crystals (30, 33), while uric

acid is not effective (30, 33). Interestingly, this effect of calcium
oxalate on calcium phosphate precipitation is not as efficient as
the reverse (31). This might be because hydroxyapatite is not the
first salt to form when calcium phosphate precipitates, but is
preceded by other phases, the nature of which is still disputed.
These initially formed phases spontaneously transform into the more
stable hydroxyapatite.

This epitactic mechanism of precipitation could give an explanation for the well known fact that most stones are formed not just
by one salt but by a mixture of different kinds of salts. It could
also explain the clinical findings that patients with calcium stones
are often hyperuricosuric (34) and that treatment with allopurinol,
which decreases the excretion of urate, reduces the formation of
calcium stones (34, 35).

INHIBITORS OF CRYSTALLIZATION

Since urine usually is supersaturated with respect to the
various stone-forming salts and contains crystals, the conditions
for crystal formation, aggregation, and thus stone formation are
satisfied. Thus, the main question is not why stones can form but
why stones do not form more generally. One explanation might be the
presence in urine of very effective inhibitors for both these processes, crystal formation and aggregation.

INHIBITORS OF THE FORMATION OF CALCIUM PHOSPHATE

Chemical estimation of inhibitory activity. One of the first
techniques testing urine's inhibitory capacity (36, 37) made use of
a system, described in the 1930's, for studying the calcification
of cartilage (38). Epiphyseal cartilage from rachitic rats was
incubated in vitro in a supersaturated salt solution, and the precipitation of calcium phosphate was studied. Urine or the substances
to be tested are added to the incubation fluid. The results, however, are difficult to interpret. Since the cartilage is enzymatic-
ally very active, it is likely to destroy certain inhibitors.
Furthermore, the test substance or urine may influence the enzyme
activity as well as the metabolic machinery of the incubating
cartilage, thereby altering the precipitation by a mechanism un-

related to crystal growth inhibitors (39). Obviously, the results
obtained can be totally unrelated to stone formation, and the technique, while possibly useful for cartilage calcification, should be
abandoned for the study of stone formation.

A theoretically correct approach of practical importance is
the determination in vitro of the minimum product of calcium and
phosphate (formation product) necessary for the formation of a solid
phase within a present time period under defined conditions. The
experimental conditions of this technique, which measures mainly
heterogenous nucleation, as demonstrated by Finlayson, are critical
and need to be carefully controlled. For example, the time of incubation has an inverse effect: the shorter the time of incubation
the greater the formation product (40). Inhibitors will increase
this product. It has been reported that inhibitors are not active
at very short incubation times, and that they increase the formation
product value only up to that obtained at short incubation times
(40). It has been suggested that this occurs because homogenous
nucleation is predominant initially, and is later replaced by heterogenous nucleation. However, it appears to be quite unlikely that
the above mentioned phenomenon of homogenous nucleation could proceed in complex solution such as urine. Moreover, the report showing
the independence of formation product from inhibitor activity has
not been confirmed (41).

It must be kept in mind that the determination of the formation
product does not measure specifically an effect on the formation of
the final phase, since in some cases there is first one salt formed
which then transforms into a salt of another type (42). Thus, an
inhibitor acting on the growth of the second salt might be missed
since the first will be formed normally. The nature of the first
phase is still disputed. Amorphous calcium phosphate (43, 44),
brushite (45, 47), octocalcium phosphate (40, 45, 48, 49) are some
of the proposed forms.

The formation product can also be determined in the presence
of a nucleating agent of some kind, such as collagen (50), elastin,
or crystals of another salt. Nucleating agents will decrease the
formation product for a given incubation time and permit the detection of inhibitory activity at shorter incubation times.

While very useful to investigate the effect of individual

compounds, the determination of the formation product presents
problems when used to test urine. In urine, one must measure the
thermodynamic product of the ion activities and not just the simple
calcium x phosphate product. Various means have been investigated
to determine this product. One is to determine chemically all the
ions and complexes involved and to calculate their activities in
the urine using the chemical stability constants described in the
literature (2). This procedure is tedious and probably not without
error, since some of the stability constants are still doubtful
(51). A simplified approach has been suggested which consists of
measuring the formation product as the ionic concentration of calcium times phosphate (Ca^{2+} x $HPO_4^=$, or Ca x P_i) at the limit of
metastability and relating it to the solubility of brushite ($CaHPO_4 \cdot$
$2H_2O$) in the individual urine (47, 52, 53). This "formation product
ratio" gives an estimate of the minimum supersaturation (given as
number of times saturated) required to induce nucleation of brushite
(9, 47).

A further problem when measuring the formation product in whole
urine is that the measurement of inhibitor activity may not be
sensitive enough. This drawback could be solved by using smaller
concentrations of urine in the test system. An amount of a few
percent is still very effective, and at these small concentrations,
the system is much more sensitive (54). This dilution of the urine
has also the advantage of making the influences of changing ionic
strength and ion complexing negligible. Unfortunately, the measurement of inhibitory activity in diluted urine may not be representative of whole urine. Since inhibitory activity is generally not
related linearly with its concentration, and since the concentration-
effect relation is different for the various inhibitors, their
relative importance will change with dilution. Furthermore, urine
itself can influence the effect of certain inhibitors so that an
effect can be different in inorganic solutions, diluted urine, and
whole urine.

Another approach which measures mainly crystal growth is to
measure the kinetics of the precipitation of the mineral after
addition of a seed to trigger the reaction. Calcium phosphate (55) or
other nucleating materials are added to a solution with a defined
supersaturation, and the decrease in the solution of calcium or
phosphate concentration or both are measured. At acid pH, where
brushite is formed, the rate of crystal growth is proportional to

the square of the supersaturation, allowing one to calculate readily
the rate of crystal formation (56). Unfortunately, at higher pH,
where salts other than brushite are formed, the precipitation curve
obtained is more complex and cannot be analyzed quantitatively,
probably because of the formation of precursor phases which latter
transform into more stable ones, especially apatite (42, 55). Unfortunately, it is these conditions which are likely to be more
relevant physiologically, since although brushite is thought by some
to be the first salt to form in calcium phosphate stones (47, 57), it
is normally not found in the formed stones. A simplified version of
this approach is to determine just the amount of calcium salt formed
within one chosen time (9, 46, 57-60).

In both these approaches it is important that the supersaturation level is well defined. As suggested by Finlayson with regard
to whole urine, the comparison of the ratio of the Ca x P_i versus
the Ca x P_i measured after incubating with brushite, is probably
the most convenient method (9, 51).

Recently, we have worked out a new method based on the determination of the amount of seed crystals needed to induce a certain
amount of precipitation in the urine specimen (133). This amount
will reflect the amount of inhibitor present in the urine. The
Ca x P_i product of the urine is adjusted to the solubility of
brushite, therefore at a constant calcium x phosphate activity product. Calcium apatite, the solubility of which is much below that of
brushite (61), is then added, and the minimum amount of apatite
necessary to initiate a precipitation of 50% in a predetermined
time is measured.

Nature of the inhibitory activity. Much work has been devoted
to the identification of the various inhibitors since they are
likely to be important not only in stone formation but also in the
process of normal and pathological calcification. It has been known
for nearly 20 years that urine contains substances capable of inhibiting calcium phosphate deposition (36, 62). The first of these
inhibitors to be isolated and identified from urine was inorganic
pyrophosphate (54). This compound has been shown to be very powerful in all of the test systems. Inorganic pyrophosphate increases
strongly the minimum Ca x P_i product necessary to form a precipitation (50, 54); it diminishes the rate of precipitation upon added
crystals of brushite (63), octocalcium phosphate (64) and apatite

(65, 66), inhibits the precipitation induced by nucleators such as collagen (46, 50, 58, 59, 67), increases the amount of apatite crystals needed to induce the precipitation of calcium phosphate (133), and inhibits the calcification of cartilage (68, 69). Finally, pyrophosphate inhibits the transformation of amorphous calcium phosphate into its crystalline form (70, 71). The amount of pyrophosphate excreted daily lies between 10 and 60 μmoles. The concentration is in the range of 10^{-5} to 10^{-4} M (72, 73), enough to exert a strong, sometimes even a maximal inhibition in the various tests used.

Other compounds such as magnesium (57, 59, 66, 69, 74), zinc (39), fluoride (57, 59, 75, 76) (although in some conditions, this ion can also activate the precipitation (75, 76)), stannous ions (75), and citrate (66, 133) are able to inhibit the precipitation of calcium phosphate under certain conditions. Besides these known substances, there are possibly other yet unidentified components present in urine which inhibit the formation of calcium phosphate (54, 66, 77, 78). For a time, these inhibitors were thought to be of peptidic nature (58, 77-79), a theory which has recently been abandoned, the compounds studied being probably contaminants (66). Recently, phosphocitrate has been implicated in inhibition of calcium posphate formation.

The question of the role played by the various compounds in the total inhibitory activity is not yet answered. One difficulty is due to the fact that the relation depends on the dilution of the urine tested, as well as on the technique used to measure the activity. The fact that pyrophosphate is the main inhibitor at a concentration of two per cent urine (54) does not necessarily mean that this compound is also the main inhibitor in whole urine. Even when using whole urine, results may vary according to whether the test utilized measures principally crystal growth or nucleation (formation product). When examined in the system where the amount of apatite needed for inducing the precipitation of this salt is determined, pyrophosphate was found to represent only about 10% of total activity while magnesium represented 20% and citrate represented about 50% (133). In this test, 20% inhibition was due to unknown compounds. This percentage increases in other assays, especially when the urine is diluted (54). While it appears undisputed that test systems using whole undiluted urine are more relevant to what occurs in urine during stone formation, it is not possible to decide at the present time which of them is the most representative one.

Recently, efforts have been devoted to finding inhibitory compounds exogenous to normal urine, but which would be excreted after oral administration. The fact that not only pyrophosphate, but also polyphosphates (50), ATP (50), phosphorylated inositols (67), as well as compounds characterized by a P-N-P bond, the imidophosphates (80), were found to be very effective inhibitors, raised the possibility that other P_i compounds, related in structure, but resistant to enzymic hydrolysis could be found and used therapeutically. The phosphonates, characterized by C-P bonds and, more specifically, the diphosphonates (P-C-P) seemed very promising. Diphosphonates are entirely resistant to breakdown in the body and were found by many authors to be very effective in a variety of test systems (41, 59, 60, 71, 81, 82).

INHIBITORS OF THE AGGREGATION OF THE CALCIUM PHOSPHATE CRYSTALS

Recently, attention has been given to a process which has been relatively ignored, that of crystal aggregation, i.e., the binding of individual crystals into larger aggregates. This approach developed from the observation that stone-formers tend to excrete in their urine large aggregates of oxalate crystals, while normal people excrete only individual crystals (19). It was then studied extensively for this salt (16, 17, 83-85).

Aggregation occurs also with calcium phosphate crystals. When individual apatite crystals are incubated in slightly metastable solution, they will tend to bind to larger clusters. This process can be measured quantitatively by counting the number of "particles" of various sizes with a Coulter counter (18). Urine inhibits this process very efficiently (18). Unfortunately, while the inhibitors can be measured in diluted urine, no technique exists yet for the measurement in whole urine. Citrate, pyrophosphate, and the diphosphonates, and interestingly, also the glycosaminoglycans, especially heparin, hyaluronic acid, and dermatan sulfate are inhibitory (18). The fact that these glycosaminoglycans are ineffective at these concentrations on crystal formation (50) shows that the two processes, crystal growth and aggregation, seem to be favored by different factors. Methylene blue, which has been suggested to be useful in prevention of renal stones (86), has no effect (18). Since the

inhibitory activity of urine is to a large part not filtrable
through a 50,000 mol wt filter (18), it is possible that the non-
filtrable part is due to glycosaminoglycans. This would bring back
the old theory of the presence in urine of so-called "colloid
protectors" (87).

Of interest is finally the fact that diphosphonates and heparin
also induce the disaggregation of already aggregated crystals (88).
This could possibly be of importance in urine leading to disruption
of already formed aggregates.

INHIBITORS OF THE FORMATION OF CALCIUM OXALATE

Chemical estimation of inhibitory activity. The quantitative
methods to determine the inhibitors of calcium oxalate precipitation
are similar to those described for calcium phosphate. They include
the measurement of the minimum Ca x oxalate product necessary for
crystal formation both in diluted (89) as well as in whole urine
(90, 91). For the latter, the ratio of Ca x oxalate formation product vs. the Ca x oxalate product after equilibration with solid
calcium oxalate has been used to determine the limit of metastability (formation product ratio) (91).

The other approach, namely to determine the rate of precipitation after addition of a seed, has been very fruitful and extensively used. Indeed, with calcium oxalate, the kinetic analysis of
the crystal growth is much more straight forward than for calcium
phosphate. The reaction is controlled by a bimolecular surface
reaction and is thus of second order. The rate of disappearance of
each ion being proportional to the square of the supersaturation
(92), the disappearance thus can be analyzed easily quantitatively
and the rate constants determined. The influence of various substances as well as of urine on the rate of crystal growth therefore
can be determined quantitatively (93). Unfortunately, the measurements up to now only have been done on diluted urine (93). Whether
the extrapolation of results obtained at various urine dilutions can
be extrapolated to full urine has not been proven yet.

A simplified version of this approach is to set the supersaturation in a straight solution or whole urine to a known level, and
to determine to what extent precipitation has occurred, either in the

solution or on a nucleator such as a glass rod or a wire after a
defined time lapse (5, 9, 90, 91, 94-98). While this technique is
acceptable in straight solutions whith defined ion concentrations,
it is valid in whole urine only if the supersaturation can be determined (9).

Nature of the inhibitory activity. As for calcium phosphate,
urine has been found to inhibit strongly the precipitation of calcium
oxalate (93, 94, 96). The same is true for pyrophosphate (32, 83,
90, 93, 95, 97, 99, 100), citrate, and magnesium (93), as well as
for a great number of substances, i.e., dyes (95, 97) (e.g;, methylene blue (102)), phytate (95), polyelectrolytes (95), urinary
macromolecules (90), and especially metals (95, 97, 101) and heparin
(103). As for calcium phosphate, it is still difficult to evaluate
what is the relative part played by the various known inhibitors.
This question is especially difficult to answer since many compounds
can have different activities when tested in simple solutions or in
urine. Thus, from the many compounds known to inhibit growth of
calcium oxalate, only very few have been found to be effective also
in whole urine (97). One of them was inorganic pyrophosphate. It
would appear, nevertheless, that at least in the system measuring
the rate of precipitation, the known inhibitors represent only a
small portion of the total inhibitory activity present (93).

In analogy to calcium phosphate, among the nonphysiological
compounds, the diphosphonates are probably the most powerful inhibitors of calcium oxalate precipitation found up to the present time
(91, 99, 104).

INHIBITORS OF THE AGGREGATION OF CALCIUM OXALATE CRYSTALS

An interesting new approach to the problem of stone formation
is the role played by the process of calcium oxalate crystal aggregation. As mentioned above, it appears that stone-formers are more
prone to excrete aggregates than normal people (16, 19), and there
is an actual correlation between the excretion of large clusters
and the rate of stone formation (4).

Techniques have been devised to measure aggregation in vitro
(16, 17, 83, 85). Disaggregated oxalate crystals are incubated in
a slightly supersaturated solution, and the development of crystal

clusters which occur spontaneously with time is measured. When
urine is added to the system, the aggregation of calcium oxalate
crystals is strongly inhibited (16, 17, 85). Unfortunately, it has
not been possible to develop the measurement of this inhibition in
whole undiluted urine, since addition of calcium oxalate induces
precipitation of other salts.

Pyrophosphate, diphosphonates, and glycosaminoglycans, especially heparin and chondroitin sulfate, were found to inhibit very
strongly the aggregation of calcium oxalate (17, 84, 85, 105).
Methylene blue had no effect (84). Diphosphonates are active inhibitors also when given in vivo (105). The effect of pyrophosphate and
the diphosphonates increases at alkaline pH, while the effect of
urine is not pH-dependent (85).

It has been difficult up to now to assess the amount of the
total inhibition played by the various parts. Some data indicate
that pyrophosphate would account for about 15% of the total activity
(106). A large part would be due to a macromolecule of large molecular weight (85, 107) which is precipitated by cetyl pyridinium
chloride, which suggests that it could be a glycosaminoglycan (106).

MECHANISM OF THE INHIBITORY ACTIVITY

Most of the work on the mode of action of the relevant inhibitors has been done on pyrophosphate and diphosphonates. These
compounds were found to delay the various processes involved in the
formation of the solid phase, namely epitactic or heterogenous
nucleation, crystal growth, crystal aggregation, and phase transformation from the amorphous to the crystalline form. These various
effects appear to be related to the binding of the inhibitors onto
the crystal surface where they inhibit induction of new crystals,
growth, and aggregation. The relation between the absorption of
substances onto the crystals and their effect on crystal growth is
a well known phenomenon. Many compounds, called crystal modifiers,
can inhibit the crystal growth of certain salts, and there is a
close correlation between the concentration of the compounds on the
crystal surface and the degree of inhibition (108). It is not necessary, however, for the whole surface to be covered by a macromolecular layer in order to obtain an inhibition. Specific binding onto
the sites of crystal growth will suffice, so that an inhibition is

obtained when less than one percent of the surface is covered.

Pyrophosphates and diphosphonates have both been shown to be bound strongly onto the crystals of hydroxyapatite (109-111). The amount bound can be very large (for pyrophosphate, about two molecules per surface unit cell (109, 111), and is accompanied by a displacement of orthophosphate (111-113). From the results obtained by Scatchard plots, it appears that the binding is not restricted to a single class of binding sites with constant affinity (111). Interestingly, the affinity to the first site goes in parallel with the inhibitory activity of crystal formation (111), suggesting it represents binding onto the sites responsible for crystal growth. It is likely that the binding occurs through a calcium ion. A further support for such a surface controlled mechanism is the fact that inhibition of the formation rate of brushite (63) and calcium oxalate (93) follows the laws of the Langmuir adsorption isotherms. Inhibition of aggregation, on the other hand, is more likely to be due to a change in the zeta potential of the surface, which will alter attraction or repulsion between crystals.

INHIBITORS IN PATIENTS WITH URINARY STONES

The question as to whether stone formers have less inhibitors in their urine is still not completely settled, although there is now strong evidence that many patients do show such a defect. This uncertainty may be attributed partly to the wide variety of techniques used, insufficient data available from study of whole urine, and the failure often seen to obtain control groups, matched with respect to age, sex, and diet.

A decrease of inhibitory activity of calcium phosphate precipitation has been described in stone-formers, using the rat cartilage system (36, 37, 68, 74). A decrease has also been found using the determination of the $Ca \times P_i$ formation product, either in presence of diluted urine (134) or in whole urine (9). For calcium oxalate, a decrease in inhibitors was detected using the determination of the formation product (9), as well as the crystallization (5, 9, 94) in whole urine. A decrease in inhibition of calcium oxalate aggregation has been described in urine of stone patients (4, 16, 114). This is especially the case in patients presenting a relatively smaller urinary calcium x oxalate supersaturation, so that when both pa-

rameters, supersaturation and inhibition of aggregation, are taken
into account,stone-formers and controls can be very clearly separated (114).

Various data indicate that pyrosphosphates are decreased in
stone-formers (73, 115, 116), although such a decrease has not been
found by all investigators (69). The observed difference seems to
occur mainly in middle-aged men (73), a group in which stone formation is actually most frequent, and in the patients without hypercalciuria (116). The lack of pyrophosphate becomes more apparent
when patients are matched to control with respect to age, sex, and
diet, in which condition a diminution of 50% has been observed in
men (134). Women, however, showed no abnormality.

Until now, only a few ways to increase urinary inhibitors have
been established. According to our present knowledge, it appears that
one of them is the oral administration of orthophosphate, which will
increase the excretion of pyrophosphate (117, 118). Cellulose
phosphate, however, does not (52). Orthophosphate administration
decreases both calcium oxalate and calcium phosphate aggregation in
urine (119). Furthermore, it is claimed by many (120-124), although
not accepted by all (125, 126), to be efficient in the treatment of
recurrent calcium stones. All these effects could be explained at
least partly by the increase in urinary pyrophosphate, although the
decrease in calciuria which occurs and which decreases in turn the
urinary saturation vs. calcium oxalate (119) is also likely to play
a role. The saturation vs. calcium phosphate, however, is not altered
(119) nor is it increased (52). Recently, thiazide, which has been
proposed in the treatment of stones because it decreases urinary
calcium (34, 127), has been shown to increase the formation product
in whole urine both of calcium oxalate (128) and brushite (129).
This effect is probably due to the increase of pyrophosphate following the increase in orthophosphate excretion. From more recent data,
it has been suggested, in addition, that renal excretion of other
inhibitors, such as zinc, may be increased as well.

The parallelism between the excretion of orthophosphate and
pyrophosphate is probably due to a direct action of orthophosphate
on the kidney. Indeed, the oral administration of orthophosphate
causes a rapid increase in the renal clearance of pyrophosphate
without a change in the blood concentration (130). Furthermore, as
shown in studies with dogs, the unilateral infusion of orthophos-

phate into one kidney causes a greater and more rapid increase in
the infused kidney than in the noninfused one (130).

The powerful effects of the diphosphonates on the formation and
aggregation of calcium phosphate and calcium oxalate crystals and
the fact that they are excreted in the urine after oral administration raised the possibility that these compounds might be useful in
the treatment of urinary stones. In animal studies they were found,
when given orally, to decrease the formation of experimentally induced
bladder stones (104). In man, the oral administration of ethane-1-
hydroxy-1, 1-diphosphonate (EHDP) decreases the amount of calcium
oxalate crystal aggregates (105), increases the formation product
of calcium phosphate when tested in diluted urine (131), and decreases the growth of brushite in full urine (132). Clinically, it
appeared to decrease the recurrence of stones (131). However, the
dosage needed to be effective was at a level where an effect on bone
turnover and mineralization was likely, so that its use does not
seem to be warranted. The development of a phosphonate which is
excreted in urine but does not enter bone might solve this problem.

In the studies of stone formation, it is necessary, however,
to keep in mind that urolithiasis is most probably a multifactorial
disease. Several predisposing factors, such as hypercalciuria,
hyperoxaluria, hyperuricosuria, urinary infection, deficiency of
inhibitors of crystal formation, and deficiency in inhibitors of
crystal aggregation can play a role. Not all are likely to be present
in the same patient, so that it will become necessary to divide the
patients into subgroups, according to their pathophysiological disturbance. The same is likely to be true for the treatments. A
specific treatment for one disturbance can be beneficial in this
group, but possibly not in others. This basic principle is frequently
disregarded which explains why the treatment is most often far from
satisfactory.

SUMMARY

The understanding of the formation of urinary stones centers
around three main mechanisms: the urinary concentration of stone-
forming ions, the role of promoters, and the role of inhibitors of
crystal formation and crystal aggregation. With respect to the
promoting activity, lately emphasis has shifted from the role of the

organic matrix to that of one salt inducing by epitaxy the precipitation of another salt. Among the inhibitors, it has become necessary
to distinguish between those affecting crystal formation and those
affecting crystal aggregation. For measuring the inhibitory activity,
the various techniques and their relevance have been reviewed. It
has been found that the main inhibitors for calcium phosphate and
calcium oxalate precipitation are citrate, pyrophosphate, and
perhaps magnesium. Those for calcium phosphate and calcium oxalate
aggregation are glycosaminoglycans, pyrophosphate, and citrate.
Among the synthetic inhibitors, the diphosphonates are the most
powerful for both processes. The role and the therapeutic indications
of these various concepts have been discussed.

ACKNOWLEDGMENT

This work was supported by the Swiss National Science Foundation
(3.725.76) and by Procter and Gamble Company, U.S.A.

REFERENCES

1. Yarbro, C.L.: Studies on the mechanism of formation of renal
    calculi: Part II. J. Urol. 80: 10–12, (1958).
2. Robertson, W.G., Peacock, M. and Nordin, B.E.C.: Activity
    products in stone-forming and non-stone forming urine. Clin.
    Sci. 34: 579–594, (1968).
3. Robertson, W.G., Peacock, M. and Nordin, B.E.C.: Calcium oxalate
    crystalluria and urine saturation in recurrent renal stone-
    formers. Clin. Sci. 40: 365–374, (1971).
4. Robertson, W.G.: Physical–chemical aspects of calcium stone-
    formation in the urinary tract. In Urolithiasis Research, edited
    by Fleisch, H., Robertson, W.G., Smith, L.H., Vahlensieck, W.,
    New York, Plenum Press, 1976, pp. 25–39.
5. Gill, W.B., Silvert, M.A. and Roma, M.J.: Supersaturation levels
    and crystallization rates from urines of normal humans and
    stone-formers determined by a ^{14}C-oxalate technique. Invest.
    Urol. 12: 203–209, (1974).
6. Elliot, J.S. and Ribeiro; M.: Calcium oxalate solubility in
    urine: The state of relative saturation. Invest. Urol. 5:
    239–243, (1967).
7. Pak, C.Y.C.: Physicochemical basis for formation of renal

stones of calcium phosphate origin: Calculation of the degree
of saturation of urine with respect to brushite. J. Clin.
Invest. 48: 1914-1922, (1969).

8. Pak, C.Y.C., Diller, E.C., Smith, G.W. and Howe, E.S.: Renal
    stones of calcium phosphate: Physicochemical basis for their
    formation. Proc. Soc. Exp. Biol. Med. 130: 753-757, (1969).

9. Pak, C.Y.C. and Holt, K.: Nucleation and growth of brushite
    and calcium oxalate in urine of stone-formers. Metabolism
    25: 665-673, (1976).

10. Marshall, R.W., Cochran, M., Robertson, W.G., Hodgkinson, A.
    and Nordin, B.E.C.: The relation between the concentration of
    calcium salts in the urine and renal stone composition in
    patients with calcium-containing renal stones. Clin. Sci. 43:
    433-441, (1972).

11. Hodgkinson, A. and Pyrah, L.N.: The urinary excretion of calcium
    and inorganic phosphate in 344 patients with calcium stone of
    renal origin. Br. J. Surg. 46: 10-18, (1958).

12. Melick, R.A. and Henneman P.H.: Clinical and laboratory studies
    of 207 consecutive patients in a kidney-stone clinic. N. Engl.
    J. Med. 259: 307-314, (1958).

13. Robertson, W.G., Peacock, M., Marshall, R.W., Speed, R. and
    Nordin, B.E.C.: Seasonal variations in the composition of urine
    in relation to calcium stone-formation. Clin. Sci. Mol. Med.
    49: 597-602, (1975).

14. Robertson, W.G., Marshall, R.W., Peacock, M. and Knowles, F:
    The saturation of urine in recurrent, idiopathic calcium stone-
    formers, in Urolithiasis Research, edited by Fleisch, H.,
    Robertson, W.G., Smith, L.H., Vahlensieck, W., New York, Plenum
    Press, 1976, pp. 335-338.

15. Pak, C.Y.C., Waters, O., Arnold, L.H., Holt, K., Cox, C. and
    Barilla, D.: Mechanism for calcium urolithiasis among patients
    with hyperuricosuria: Supersaturation of urine with respect
    to monosodium urate. J. Clin. Invest. 59: 426-431, (1977).

16. Robertson, W.G. and Peacock, M.: Calcium oxalate crystalluria
    and inhibitors of crystallization in recurrent renal stone-
    formers. Clin. Sci. 43: 499-506, (1972).

17. Fleisch, H. and Monod, A.: A new technique for measuring
    aggregation of calcium oxalate crystals in vitro: Effect of
    urine, magnesium, pyrophosphate and diphosphonates, in Urinary
    Calculi, edited by Cifuentes Delatte, L., Rapado, A., Hodgkinson,
    A., Basel, S. Karger, 1973, pp. 53-56.

18. Hansen, N.M., Felix, R., Bisaz, S. and Fleisch, H.: Aggrega-

tion of hydroxyapatite crystals. Biochim. Biophys. Acta. 451:
549-559, (1976).

19. Robertson, W.G., Peacock, M. and Nordin, B.E.C.: Calcium
     crystalluria in recurrent renal-stone formers. Lancet. 2: 21-
     24, (1969).

20. Boyce, W.H. and Garvey, F.K.: The amount and nature of the
     organic matrix in urinary calculi: A review. J. Urol. 76: 213-
     227, (1956).

21. Boyce, W.H., Garvey, F.K. and Norfleet, C.M.: The metal chelate
     compounds of urine. Am. J. Med. 18: 87-95, (1955).

22. Boyce, W.H., Garvey, F.K. and Norfleet, C.M.: Ion-binding
     properties of electrophoretically homogeneous mucoproteins of
     urine in normal subjects in patients with renal calculous dis-
     ease. J. Urol. 72: 1019-1031,(1954).

23. Boyce, W.H. and Swanson, M: Biocolloids of urine in health and
     in calculous disease: II. Electrophoretic and biochemical
     studies·of a mucoprotein insoluble in molar sodium chloride.
     J. Clin. Invest. 34: 1581-1589, (1955).

24. Boyce, W.H., King, J.S. and Fielden, M.L.: Total nondialyzable
     solids (TNDS) in human urine: XIII. Immunological detection
     of a component peculiar to renal calculous matrix and to urine
     of calculous patients. J. Clin. Invest. 41: 1180-1189, (1962).

25. King, J.S. and Boyce, W.H.: Immunological studies on serum and
     urinary proteins in urolith matrix in ·man. Ann. N.Y. Acad. Sci.
     104: 579-591, (1963).

26. Boyce, W.H. and King, J.S.: Crystal-matrix interrelations in
     calculi. J. Urol. 81: 351-365, (1959).

27. Boyce, W.H., Pool, C.S., Meschan, I. and King, J.S.: Organic
     matrix of urinary calculi. Acta Radiol. 50: 543-560, (1958).

28. Vermeulen, C.W., Lyon, E.S. and Gill, W.B.: Artificial urinary
     concretions. Invest. Urol. 1: 370-386, (1964).

29. Lonsdale, K.: Epitaxy as a growth factor in urinary calculi
     and gallstones. Nature 217: 56-58, (1968).

30. Pak, C.Y.C., Hayashi, Y. and Arnold, L.H.: Heterogeneous nu-
     cleation with urate, calcium phosphate and calcium oxalate.
     Proc. Soc. Exp. Biol. Med. 153: 83-87, (1976).

31. Meyer, J.L., Bergert,J.H. and Smith, L.H.: Epitaxial relation-
     ships in urolithiasis: The calcium oxalate monohydrate-hydroxy-
     apatite system. Clin. Sci. Mol. Med. 49: 369-374, (1975).

32. Coe, F.L., Lawton, R.L., Goldstein, R.B. and Tembe, V.: Sodium
     urate accelerates precipitation of calcium oxalate in vitro.
     Proc. Soc. Exp. Biol. Med. 149: 926-929, (1975).

33. Pak, C.Y.C. and Arnold, L.H.: Heterogeneous nucleation of cal-
    cium oxalate by seeds of monosodium urate. Proc. Soc. Exp. Biol.
    Med. 149: 930-932, (1975).
34. Coe, F.L. and Kavalach, A.G.: Hypercalciuria and hyperuricosuria
    in patients with calcium nephrolithiasis. N. Engl. J. Med.
    291: 1344-1350, (1974).
35. Coe, F.L. and Raisen, L.: Allopurinol treatment of uric-acid
    disorders in calcium-stone formers. Lancet 1: 129-131, (1973).
36. Howard, J.E. and Thomas W.C.: Some observations on rachitic
    rat cartilage of probable significance in the etiology of renal
    calculi. Trans. Am. Clin. Clim. Assoc. 70: 94-102, (1958).
37. Thomas, W.C. and Howard, J.E.: Studies on the mineralizing
    propensity of urine from patients with and without renal cal-
    culi. Trans. Assoc. Am. Physicians 72:181-187, (1959).
38. Gutman, A.B. and Yu, T.F.: Further studies of the relation
    between glycogenolysis and calcification in cartilage, in Meta-
    bolic Interrelations, edited by Reifenstein, E.C., New York,
    Josiah Macy Jr. Foundation, 1949, pp. 11-26.
39. Bird, E.D. and Thomas, W.C.: Effect of various metals on min-
    eralization in vitro. Soc. Exp. Biol. Med. 112: 640-643, (1963).
40. Robertson, W.G.: Factors affecting the precipitation of calcium
    phosphate in vitro. Calc. Tiss. Res. 11: 311-322, (1973).
41. Pak, C.Y.C.: Disorders of stone formation, in The Kidney,
    edited by Brenner, B.M., Rector, F.C. Jr., Saunders, W.B., 1976,
    pp. 1326-1354.
42. Nancollas, G.H. and Tomazic, B.: Growth of calcium phosphate
    on hydroxyapatite crystals: Effect of supersaturation and ionic
    medium. J. Phys. Chem. 78: 2218-2225, (1974).
43. Eanes, E.D., Gillessen, I.H. and Posner, A.S.: Intermediate
    states in the precipitation of hydroxyapatite. Nature 208:
    365-367, (1965).
44. Eanes, E.D. and Posner, A.S.: Intermediate phases in the basic
    solution preparation of alkaline earth phosphates. Calc.Tiss.
    Res. 2: 38-48, (1968).
45. Füredi-Milhofer, H., Purgaric, B., Brecević, L. and Pavković,
    N.: Precipitation of calcium phosphates from electrolyte solu-
    tions: I. A study of the precipitates in the physiological pH
    region. Calc. Tiss. Res. 8: 142-153, (1971).
46. Pak, C.Y.C. and Ruskin, B.: Calcification of collagen by urine
    in vitro: Dependence on the degree of saturation of urine with
    respect to brushite. J. Clin. Invest. 49: 2353-2361, (1970).
47. Pak, C.Y.C., Eanes, E.D. and Ruskin, B.: Spontaneous precipi-

tation of brushite in urine: Evidence that brushite is the
nidus of renal stones originating as calcium phosphate. Proc.
Nat. Acad. Sci. 68: 1456–1460, (1971).

48. Brown, W.F.: Crystal growth of bone mineral. Clin. Orthop.
     44: 205–220, (1966).

49. McGregor, J. and Brown, W.F.: Blood: bone equilibrium in cal-
     cium homeostasis. Nature 205: 359–361, (1965).

50. Fleisch, H. and Neuman, W.F.: Mechanisms of calcification:
     Role of collagen, polyphosphates,and phosphatase. Am. J.
     Physiol. 200: 1296–1300, (1961).

51. Pak, C.Y.C., Hayashi, Y., Finlayson, B. and Chu, S.: Estimation
     of the state of saturation of brushite and calcium oxalate in
     urine: A comparison of three methods. J. Lab. Clin. Med. 89:
     891–901, (1977).

52. Pak, C.Y.C.: Effects of cellulose phosphate and sodium phos-
     phate on formation product and activity product of brushite in
     urine. Metabolism 21: 447–455, (1972).

53. Pak, C.Y.C. and Chu, S.: A simple technique for the determina-
     tion of urinary state of saturation with respect to brushite.
     Invest. Urol. 11: 211–215, (1973).

54. Fleisch, H. and Bisaz, S.: Isolation from urine of pyrophos-
     phate, a calcification inhibitor. Am. J. Physiol. 203: 671–675,
     (1962).

55. Nancollas, G.H. and Mohan, M.S.: The growth of hydroxyapatite
     crystals. Arch. Oral Biol. 15: 731–745, (1970).

56. Nancollas, G.H.: The kinetics of crystal growth and renal stone-
     formation, in Urolithiasis Research, edited by Fleisch, H.,
     Robertson, W.G., Smith, L.H., Vahlensieck, W., New York, Plenum
     Press, 1976, pp. 5–23.

57. Bachra, B.N. and Fischer, H.R.A.: The effect of some inhibitors
     on the nucleation and crystal growth of apatite. Calc. Tiss.
     Res. 3: 348–357, (1969).

58. Wadkins, C.L.: Experimental factors that influence collagen
     calcification in vitro. Calc. Tiss. Res. 2: 214–228, (1968).

59. Jethi, R.K. and Wadkins, C.L.: Studies of the mechanism of
     biological calcification by tendon matrix. Calc. Tiss. Res.
     7: 277–289, (1971).

60. Ohata, M. and Pak, C.Y.C.: The effect of diphosphonates on
     calcium phosphate crystallization in urine in vitro. Kidney
     Int. 4: 401–406, (1973).

61. Neuman, W.F. and Neuman, M.W.: The Chemical Dynamics of Bone
     Mineral. Chicago, The University of Chicago Press,(1958).

62. Vermeulen, C.W., Lyon, E.S. and Miller, G.H.: Calcium phos-
     phate solubility in urine as measured by a precipitation test:
     Experimental urolithiasis: Part XIII. J. Urol. 79: 596-606,
     (1958).

63. Marshall, R.W. and Nancollas, G.H.: The kinetics of crystal
     growth of dicalcium phosphate dihydrate. J. Phys. Chem. 73:
     3838-3844, (1969).

64. Le Geros, R.Z. and Morales, P.: Renal stone crystals grown in
     gel systems. Invest. Urol. 11: 12-20, (1973).

65. Fleisch, H., Russell, R.G.G. and Straumann, F.: Effect of
     pyrophosphate on hydroxyapatite and its implications in cal-
     cium homeostasis. Nature 212: 901-903, (1966).

66. Smith, L.H., Meyer, J.L. and McCall, J.T.: Chemical nature of
     crystal inhibitors isolated from human urine, in Urinary cal-
     culi, edited by Cifuentes Delatte, L., Rapado, A., Hodgkinson,
     A., Basel, S. Karger, 1973, pp. 318-327.

67. Thomas, W.C. and Tilden, M.T.: Inhibition of mineralization
     by hydrolysis of phytic acid. Johns Hopkins Med. J. 131: 133-
     142, (1972).

68. Thomas, W.C., Bird, E.D. and Tomita, A.: Some concepts con-
     cerning the genesis of urinary calculi. J. Urol. 90: 521-526,
     (1963).

69. Lewis, A.M., Thomas, W.C. and Tomita, A.: Pyrophosphate and
     the mineralizing potential of urine. Clin. Sci. 30: 389-397,
     (1966).

70. Fleisch, H., Russell, R.G.G., Bisaz, S., Termine, J.D. and
     Posner, A.S.: Influence of pyrophosphate on the transformation
     of amorphous to crystalline calcium phosphate. Calc. Tiss. Res.
     2: 49-59, (1968).

71. Francis, M.D.: The inhibition of calcium hydroxyapatite crystal
     growth by polyphosphonates and polyphosphates. Calc. Tiss. Res.
     3: 151-162, (1969).

72. Fleisch, H. and Bisaz, S.: Die Pyrophosphatausscheidung im
     Harn beim gesunden Menschen. Helv. Physiol. Pharm. Acta 21:
     88-94, (1963).

73. Russell, R.G.G. and Hodgkinson, A.: Urinary excretion of inor-
     ganic pyrophosphate by normal subjects and patients with renal
     calculus. Clin. Sci. 31: 51-62, (1966).

74. Mukai, T. and Howard, J.E.: Some observations on the calcifi-
     cation of rachitic cartilage by urine. Bull Johns Hopkins
     Hospital 112: 279-290, (1963).

75. Meyer, J.L. and Nancollas, G.H.: Effect of stannous and fluoride

ions on the rate of crystal growth of hydroxyapatite. J. Dent. Res. 51: 1443–1450, (1972).

76. Taves, D.R. and Neuman, W.F.: Factors controlling calcification in vitro: Fluoride and magnesium. Arch. Biochem. Biophys. 108: 390–397, (1964).

77. Howard, J.E., Thomas, W.C., Barker, L.M., Smith, L.H. and Wadkins, C.L.: The recognition and isolation from urine and serum of a peptide inhibitor to calcification. Johns Hopkins Med. J. 120: 119–136, (1967).

78. Smith, L.H. and McCall, J.T.: Chemical nature of peptide inhibitors isolated from urine, in Renal Stone Research Symposium, Churchill, 1969, pp. 153–163.

79. Robertson, W.G., Hambleton, J. and Hodgkinson, A.: Peptide inhibitors of calcium phosphate precipitation in the urine of normal and stone-forming men. Clin. Chim. Acta 25: 247–253, (1969).

80. Robertson, W.G. and Fleisch, H.: The effect of imidodiphosphate (P–N–P) on the precipitation and dissolution of calcium phosphate in vitro. Biochim. Biophys. Acta 222: 677–680, (1970).

81. Meyer, J.L. and Nancollas, G.H.: The influence of multidentate organic phosphonates on the crystal growth of hydroxyapatite. Calc. Tiss. Res. 13: 295–303, (1973).

82. Fleisch, H., Russell, R.G.G., Bisaz, S., Muhlbauer, R.C. and Williams, D.A.: The inhibitory effect of phosphonates on the formation of calcium phosphate crystals in vitro and on aortic and kidney calcification in vivo. Eur. J. Clin. Invest. 1: 12–18, (1970).

83. Robertson, W.G.: A method for measuring calcium crystalluria. Clin Chim. Acta 26: 105–110, (1969).

84. Robertson, W.G., Peacock, M. and Nordin, B.E.C.: Inhibitors of the growth and aggregation of calcium oxalate crystals in vitro. Clin. Chim. Acta 43: 31–37, (1973).

85. Felix, R., Monod, A., Broge, L., Hansen, N.M. and Fleisch, H.: Aggregation of calcium oxalate crystals: Effect of urine and various inhibitors. Urol. Res. 5: 21–28, (1977).

86. Boyce, W.H., McKinney, W.M., Long, T.T. and Drach, G.W.: Oral administration of methylene blue to patients with renal calculi. J. Urol. 97: 783–789, (1967).

87. Butt, A.J.: Etiologic Factors in Renal Lithiasis. Springfield, Ill., Thomas, C.C., 1956.

88. Bisaz, S., Felix, R., Hansen, N.M. and Fleisch, H.: Disaggregation of hydroxyapatite crystals. Biochim. Biophys. Acta

451: 560-566, (1976).

89. Fleisch, H. and Bisaz, S.: The inhibitory effect of pyrophos-
     phate on calcium oxalate precipitation and its relation to
     urolithiasis. Experientia 20: 276-280, (1964).

90. Gill, W.B. and Karesh, J.W.: Demonstration of protective (in-
     hibitory) effects of urinary macromolecules on the crystalli-
     zation of calcium oxalate, in Urolithiasis Research, edited
     by Fleisch, H., Robertson, W.G., Smith, L.H., Vahlensieck, W.,
     New York, Plenum Press, 1976, pp. 277-280.

91. Pak, C.Y.C., Ohata, M. and Holt, K.: Effect of diphosphonates
     on crystallization of calcium oxalate in vitro. Kidney Int.
     7: 154-160, (1975).

92. Meyer, J.L. and Smith, L.H.: Growth of calcium oxalate crystals:
     I. A model for urinary stone growth. Invest. Urol. 13: 31-35,
     (1975).

93. Meyer, J.L. and Smith, L.H.: Growth of calcium oxalate crystals:
     II. Inhibition by natural urinary crystal growth inhibitors.
     Invest. Urol. 13: 36-39, (1975).

94. Dent, C.E. and Sutor, D.J.: Presence or absence of inhibitor
     of calcium-oxalate crystal growth in urine of normals and of
     stone-formers. Lancet 775-778, (1971).

95. Sutor, D.J.: Growth studies of calcium oxalate in the presence
     of various ions and compounds. Br. J. Urol. 41: 171-178, (1969).

96. Rose, M.B.: Renal stone formation: The inhibitory effect of
     urine on calcium oxalate precipitation. Invest. Urol. 12: 428-
     433, (1975).

97. Welshman, S.G. and McGeown, M.G.: A quantitative investigation
     of the effects on the growth of calcium oxalate crystals on
     potential inhibitors. Br. J. Urol. 44: 677-680, (1972).

98. Lyon, E.S. and Vermeulen, C.W.: Crystallization concepts and
     calculogenesis: Observations on artificial oxalate concretions.
     Invest. Urol. 3: 309-320, (1965).

99. Will, E.J., Bijvoet, O.L.M. and te Brakevan der Linden, H.:
     Inhibition of calcium oxalate crystal growth: A simple method
     of measurement and preliminary results, in Urolithiasis Re-
     search, edited by Fleisch, H., Robertson, W.G., Smith, L.H.,
     Vahlensieck, W., New York, Plenum Press, 1976, pp. 367-370.

100. Nancollas, G.H. and Gardner, G.L.: Kinetics of crystal growth
     of calcium oxalate monohydrate. J. Crystal Growth 21: 267-276,
     (1974).

101. Eusebio, E., Elliot, J.S.: Effect of trace metals on the crys-
     tallization of calcium oxalate. Invest. Urol. 4: 431-435, (1967).

102. Rollins, R. and Finlayson, B.: Mechanism of prevention of
      calcium oxalate encrustation by methylene blue and demonstra-
      tion of the concentration dependence of its action. J. Urol.
      110: 459–463, (1973).

103. Crawford, J.E., Crematy, E.P. and Alexander, A.E.: The effect
      of natural and synthetic polyelectrolytes on the crystalliza-
      tion of calcium oxalate. Aust. J. Chem. 21: 1067–1072, (1968).

104. Fraser, D., Russel, R.G.G., Pohler, O., Robertson, W.G., and
      Fleisch, H.: The influence of disodium ethane-1-hydroxy-1,
      1-diphosphonate (EHDP) on the development of experimentally
      induced urinary stones in rats. Clin. Sci. 42: 197–207, (1972).

105. Robertson, W.G., Peacock, M., Marshall, W.R. and Knowles, F.:
      The effect of ethane-1-hydroxy-1, 1-diphosphonate (EHDP) on
      calcium oxalate crystalluria in recurrent renal stone-formers.
      Clin. Sci. Mol. Med. 47: 13–22, (1974).

106. Robertson, W.G., Knowles, F. and Peacock, M.: Urinary acid
      mucopolysaccharide inhibitors of calcium oxalate crystalliza-
      tion, in Urolithiasis Research, edited by Fleisch, H.,
      Robertson, W.G., Smith, L.H., Vahlensieck, W., New York,
      Plenum Press, 1976, pp. 331–334.

107. Robertson, W.G., Peacock, M. and Knowles, C.F.: Calcium oxa-
      late crystalluria and inhibitors of crystallization in recur-
      rent renal stone formers, in Urinary Calculi, edited by Ci-
      fuentes Delatte, L., Rapado, A., Hodgkinson, A., Basel, S.
      Karger, 1973, pp. 302–306.

108. Bliznakov, G.: Sur le mécanisme de l'action des additifs
      adsorbants dans la croissance cristalline, in Adsorption et
      Croissance Cristalline, Editions du Centre National de la
      Recherche Scientifique, Paris, 1965, pp. 291–301.

109. Krane, S.M. and Glimcher, M.J.: Transphosphorylation from
      nucleoside di-and triphosphates by apatite crystals. J. Biol.
      Chem. 237: 2991–2998, (1962).

110. Burton, F.G., Neuman, MW and Neuman, W.F.: On the possible
      role of crystals in the origin of life: I. The absorption of
      nucleosides, nucleotides and pyrophosphate by apatite crys-
      tals. Curr. Mod. Biol. 3: 20–26, (1969).

111. Jung, A., Bisaz, S. and Fleisch, H.: The binding of pyrophos-
      phate and two diphosphonates by hydroxyapatite crystals. Calc.
      Tiss. Res. 11: 269–280, (1973).

112. Robertson, W.G. and Morgan, D.B.: Effect of pyrophosphate on
      the exchangeable calcium pool of hydroxyapatite crystals.
      Biochim. Biophys. Acta 230: 495–503, (1971).

113. Jung, A., Bisaz, S., Bartholdi, P. and Fleisch, H.: Influence of pyrophosphate on the exchange of calcium phosphate ions on hydroxyapatite. Calc. Tiss. Res. 13: 27-40, (1973).

114. Robertson, W.G., Peacock, M., Marshall, R.W., Marshall, D.H. and Nordin, B.E.C.: Saturation-inhibition index as a measure of the risk of calcium oxalate stone formation in the urinary tract. N. Engl. J. Med. 294: 249-252, (1976).

115. Valyasevi, A. and Van Reen, R.: Pediatric bladder stone disease: Current status of research. J. Pediat. 72: 546-553, (1968).

116. Russell, R.G.G. and Fleisch, H.: Pyrophosphate and stone formation, in Renal Stone Research Symposium, Churchill, 1969, pp. 165-180.

117. Fleisch, H. Bisaz, S. and Care, A.D.: Effect of orthophosphate on urinary pyrophosphate excretion and the prevention of urolithiasis. Lancet 1: 1065-1067, (1964).

118. Russell, R.G.G., Edward, N.A. and Hodgkinson, A.: Urinary pyrophosphate and urolithiasis. Lancet 1446, (1964).

119. Robertson, W.G., Peacock, M., Marshall, R.W., Varnavides, C.K., Heyburn, P.J. and Nordin, B.E.C.: Effect of oral orthophosphate on calcium crystalluria in stone-formers, in Urolithiasis Research edited by Fleisch, H., Robertson, W.G., Smith, L.H., Vahlensieck, W., New York, Plenum press, 1976, pp. 339-342.

120. Howard, J.E.: Urinary stone. Can Med. Assoc. J. 86: 1001-1007, (1962).

•121. Howard, J.E., Thomas, W.C., Mukai, T., Johnston, R.A. and Pascoe, B.J.: The calcification of cartilage by urine, and a suggestion for therapy in patients with certain kinds of calculi. Trans. Assoc. Am. Physicians. 75: 301-306, (1962).

122. Thomas, W.C. and Miller, G.H.: Inorganic phosphates in the treatment of renal calculi. Modern Treatment 4: 494-504, (1967).

123. Smith, L.H., Thomas, W.C. and Arnaud, C.D.: Orthophosphate therapy in calcium renal lithiasis, in Urinary Calculi, edited by Cifuentes Delatte, L. Rapado, A., Hodgkinson, A., Basel, S. Karger, 1973, pp. 188-197.

124. Bernstein, D.S. and Newton, R.: The effect of oral sodium phosphate on the formation of renal calculi and on idiopathic hypercalciuria. Lancet 1105-1107, (1966).

125. Ettinger, B.: Recurrent nephrolithiasis: Natural history and effect of phosphate therapy. Am. J. Med. 61: 200-206, (1976).

126. Ettinger, B. and Kolb, F.O.: Inorganic phosphate treatment of
     nephrolithiasis. Am. J. Med. 55: 32-37, (1973).
127. Yendt, E.R.: Renal calculi. Can. Med. Assoc. J. 102: 479-489,
     (1970).
128. Woelfel, A., Kaplan, R.A. and Pak, C.Y.C.: Effect of hydrochlo-
     rothiazide therapy on the crystallization of calcium oxalate
     in urine. Metabolism 26: 201-205, (1977).
129. Pak, C.Y.C.: Hydrochlorothiazide therapy in nephrolithiasis:
     Effect on the urinary activity product and formation product
     of brushite. Clin. Pharmacol. Ther. 13: 209-217, (1973).
130. Russell, R.G.G., Bisaz, S. and Fleisch, H.: The influence of
     orthophosphate on the renal handling of inorganic pyrophos-
     phate in man and dog. Clin. Sci. Mol. Med. 51: 435-443, (1976).
131. Baumann, J.M., Ganz, U., Bisaz, S., Fleisch, H. and Rutishau-
     ser, G.: Verabreichung eines Diphosphonates zur Steinprophy-
     laxe.Helv. Chir. Acta 41: 421-424, (1974).
132. Ohata, M. and Pak, C.Y.C.: Preliminary study of the treatment
     of nephrolithiasis (calcium stones) with diphosphonate. Metab-
     olism 23: 1167-1173, (1974).
133. Bisaz, S., Felix, R., Neuman, W.F. and Fleisch, H.: Quantita-
     tive determination of inhibitors of calcium phosphate precipi-
     tation in whole urine. Mineral Electrolyte Metabol. 1: 74-83,
     (1978).
134. Bauman, J.M., Bisaz, S., Felix, R., Fleisch, H., Ganz, U. and
     Russell, R.G.G.: The role of inhibitors and other factors in
     the pathogenesis of recurrent calcium-containing renal stones.
     Clin. Sci. Mol. Med. 53: 141-148, (1977).

RENAL STONE INHIBITORS

F. Corrado, M. Fini and A. Ligabue

Department of Urology
M. Malpighi Hospital
Bologna (Italy)

ABSTRACT

The size and the degree of aggregation of crystals, dependent on salt saturation and crystal inhibitor concentration, seem to be the main differences between stone-formers and healthy subjects.

Citrate, pyrophosphate and magnesium inhibit calcium-oxalate and calcium-phosphate crystal formation and growth, while glycosaminoglycans and diphosphonates are crystal aggregation inhibitors.

The role of uric acid in the pathogenesis of calcium-oxalate lithiasis is reviewed, as well as the methods employed for identification and measurement of urinary inhibitors. We present a simple new method to evaluate the inhibitory capacity of urine with regard to calcium-oxalate crystal growth.

———

One of the many unclear aspects of nephrolithiasis is that it affects only a small part of the population, while physico-chemical conditions exist which should indicate a wider diffusion of the disease. As demonstrated in figure 1, if we study the urinary saturation of the principal lithogenic salts in a group of recurrent stone formers versus that in a control group, we observe that saturation values in both groups are often very similar in free diet

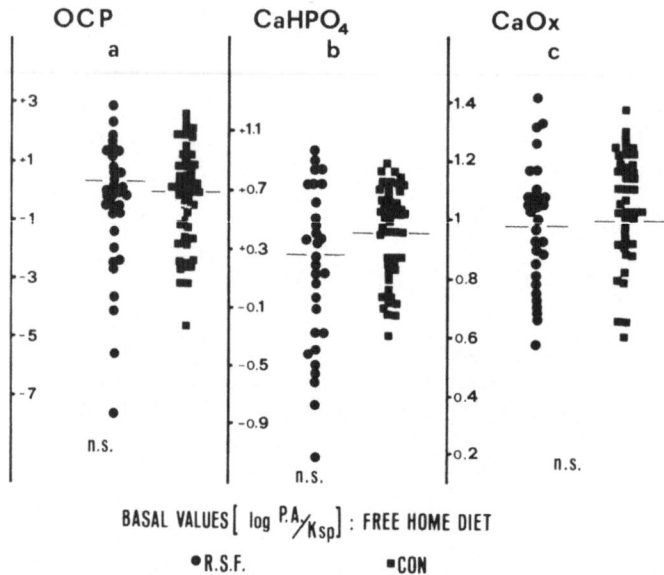

Fig. 1.   Urinary saturation of octocalcium phosphate (OCP),
           calcium phosphate (CaHPO$_4$) and calcium oxalate salts
           in recurrent stone formers.

conditions. Furthermore the urine of normal subjects who do not
usually drink large amount of fluids is at times more supersaturated
than that of a recurrent stone former.

Despite these observations, most Authors (1) are of the
opinion that supersaturation must be considered a condition which is
necessary for crystallisation, even if in itself it is insufficient
to explain stone formation.

Therefore, it must be admitted that substances are present in
the urine which are either capable of facilitating stone formation
or of preventing a supersaturated solution from forming stones.

Among the so-called "facilitating" substances which can favour
the formation of renal calculi, even when solutions are not highly
supersaturated or are below the formation product, considerable
importance was attributed, several years ago (2), to the activating
role of the matrix.

This hypothesis, particularly supported by Boyce, was confirmed both by experimental and clinical results. Experimental data indicated that the proteins incorporated in the matrix may combine with calcium, thus inducing in vitro calcification. Clinical results showed that the patients with stones excreted greater quantities of proteins than the controls.

More recently, Lonsdale (3) has stressed the importance that the crystals of a given salt can have in inducing the crystallisation of a salt of a different nature. This phenomenon is called epitaxy; it is based on the possibility that the growth of a salt may occur on the surface of a crystal of different chemical composition, but having similarities in lattice dimensions.

Such structural affinities exist, for example, between crystals of calcium oxalate, uric acid and calcium phosphate. Therefore, urate precipitation may be favoured by hydroxyapatite as well as by calcium oxalate, while calcium oxalate precipitation may be induced by the presence of hydroxyapatite crystals as well as by those of brushite and urate.

These mechanisms of precipitation explain why the composition of most stones is mixed and why hyperuricosuria may, in some cases, be considered responsible for calcium-oxalate stone formation.

In fact, Pak and Arnold (4), and Coe et al. (5), have recorded how some normocalciuric subjects, with no alterations in their phospho-calcium, metabolism can produce oxalate or calcium-phosphate stones if they are hyperuricosuric and with a urinary pH higher than 5.7. The explanation can be found in the fact that the aforementioned conditions create a physico-chemical state which favours the nucleation of monosodium urate, since the urines are supersaturated with respect to this salt. If urines are also supersaturated with respect to oxalate and phosphate, urate may constitute the nucleus for the precipitation of these salts.

Particular mention is made here of two of the methods used to evaluate the inhibitory activity of urine: the first allows determination of the inhibitory power of whole urine, the second of diluted urine.

Concerning whole urines, one can establish the formation prod-

uct of a certain salt; since this value represents the minimum concentration necessary for the spontaneous precipitation of a salt,
it is clear that the greater the inhibitory activity, the higher
this product. This method has the advantage of measuring inhibitory
activity under conditions which are very similar to those existing
"in vivo", provided that the saturation values can be strictly
checked.

On the other hand the method employing diluted urines consists
of seeding crystals in metastable solutions containing small quantities of urine (1-2%). From the kinetics of the precipitation of
the mineral, we obtain an evaluation of inhibitory activity. The
amount of precipitated salt will, of course, be inversely proportional to inhibition.

An objection to the use of the latter system is that the first
stages of crystallisation are, as yet, unclear, as is the action of
inhibitors on the growth of the first or successive salts: therefore
crystals of a different nature from those first precipitated can be
employed for seeding.

The methods employing diluted urines allow the detection of
even small differences between the inhibitors. However, while it
is true that the influence of the ionic strenght of salts and pH can
be minimized, it is also true that this method adopted "in vitro"
shows situations far removed from those "in vivo".

In man, in fact, we can observe that some inhibitors which are
active in whole urine, may be ineffective in diluted urine. Also the
activity of other inhibitors might be favoured by the altered conditions of pH and ionic concentrations.

We have recently proposed a method allowing the measurement
of the degree of urinary inhibitory activity (6); this involves
incubating crystals of calcium oxalate in a standard metastable
solution of calcium oxalate in the presence of a known amount of
oxalate labelled with C^{14} as a tracer. Since oxalate will ·deposit
on crystals, the growth of the latter will be inversely proportional
to the residual radioactivity.

The Inhibition Index (I.I.) is obtained by comparing a solution
in which inhibitors are present with a non-inhibited reference so-

lution.

Concerning the nature of single inhibitors, it is useful to
point out that the inhibitors of the formation and growth of calcium
phosphate and calcium oxalate crystals are to be distinguished from
those acting upon crystal aggregation.

Figure 2 sums up the principal inhibitors belonging to the
groups mentioned.

We do not intend to deal with all the substances listed here,
but will mention the properties of those which are most important.

Inorganic pyrophosphate is one of the main inhibitors in the
formation of calcium oxalate and calcium phosphate crystals.

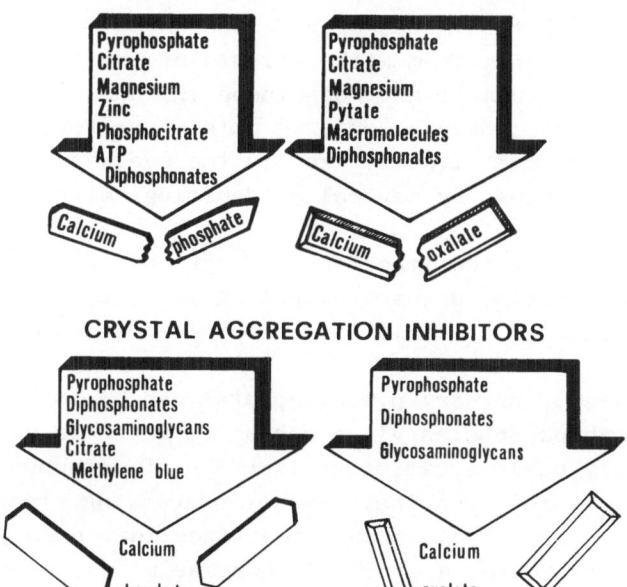

Fig. 2. Crystal formation, growth and aggregation inhibitors.

According to Fleisch (7), it possesses, in vitro, the following properties:

1) it increases the formation product of phosphate and calcium oxalate;

2) it delays the transformation of amorphous calcium phosphate into hydroxyapatite;

3) it absorbs apatite crystals and slows down both the growth and the dissolution of these crystals;

4) it prevents the aggregation of calcium oxalate crystals;

5) it delays calcification processes in animal soft tissues.

Pyrophosphates excreted in urine at a concentration varying between 10^{-4} to 10^{-5} moles, have often been the object of research. However, Russell et al. (8) have observed a decrease in pyrophosphates in the urines of stone formers compared with controls.

Having studied the values of pyrophosphate in patients with lithiasis and comparing them with patients of the same age and sex taken as controls, Russell and Hodgkinson (9) has pointed out that differences could be observed only in male stone formers between 40 and 50 years of age, i.e. at exactly the age when lithiasis is most frequent, as shown by several epidemiological studies.

Another proof of the importance of pyrophosphates lies in the fact that by increasing urinary excretion in various ways, some authors have obtained a significant reduction of stone recurrences.

An increase of urinary pyrophosphates can be obtained by administering orthophosphates via a mechanism which is not as yet clear. The daily administration of 1 or 2 g of orthophosphates is considered one of the principal ways of preventing stone formation (10), since it not only diminishes the intestinal absorption of calcium, but also causes a striking increase of pyrophosphaturia.

The excretion of pyrophosphates may also increase after the administration of hydrochlorothiazide. Its action on stone formation could be explained by an increase in urine inhibitors as well as by a decrease of urinary calcium excretion.

Citrate is another substance seemingly provided with inhibitory activity. Smith et al. (11) have studied the effect of various concentrations of citrate on the growth of calcium phosphate crystals. Citric acid acts as a powerful inhibitor of crystal growth, with complete inhibition at a concentration of 1×10^{-3} moles.

We must point out that if citric acid had only a chelating action, a marked inhibitory effect would result at concentrations quite different from those mentioned above.

Among synthetic inhibitors, we must mention the role of diphosphonates, (12), substances which are characterized by a P-C-P group making their structure very similar to pyrophosphates, which instead bear P-O-P groups.

Diphosphonates might be useful in the prevention of renal stones in so far as they can act through one of these mechanisms:

1) inhibition of the precipitation of calcium phosphate and calcium oxalate in metastable urine;

2) prevention of transformation of calcium phosphate from the amorphous form into a much less soluble crystalline form.

3) inhibition of growth and aggregation of calcium oxalate crystals.

Some diphosphonates at high concentrations, perhaps owing to their strong affinity with calcium salts, can determine the transformation of calcium phosphate from a crystalline form into the more soluble amorphous one, probably by splitting crystalline aggregations. It has been observed that the administration of these substances to stone formers may cause increased oxaluria, although, paradoxically, a reduction of oxalate crystal formation may be observed.

Unfortunately the doses required to modify crystalluria may also be responsible for severe osseous lesions and this drug is therefore of little practical use in the prevention of stones.

Another substance which seems to possess inhibitory activity
is magnesium, but it is rarely used nowadays. Its use was justified
by the fact that Lyon et al. (13) had observed that this ion increased the solubility of calcium oxalate and by the discovery that
administration of a magnesium-free diet induced Ca oxalate stones
in rats.

The efficacy of magnesium has been questioned by Smith et al.
(11) and we also have observed that the drug causes little reduction
in the number of recurrences. In our experience, no dissolution of
oxalate stones can be obtained by incubating them in vitro with
urines containing magnesium in high concentrations. Furthermore,
magnesium-rich urines in no way prevented the tendency towards
aggregation of crystals, which by some authors is considered as one
of the main characteristics of renal stone formers.

Crystal aggregation appears to be considerably influenced by
a group of substances, including hyaluronic acid, heparin sulfate,
dermatan sulfate and chondroitin sulfate, known as glycosoaminoglycans. Robertson et al.(14) were among the first to note the efficacy of some glycosoaminoglycans, after incubating crystals of Ca
oxalate in a standard metastable solution of Ca oxalate.They observed
variation in the volume of these crystals as a result of the addition
of various substances with inhibitory power.

Such experiments suggest that both magnesium and methylene blue
have no effect on the growth and aggregation of crystals while, on
the other hand, diphosphonates and pyrophosphates totally inhibit
such processes at a concentration higher than that of heparin and
chondroitin sulfate (0,1-1 micromoles/l against 0,001 micromoles/l).
The latter substances act, therefore, as the must powerful inhibitors
of crystal aggregation.

The study of glycosoaminoglycans was subsequently developed by
Sallis and Lumley (15) who proved that the highest inhibitory power
lies in unknown substances, having the same characteristics of
precipitation as glycosoaminoglycans or precipitating with them.

Foye et al. (16) have also investigated to what extent the
degree of sulfatation of glycosoaminoglycans is important in favouring or preventing the formation of stones.

It has lately been reported that the total inhibitory activity
of glycosoaminoglycans (which represents 80% of the total inhibitory
activity) can be abolished by uric acid and by monosodium urate.
This may be due to binding and formation of colloidal products
lacking any inhibitory power.

The experimental basis of this hypothesis is, above all, due
to Robertson et al (14). In their effort to pin-point the role of
uric acid in recurrent stone formation, they observed that there
was no marked difference in the excretion of acid mucopolysaccharides
(AMPS) between recurrent stone formers and controls. It was found,
however, that the inhibitory activity of urine was highly influenced
by the AMPS-urates ratio. It would seem that the higher the urinary
concentration of uric acid and of sodium urate, the smaller the
capacity of AMPS to inhibit crystal aggregation. This hypothesis was
later confirmed by Pak et al.(17) who proved that both heparin and,
to a lesser extent, chondroitin sulfate considerably increase the
formation product of Ca-oxalate. Such action is far less pronounced
if the solution containing mucopolysaccharide is preincubated with
crystalline sodium urate and is secondary to the close binding that
urate forms with heparin. With regard to the mechanism of action of
these inhibitor substances, it has been postulated that since most
of them consist of negatively-charged molecules, they may combine
with Ca and form soluble, more stable compounds. In addition to
their action by sequestering Ca ions from precipitation, inhibitors
owing to their electrostatic charge can become absorbed on the
activated growing points of crystals, preventing them from increasing in size.

REFERENCES

1. Robertson, W.G. and Nordin,B.E.C.: Physicochemical factors
    governing stone-formation. In Scientific Fundations in Urology,
    edited by Williams, D.I. and Chisholm, G.D. pp. 254-267. London,
    Heinemann (1976).
2. Boyce, W.H.: Organic matrix of native human urinary concretions.
    In Renal Stone Research Symp. edited by Hodgkinson, A. and
    Nordin, B.E.C. pp. 93-104. London, Churchill, (1969).
3. Lonsdale, K.: Epitaxy as a growth factor in urinary calculi
    and gallstones. Nature 217: 56, (1968).
4. Pak, C.Y.C. and Arnold, L.H.: Heterogeneous nucleation of cal-

cium oxalate by seeds of monosodium urate. Proc. Soc. Exp. Biol.
Med. 149: 930,(1975).

5. Coe, F.L., Lawton, R.L., Goldstein, R.B. and Temple, V.: Sodium
    urate accelerates precipitation of calcium oxalate in vitro.
    Proc. Soc. Exp. Biol. Med. 149: 926, (1975).

6. Ligabue, A., Fini, M. and Robertson, W.G.: Influence of urine
    on "in vitro" crystallization rate of calcium oxalate: deter-
    mination of inhibitory activity by a 14C oxalate technique.
    Clinica Chimica Acta 98: 39,(1979).

7. Russell, R.G.G. and Fleisch, H.: Inhibitors in urinary stone
    disease: Role of pyrophosphate. Urinary Calculi. Int. Symp.
    Renal Stone Res. Madrid 1972, pp. 307-312 (Karger, Basel 1973).

8. Russell, R.G.G., Edwards, B.A. and Hodgkinson, A.: Urinary
    pyrophosphate and urolithiasis. Lancet 1: 1446, (1964).

9. Russell,R.G.G. and Hodgkinson, A.: Urinary excretion of inor-
    ganic pyrophosphate by normal subjects and patients with renal
    calculus. Clin. Sci. 31: 51, (1966).

10. Thomas, W.C. and Miller, G.M.: Inorganic phosphates in the
    treatment of renal calculi. Mod. Treatment 4: 494, (1967).

11. Smith, L.H., Meyer, J.L. and McCall, J.T.: Chemical nature of
    crystal inhibitors isolated from human urine. Urinary Calculi.
    Int. Symp. Renal Stone Res. Madrid 1972, pp. 318-327 (Karger,
    Basel 1973).

12. Fleisch, H. and Russell,R.G.G.: Inhibitors in urinary stone
    disease: role of diphosphonates. Urinary Calculi. Int. Symp.
    Renal Stone Res. Madrid 1972, pp. 296-301 (Karger, Basel 1973).

13. Lyon, E.S., Borden, T.A., Ellis, J.E. and Vermeulen, C.W.:
    Calcium oxalate lithiasis produced by pyridoxine deficiency and
    inhibition with high magnesium diets. Invest. Urol. 4: 133,
    (1966).

14. Robertson, W.G., Knowles, F. and Peacock, M.: Urinary acid
    mucopolysaccharide inhibitors of calcium oxalate crystalliza-
    tion. In Urolithiasis Research, pp. 331-338, edited by Fleisch,
    H., Robertson, W.G., Smith, L.H. and Vahlensieck, W. Plenum
    Publishing Corp. New York, 1979.

15. Sallis, J.D. and Lumley, M.F.: On the possible role of glyso-
    aminoglycans as natural inhibitors of calcium oxalate stones.
    Invest. Urol. 16: 296, (1978).

16. Foye, W.O., Hong, H.S., Kim, C.M. and Prien, E.L. Sr.: Degree
    of sulfation in mucopolysaccharide sulfates in normal and stone-
    forming urines. Invest. Urol. 14: 33, (1976).

17. Pak, C.Y.C., Holt, K. and Zerwekh, J.E.: Attenuation by mono-
     sodium urate of the inhibitory effect of glycosaminoglycans
     on calcium oxalate nucleation. Invest. Urol. 17: 138, (1979).

PHYSIOPATHOLOGY AND DIAGNOSIS OF HYPERPARATHYROIDISM

L. Giuliani

Urology Clinic
University of Genova
Viale Benedetto XV
16132 Genova (Italy)

ABSTRACT

The following points of interest are discussed: the frequency, classification, pathological patterns, clinical aspects and the role of hyperparathyroidism (HPT) in the genesis of calcium lithiasis. The steps of the diagnostic screening adopted at the Urological Clinic of the University of Genova are described. The problems of pathogenesis and classification of HPT are however far from a uniform, definitive and satisfactory solution. We still distinguish, for the sake of simplicity, between primary and secondary HPT. The latter form may become autonomous (tertiary HPT) and unable to respond to correction of the basic defect. From a histopathological standpoint, the distinction between adenoma and hyperplasia on the one part and between mild hyperplasia and normal glands on the other, is still fraught with difficulties. On clinical grounds, the work-up for screening, diagnosis of nature and location and, closely related, indications for surgery, suffers from these drawbacks. Fortunately, however, diagnostic methods today are sufficiently reliable to lead to a correct diagnosis in most cases.

A brief survey of personal experience is presented, consisting of over 200 evaluable patients. Surgery was performed in 38, including 24 orthotopic adenomas, 4 ectopic adenomas, and 10 diffuse

hyperplasias. Six of the latter cases were treated with total parathyroidectomy and autotransplantation in the forearm.

Follow-up of operated patients ranges between a few months
and 8 years. No recurrent HPT and no recurrence of urolithiasis
have been observed in our series.

INTRODUCTION: FREQUENCY AND POINTS OF INTEREST

1) The nephro-urologic patient has a high probability of
suffering from hyperparathyroidism (HPT).

2) The urologist more than anyone else has the opportunity to
find patients affected by hyperparathyroidism.

Dealing with hyperparathyroidism in urology, I would limit
myself to the following observations:

- there are still some doubts about classification and etio-
pathogenesis of this disease;

- from the clinical standpoint the most important problem is
screening.

Less important is the choice of therapy. Technical aspects
of surgery are, however, no longer a problem.

In spite of the interdisciplinary nature of this disease, a
new specialist has emerged in recent years who concerns himself
with all aspects of HPT, regardless of his clinical background.

The reason why HPT is found quite frequently in urologic patients, is associated with the high incidence of recurrent lithiasis
due to HPT (6, 9). Furthermore, 65% of patients with HPT have a
secondary involvement of the urinary tract (7). It is therefore
usual for the urologist to be particularly involved in the treatment of these patients. I would like also to emphasize that the
urologist is in the best position to treat surgically the urinary
as well as the parathyroid conditions.

In view of the latest research concerning the relationship

between lithiasis and HPT, a total view of the disease would be useful. Patients with HPT should be studied, treated and followed in a highly specialized center.

We cannot forget that a great deal of knowledge in HPT has recently come from progress in the field of lithiasis.

CLASSIFICATION

HPT is classically divided into primary hyperparathyroidism (PHPT), caused by adenoma or idiopathic cell hyperplasia, and secondary hyperparathyroidism (SHPT), found in patients with advanced chronic renal failure, especially if dialyzed. The cause of SHPT is an alteration in vitamin D metabolism with consequent malabsorption of calcium. HPT develops as a compensatory mechanism to correct the altered Ca-P metabolism. In the initial phase this form is reversible once the primary cause is eliminated.

If the disease goes on, the process becomes irreversible and independent of the primary cause (so-called tertiary HPT).

PATHOLOGY

HPT can be caused by adenoma, hyperplasia or, rarely, functional carcinoma. The histological differentiation between adenoma and hyperplasia or between mild hyperplasia and normal gland is not always easy (14, 24). It is, however, very important in view of the choice of surgery: removal of one gland in a patient with an adenoma, and total or subtotal parathyroidectomy in patients with hyperplasia. According to some Authors the routine use of electron microscopy would improve the diagnostic accuracy (3).

CLINICAL FINDINGS

The clinical features of HPT allow a classification into three forms: renal, skeletal and gastro-intestinal.

The renal form is characterized by recurrent renal calculi. 65% of patients with HPT have at least one episode of nephrolithiasis (7). The findings of nephrocalcinosis and staghorn stones are quite rare. The stones are mostly composed of calcium. In PHPT, because of the high serum calcium level, the excretion of calcium is enhanced, not compensated by increased tubular calcium absorption. In SHPT a tubular defect is the cause of the hypercalciuria.

The depletion of intracellular phosphates (HPT causes hyper-phosphaturia) leads to increased pH, decreased tubular bicarbonate reabsorption and renal tubular acidosis, favouring stone formation.

The deterioration of renal function is very slow and for the most part related to the complications of renal calculi rather than to the HPT itself.

The skeletal form is less frequent than the renal form, although bone involvement can be observed in 20-40% of cases. The lesions vary from mild asymptomatic subperiosteal reabsorption of the phalanges to advanced bone disease (Recklinghausen's disease).

The gastrointestinal form is characterized by the presence of peptic ulcer which is rather frequent in patients with HPT (10-15%) (5).

In these cases, a finding of peptic ulcer, alone or associated with acute or chronic pancreatitis, is the only feature of HPT.

Other Syndromes

In addition to pancreatitis and peptic ulcer, several other diseases may be associated with PHPT. Psychological alterations are frequent and can sometimes be the only symptoms (15). Hypertension occurs in up to 50% of cases. There can also be an increased incidence of ill-defined arthralgia as well as of gout and pseudo-gout. The latter is associated with intra-articular deposition of calcium pyrophosphate crystals.

LABORATORY DIAGNOSIS

Hypercalcemia with hypophosphatemia, although not specific, is highly indicative of PHPT.

Hypercalcemia

The presence of hypercalcemia is still the most important finding for a diagnosis of PHPT. The hypercalcemia is not always present. Sometimes the serum calcium levels are only slightly above the normal limits and sometimes periods of hypercalcemia alternate with periods of normocalcemia.

It is most important to detect even slight elevations of the serum calcium. In order to accomplish this, it is necessary to use a correct blood sampling technique, to have a reliable method of blood calcium assay and above all an accurate interpretation of the laboratory data (7).

It has therefore become very important to measure the ionized serum calcium directly, which can now be easily done with a highly sensitive electrode.

Hypophosphatemia

Hypophosphatemia is present in 50% of cases of HPT and this finding is therefore less important than hypercalcemia. The serum phosphorus level is strictly related to the renal function. As the renal function worsens, a retention of phosphates occurs and the previous low serum phosphorus level might become normal.

cAMP

The determination of the urinary excretion of cyclic adenosine monophosphate (cAMP) which represents the effect of PTH on the renal tubule cells seems to be a test of great value. This is easily accomplished since the introduction of a reliable radioimmunoassay method.

Comparing the urinary cAMP with the creatinine clearance (cAMP/g of urinary creatinine) makes this test even more sensitive (23).

PTH Dosage (PTH Radioimmunoassay)

PTH dosage represents the most important advancement in the
study of HPT but its diagnostic value is still controversial. As a
matter of fact the possibility of detecting the increased activity
of the parathyroid glands, by measuring the PTH, would be the most
reasonable way to make an unquestionable diagnosis. Unfortunately
the presence of different hormone fractions in the blood (26), the
unreliable titration and the high cost made this determination quite
difficult in clinical practice up to few years ago.

Two main techniques for PTH assay are available: one measures
the PTH concentration in the peripheral blood; and in the other
the assay is performed on blood obtained by selective or superselective catheterization from the cervical and mediastinal veins or
from the thyroidal veins (11, 16, 17).

The peripheral blood assay is useful as an initial screening
on a large number of patients. Unfortunately it is not always
capable of showing slight elevations of PTH concentrations. This is
mainly due to the fact that small amounts of hormone, released in
excess from the parathyroid glands, become highly diluted in the
peripheral blood. In order to increase the sensitivity of the test
it could be helpful to compare the PTH assay of the peripheral blood
with the calcium and phosphorus determinations. In our experience
the screening test using peripheral blood has been unsuccessful in
detecting HPT in 55% of cases. This has been especially true in the
cases secondary to renal failure where the PTH concentration is
only slightly increased. It is therefore our practive to measure
the PTH concentration in venous blood obtained with selective
angiography techniques. Following Reitz (7) and Doppman (11, 12)
we have studied about 100 patients since 1974 without any complications. The method has been previously reported (17).

Two main advantages can be obtained with the superselective
technique: 1) a slight increase of PTH concentration can be detected
even when its concentration is normal in peripheral blood;. 2) the
analysis of PTH concentration at different levels allows one to
establish in most cases whether the adenoma involves only one or
several parathyroid glands. Our own experience has been quite
significant in this regard. With this technique it was possible to
diagnose unilateral adenoma preoperatively in 80% of cases and

parathyroid hyperplasia in 100% of cases.

In one case, in which the adenoma lay in the mediastinum and in which sternotomy was performed, the highest concentration of PTH was found in the superior vena cava. The technique of selective sampling has also been found useful in cases requiring re-exploration, although the altered venous return can make the interpretation of blood samples confusing. The PTH assay from selective sampling can give false positive results in a few cases. This should not be a deterrent to using this technique, as superselective sampling for PTH assay is the test to be performed after the clinical criteria and the laboratory tests have already pointed to the diagnosis. In this way PTH assay from selective venous sampling is useful in improving the chance of making a definitive diagnosis.

This technique has also proved to be useful in detecting other forms of hypercalcemia (not PTH dependent) particularly those of neoplastic origin. In these cases the high blood calcium and the normal or low PTH concentration make the diagnosis easy. Also in cases of abnormal production of PTH by some tumours, a normal PTH concentration at the cervical level and a high concentrations at the sub-diaphragmatic level, as may occur in presence of a renal tumour, make the diagnosis quite clear.

EVALUATION OF BONE INVOLVEMENT

Some degree of bone involvement is relatively frequent and, if present, is of diagnostic value. The typical bone lesions are characterized by fraying of the distal phalanges of the hands, subperiosteal reabsorption in the long bones and skull and disappearance of the lamina dura. Cysts in the bones are highly indicative of HPT but unfortunately their presence is rare.

VISUALIZATION OF THE INCREASED VOLUME OF THE AFFECTED GLANDS

The direct visualization of the affected glands is still very difficult. We would like here to mention only those techniques most suitable for clinical practice.

Angiography. Arteriography of the parathyroid glands is quite

a delicate technique and not free from danger. Only modest results
are obtainable with this technique and it is not always possible
to distinguish between a parathyroid adenoma and a thyroid nodule.
In our opinion this technique should be reserved for use in those
patients already previously operated upon.

Scintigraphy. The scintigraphic examination of the parathyroid
glands with Se[75], widely used up to a few years ago, has now been
almost completely abandoned because of its poor results.

Recently the use of new isotopes and of very sophisticated
gamma-cameras, seems to have led to better results.

Echotomography. The use of ultrasound seldom gives good results.
This technique has approximately the same limitations as angiography;
the main advantage is that of not being invasive.

C.A.T. Scan. Only a few cases of parathyroid adenomas have
been localized preoperatively (13) thus far with this technique. A
better diagnostic accuracy is needed before this method could be
used routinely.

Thermography. Some Authors (22) have proposed this technique
for the localization of parathyroid adenomas but the results have
been unsatisfactory.

Lymphography. With this procedure the thyroid gland becomes
radio-opaque following the injection of iodinated oil (18). An
increase in size of a parathyroid gland will show up as a defect
on the thyroid surface. In our opinion however, this procedure is
not particularly accurate and is of little clinical value.

In summary we can state that no universally valid methods are
yet available to allow a definitive preoperative localization of
the affected parathyroid glands. Therefore we must still depend on
a careful and complete surgical exploration of the cervico-jugular
region.

THERAPY

The treatment of HPT is essentially surgical although medical
therapy does play a role in controlling severe hypercalcemia in
HPT crisis (2). The initial choice is wheter the hyperparathyroidism

or renal disorder should be treated first.

Usually the parathyroid lesions, are treated first (1), but if
renal obstruction is present this should be given priority. However,
the simultaneous treatment of both conditions has been recently
accepted when the patient's condition allows it.

The essential surgical problems are: 1) localization of the
parathyroid glands; 2) identification of the affected glands;
3) determination of the extent of resection.

1) The parathyroid glands, because of their small size and
multiple location, are not always easy to find (8). The superior
glands are most often found on the posterior face of the thyroid
lobes.

The glands are firmly in contact with the thyroid tissue but,
once the capsule is opened, their enucleation is simple.

The inferior glands are usually found behind the lower pole
of each thyroid lobe. Sometimes they are ectopic and can be found
near the upper part of the thymus gland. They can also be found in
the mediastinum, in which case a sternotomy is needed for their
removal.

The parathyroid glands are easily confused with limph nodes,
fat, lobules and small thyroid nodules. A frozen section examination
or the "imprint technique" (25) can be utilized for their identification.

2) The presence of an adenoma, when the remaining glands are
normal or atrophic, does not give rise to any diagnostic problem.
But when one must differentiate between multiple adenomas and
diffuse hyperplasia or between normal glands and mild hyperplasia,
the diagnosis may become difficult (21). Therefore the identification
of all four glands is mandatory and in doubtful cases a "density
test" (27) can be used, since histology is quite often unsatisfactory.

3) In presence of an adenoma, the affected gland is removed.
When dealing with multiple adenomas or diffuse hyperplasia it is
often difficult to judge the amount of tissue to resect. An insuf-

ficient resection may allow recurrent HPT and, on the other hand,
too radical an operation leads to irreversible hypoparathyroidism.

The correct approach is to resect as much as possible of the
parathyroid glands leaving behind the minimal amount of tissue
necessary to maintain a normal calcium homeostasis.

Some situations deserve a specific mention:

A) Two enlarged glands and two normal ones (5% of cases): the
possibility of two separate adenomas should be considered, although
some authors feel that an occult hyperplasia of the two normal
glands could be present in these cases (20). The removal of the
two enlarged glands only, seems to us the best approach.

B) Three or more enlarged glands: a diffuse hyperplasia is
present. Most authors perform a subtotal resection (so-called 7/8
parathyroidectomy) (19, 20), by removing three and a half glands
and leaving just half a gland with a good blood supply. This procedure is usually not completely satisfactory. There is the possibility of leaving too much tissue, which may lead to recurrent
HPT and to a risky re-operation. On the other hand too radical
a resection or damage to the blood supply of the residual gland
may bring about permanent hypoparathyroidism.

Wells (1974) (28) developed an interesting procedure which
consists of a total parathyroidectomy with autotransplantation of
the residual normal gland in the forearm. We successfully used this
technique in 6 patients, three with secondary HPT and three with
idiopathic hyperplasia. In these patients a higher serum PTH concentration was found in the arm containing the transplanted gland as
compared with the contralateral arm (29).

C) Three normal glands: two possibilities should be considered:
an intrathyroid adenoma or an ectopic adenoma. Real intrathyroid
adenomas are extremely rare. In 20% of cases adenomas of the upper
glands are covered with thyroid capsule simulating intrathyroid
adenomas. Once the capsule is opened, however, the adenoma clearly
shows itself to lie outside the thyroid tissue.

Ectopic adenomas on the other hand are very frequent. In these
cases it is necessary to extend the incision to the jugular notch

in order to perform a transcervical thymectomy. Because of their identical embryological origin, it is quite frequent to find parathyroid glands inside the thymus. The thymectomy must be performed together with a thorough exploration in order to remove even deeply located adenomas. Sternotomy is needed if the adenoma is located anteriorly, in the lower mediastinum, but must be limited to selected cases because of its high surgical risk. We have performed sternotomy in two out of 32 cases.

D) <u>Four normal glands</u>: there are two possibilities: wrong diagnosis or adenoma of a fifth parathyroid gland (5% of cases).

COMPLICATIONS

Parathyroidectomy is usually well tolerated. The worst complication is an injury to the recurrent laryngeal nerve by cutting it or by electrocoagulation performed near the nerve. This leads to permanent or reversible paralysis. The surgeon must carefully dissect the nerve in order to avoid any injury. We have had only one case of temporary dysphonia which returned to normal within three months.

RE-OPERATIONS

Re-operation should be avoided because of a high risk of inducing permanent hypoparathyroidism or nerve lesion. According to the Mayo Clinic report (4), the first operation carries an 86% success rate while that of re-operation is only 3%. Davies reports (10) permanent hypoparathyroidism in 1% of cases after the first operation, but in 23% after re-operation. Two out of 35 patients suffered from permanent bilateral laryngeal paralysis in Bearzley's report (4).

Intraoperative death from bleeding due to a large vessel lesion and damage to the cervical sympathetic nerves have been reported during re-operation.

In order to avoid these complications one should perform an accurate dissection of the cervical-jugular region and of the four parathyroid glands and carry out an adequate resection. In our opinion diffuse hyperplasia should be treated with total para-

thyroidectomy with autotransplantation. We have had to re-operate
in only one patient in whom an adenoma was found in the residual
thymus.

POST-OPERATIVE TREATMENT

Hypocalcemia is often found in patients who have undergone a
parathyroidectomy. In most cases this is due to "bone hunger" followed
by deposition of large amounts of calcium in demineralized bone.

A routine post-operative PTH assay is most valuable. If it is
normal, a reversible condition is foreseen; if close to zero, permanent hypoparathyroidism should be suspected.

The administration of calcium-lactobionate and dihydrotachy-
sterol orally (3-8 mg/daily) is advisable.

In severe case of hypocalcemia intravenous infusion of calcium
gluconate may be necessary.

We still have not had the opportunity to use 1-alfa-dihydroxy-
cholecalciferol.

Patients with parathyroidectomy and forearm autotransplantation
have been able to discontinue the treatment two or three months
after surgery.

FOLLOW-UP

The follow-up should be evaluated using the following parameters

- normal calcium homeostasis;
- lack of recurrent urolithiasis;
- disappearance of bone pain;
- improvement of damaged renal function.

Out of 38 patients operated on in our Clinic (followed from 4
months to 8 years), none presented with recurrent HPT or urolithiasis

REFERENCES

1. Albright, F. and Reifenstein, E.C.: The parathyroid glands
   and metabolism bone disease. Williams and Wilkins Co., Baltimore, 1948.
2. Barilla, E.D. and Pak, C.Y.C.: Pitfalls in parathyroid evaluation in patients with calcium urolithiasis. Urol. Res. 7: 177,
   (1979).
3. Black, W.C., III: Correlative light and electron microscopy in
   primary hyperparathyroidism. Arch. Pathol. 88: 225, (1969).
4. Beazley, R.M., Costa, J. and Ketcham, A.S.: Re-operative parathyroid surgery. Am. J. Sug. 130: 427, (1975).
5. Bone, G.H. III, Snyder, W.H. III and Pak, C.Y.C.: Diagnosis
   of hyperparathyroidism. Ann. Rev. Med. 28: 111, (1977).
6. Bracci, U. and Basso, A.: L'iperparatiroidismo in Urologia.
   Chir. Urol. 4: 13, (1962).
7. Carmignani, G., Belgrano, E. and Puppo, P.: L'iperparatiroidismo in Urologia. Urologia XLIV: 3, (1977).
8. Chigot, P.L.: Chirurgie de l'hyperparathyroidie. J. Chir. 93:
   579, (1967).
9. Coe, F.L. and Kavalach, A.G.: Hypercalciuria and hyperparathyroidism in patients with calcium nephrolithiasis. N. Engl.
   J. Med. 291: 1344, (1974).
10. Davies, D.R.: The surgery of primary hyperparathyroidism.
    Clin. Endocrin. Metabol. 3: 253, (1974).
11. Doppman, J.L., Nelson, G.L., Evens, R.G. and Hammond, W.G.:
    Selective superior and inferior thyroid vein catheterization:
    venographic anatomy and potential application. Invest. Radiol.
    4: 97, (1969).
12. Doppman, J.L., Mallette, L.E., Marx, S.J., Monchick, J.M.,
    Broadus, A., Spiegel, A.M., Beazley, R. and Aurbach, G.D.:
    The localization of abnormal mediastinal parathyroid glands.
    Radiology 115: 31, (1975).
13. Doppman, J.L., Murray, F.B., Koehler, J.O. and Marx, S.J.:
    Computed assisted tomography. J. Computed Assisted Tomography
    1: 30, (1977).
14. Esselstyn, C.B.: Parathyroid pathology: its relation to choice
    of operation for hyperparathyroidism. Surg. Clin. North Am.
    59: 77, (1979).
15. Giuliani, L., Carmignani, G., Belgrano, E., Martorana, G. and
    Puppo, P.: Doppio adenoma paratiroideo: insufficienza renale
    e turbe psichiche. Presentazione di un caso. Recenti acquisi-

zioni diagnostico-terapeutiche. Arch. It. Urol. 46: 259, (1975).

16. Giuliani, L., Carmignani, G., Belgrano, E. and Puppo, P.: Die
    selektive Parathormon-bestimmung als Lokalisationsmittel der
    Nebenschilddrüsenadenoma beim primären Hyperparathyroidismus.
    Urologe A 17: 23, (1978).

17. Giuliani, L., Carmignani, G., Belgrano, E. and Puppo, P.:
    Value of selective parathormone radioimmunoassay in primary
    hyperparathyroidism. Urology 13: 156, (1979).

18. Kato, T., Hattori, T., Miura, K. and Sato, M.: Application of
    thyroid lymphography to preoperative localization of hyper-
    functioning parathyroid adenomas. Ann. Surg. 179: 387, (1974).

19. Johansson, H., Thoren, L., Werner, I. and Grimelius, L.:
    Normocalcemic hyperparathyroidism; kidney stones and idiopathic
    hypercalciuria. Surgery 77: 691, (1975).

20. Paloyan, E., Lawrence, A.M. and Baker, W.H.: Near total para-
    thyroidectomy. Surg. Clin. North. Am. 49: 43, (1969).

21. Roth, S.I., Wang, C. and Potts, J.T.: The team approach to
    primary hyperparathyroidism. Hum. Pathol. 6: 645, (1975).

22. Samuels, B.I., Dowdy, A.H. and Lecky, J.W.: Parathyroid thermo-
    graphy. Radiology 104: 575, (1972).

23. Schmidt-Gayk H. and Rohen, H.D.: Urinary excretion of cyclic
    adenosine monophosphate in the detection and diagnosis of
    primary hyperparathyroidism. Surg. Gynec. Obstet. 137: 439,
    (1973).

24. Scholz, D.A., Purnell, D.C., Edis, A.J., Van Heedern, J.A. and
    Woolner, L.B.: Primary hyperparathyroidism with multiple para-
    thyroid gland enlargement. Review of 53 cases. Mayo Clin. Proc.
    53: 792, (1978).

25. Silverberg, S.G.: Imprints in the intraoperative evaluation of
    parathyroid disease. Arch. Pathol. 99: 378, (1978).

26. Silverman, R. and Yalow, R.S.: Heterogeneity of parathyroid
    hormone: clinical and physiologic implications. J. Clin. Invest.
    52: 1958, (1973).

27. Wang, C. and Rieder, S.U.: A density test for the intraoperative
    differentiation of parathyroid hyperplasia from neoplasia.
    Ann. Surg. 187: 63, (1978).

28. Wells, A.S., Gunnells, J.C. and Shelburne, J.D.: Transplantation
    of the parathyroid glands in man: clinical indication and
    results. Surgery 78: 34, (1975).

29. Wells, A.S., Ross, A.J., Dale, J.K. and Gray, R.S.: Trans-
    plantation of the parathyroid glands: current status. Surg.
    Clin. North Am. 59: 167, (1979).

CONTROVERSIES IN THE PATHOPHYSIOLOGY OF CALCIUM NEPHROLITHIASIS

G. Maschio, N. Tessitore, A. D'Angelo, E. Bonucci, C.
Lo Schiavo, E. Valvo, A. Lupo, A. Fabris, L. Oldrizzi,
and G. Previato

- Cattedra e Divisione di Nefrologia,
  Istituti Ospitalieri, 37100 Verona (Italy)
- I Clinica Medica, Università di Padova (Italy)
- I Cattedra di Anatomia e Istologia Patologica
  Università di Roma (Italy)

ABSTRACT

The main risk factors in the pathogenesis of calcium nephrolithiasis are low urine volume, high urine pH, reduced excretion
of inhibitors of crystal formation, and increased excretion of calcium, uric acid and oxalate. These risk factors may be influenced
by some physiological and pathological conditions. The relative
importance and the interrelationships of oral fluid intake, dietary
intake of animal protein and salt, intestinal calcium absorption,
renal handling of calcium and uric acid, and parathyroid function
are discussed.

INTRODUCTION

The main risk factors of calcium stone-formation are low urine
volume, high urine pH, impaired excretion of inhibitors of crystal
formation, and increased urinary excretion of calcium, oxalate and
uric acid. Apart from the inhibitors, these risk factors may be
influenced by a number of physiological and pathological conditions.

1. Oral fluid intake. An important etiologic role of oral fluid intake in the formation and prevention of renal stones has long been suggested (36). Evidence has been obtained that a low fluid intake, which results in a low urine volume, is a risk factor for calcium stone disease (32). Furthermore, recent data suggest that urinary dilution due to high fluid intake significantly reduces the activity product ratio of calcium phosphate, calcium oxalate and monosodium urate in calcium stone-formers (27).

2. Diet. The dietary intake of calcium, phosphate, oxalate, animal protein and salt may adversely affect some of the six urinary risk factors for nephrolithiasis. In our experience, the incidence of hypercalciuria decreased from 56% to 43% of a large population of stone-formers when their calcium intake was lowered to about 400 mg/day for a week. In clinical practice, conflicting results were obtained concerning the patients' adherence to low-calcium diets, which was good in Italy (31) but very poor in England (4). On the other hand, unless patients are selected and studied careful-ly, the possibility of producing chronic negative calcium balance and increased urinary oxalate is a potential risk for this diet. Urinary calcium excretion may rise as a consequence of extreme dietary phosphate deprivation, which is unlikely to occur in man (21). It is well documented that increasing the level of dietary animal protein intake results in a concomitant rise in urinary ex-cretion of three risk factors for nephrolithiasis, namely calcium, oxalate and uric acid (33). A decrease in the fractional reabsorption of calcium by the kidney, probably mediated by an increased net acid excretion, seems to be the most likely cause of the protein-induced hypercalciuria (2), although some additional mechanisms, such as increased renal calcium load and stimulation of bone resorption, might also play a role (1, 21). The increase in urinary oxalate may derive from the increased production of oxalate, resulting from the metabolism of some amino acids which are highly concentrated in animal protein (28). As far as hyper-uricosuria is concerned, an excessive dietary purine intake is the most likely pathogenetic mechanism, though uric acid overproduction from endogenous purine metabolism may also be regarded as a contributory factor (14). The possible importance of sodium intake is yet to be fully assessed: some data suggest that urinary calcium significantly increases when dietary sodium is maintained above 150 mEq/day (23, 24).

3. Intestinal absorption of calcium and oxalate. In many pa-

tients with calcium nephrolithiasis, an enhanced intestinal calcium absorption was shown by several techniques, including fractional calcium absorption, forearm counting bone-densitometry techniques, calciuric response to an oral calcium load, and selective intestinal perfusion (26). According to these data, in patients with idiopathic hypercalciuria an "intestinal calcium hyperabsorption theory" was suggested, which led to a separation of those patients into two groups, one with primary "absorptive hypercalciuria" and the other with primary "renal leak hypercalciuria" (13, 25, 29). Separation, however, is a rather euphemistic term for patients with idiopathic hypercalciuria.

Yet, if intestinal calcium absorption is invariably elevated in those supposed to have absorptive hypercalciuria, it is also increased in a significant percentage of those having renal leak hypercalciuria, probably as a consequence of secondary hyperparathyroidism (26). Evidence was obtained that serum concentrations of 1,25 (OH)2 vitamin D are, on the average, elevated among patients with calcium nephrolithiasis (12). In renal leak hypercalciuria, the fractional calcium absorption was directly correlated with the serum levels of 1, 25 (OH)2 vitamin D, as it is usually observed in primary hyperparathyroidism (18, 26). Therefore, these results support the hypothesis that the increased intestinal calcium absorption in renal hypercalciuria is due to an enhanced renal synthesis of 1, 25 (OH)2 vitamin D associated with secondary hyperparathyroidism. In absorptive hypercalciuria, no correlation was found between fractional calcium absorption and serum 1, 25 (OH)2 vitamin D values. Since plasma parathyroid hormone (PTH) values are usually normal or low in absorptive hypercalciuria, some other factors must be responsible for the high serum values of 1, 25 (OH)2 vitamin D which are observed in 30-50% of the patients (18). A possible role of hypophosphatemia, which would stimulate the renal synthesis of 1,25 (OH)2 vitamin D, as a consequence of a primary renal leak of phosphate was suggested (9, 18). However, conflicting results regarding serum PO4 values and renal tubular threshold for PO4 and their correlations with serum 1,25 (OH)2 vitamin D levels were obtained, so that the possible importance of phosphate depletion in the pathogenesis of absorptive hypercalciuria is far from being established (7, 34). On the other hand, studies of intestinal perfusion did not support an important etiologic role of 1,25 (OH)2 vitamin D and suggested that hyperabsorption of calcium may be vitamin D-indipendent in patients with the absorptive form of hypercalciuria (10).

An increased intestinal absorption of oxalate may be regarded as a primary mechanism for hyperoxaluria in different conditions, such as oxalate overingestion, cellulose phosphate ingestion, ileal resection, and small-bowel bypass (15). Apart from the two latter conditions - where oxalate is over-absorbed as a consequence of fat malabsorption - the majority of patients with calcium stones have a normal or only slightly raised urinary oxalate excretion. It was shown that this mild hyperoxaluria - which may be observed in about 15% of the calcium stone-formers (17) - is due to increased intestinal absorption of oxalate, probably due to a reduction in intraluminal calcium concentration (19).

4. <u>Renal handling of calcium and uric acid</u>. The "renal calcium leak theory" in the pathogenesis of idiopathic hypercalciuria is supported by the findings of high fasting urinary calcium excretion, impaired renal tubular reabsorption during calcium load, high fractional excretion of filtered ionized calcium, and exaggerated calciuric response to carbohydrate ingestion (5, 11, 13, 24, 35). In the absence of hypercalcemia or increased bone resorption, a high fasting urinary calcium may indicate that a primary renal leak is present. A more direct approach to the diagnosis of renal leak hypercalciuria, based on studies of renal calcium clearance during an intravenous calcium load, gave conflicting results. In fact, there was evidence of an impaired (35) as well as of a normal (30) tubular reabsorption of calcium. On the other hand, recent studies showed a significant increase of the fractional excretion of filtered ionized calcium, both in fasting and post-load conditions, in patients with idiopathic hypercalciuria (24). Moreover, an exaggerated calciuric response to carbohydrate ingestion - which is believed to reflect an altered function of renal tubular cells, consequent upon insulin-dependent glucose metabolism - was also observed in patients with idiopathic hypercalciuria (5). Finally, a high frequency of renal tubular defects - including impaired acidification capacity and tubular proteinuria - was recently reported to be associated with hypercalciuria in patients with recurrent nephrolithiasis (3). If the hypothesis of a primary renal calcium leak is tenable, this would eventually be associated with a fall in serum calcium, a rise in serum PTH and an increased renal synthesis of 1,25 (OH)2 vitamin D (13, 18). Obviously, this metabolic sequence makes it difficult, at best, to determine the true incidence of

renal hypercalciuria versus resorptive or absorptive hypercalciuria in stone-forming patients. Yet, the incidence of renal hypercalciuria varied from 15% (26) to 100% (24) of the patients in two different reports. That a primary renal calcium leak exists, however, is supported by the response to medical and surgical treatment. Thiazide administration for instance, by correcting the renal calcium leak, actually restores normal PTH values, serum 1,25 (OH)2 vitamin D levels and fractional calcium absorption in patients supposed to have renal hypercalciuria, thus providing a useful diagnostic tool (13, 37). Furthermore the persistence of renal hypercalciuria after removal of a parathyroid adenoma (9, 22) supports the view that a primary renal calcium leak is a distinctive entity in stone-forming patients. A disturbance in the renal handling of uric acid, such as enhanced secretion or reduced postsecretory reabsorption, could be associated with renal hypercalciuria in those patients with calcium-uric acid stones who have normal serum uric acid values (15). So far, no extended studies on this intriguing possibility have been performed.

5. Parathyroid function. Primary hyperparathyroidism is recognized as one of the major causes of calcium nephrolithiasis and, when classically expressed, its diagnosis is usually easy. Increased serum PTH values in patients with idiopathic hypercalciuria are believed to reflect the chronic parathyroid gland stimulation due to calcium deficiency as a consequence of primary renal calcium leak. Unfortunately, this condition is not always easily discriminated from a mild form of primary hyperparathyroidism. The suppression test with calcium infusion, for instance, may decrease plasma PTH values in both conditions (24). Moreover, in certain patients with renal hypercalciuria, some degree of parathyroid autonomy must be expected from a long-term stimulation of parathyroid glands (9). This, in turn, would increase the diagnostic difficulties, with important medical and surgical implications (6, 22). In addition to increased plasma PTH values, patients with renal hypercalciuria usually have high levels of total urinary cyclic AMP as well as nephrogenous cyclic AMP, and these features clearly separate them from patients with absorptive hypercalciuria (26). Moreover, it has recently been shown that the bone mineral content is significantly lower in patients with renal hypercalciuria and secondary hyperparathyroidism than in those with absorptive hypercalciuria (20). In addition, our preliminary results of bone histology in recurrent calcium nephrolithiasis indicate that

Table 1. Bone morphology in calcium stone-formers.

Number of patients   12
Normal bone   7

DEFECTIVE MINERALIZATION

    2 (renal leak hypercalciuria)

DEFECTIVE MINERALIZATION AND INCREASED BONE RESORPTION

    3 - Renal leak hypercalciuria with secondary HPTH
      - Renal leak hypercalciuria with secondary HPTH
      - Renal leak hypercalciuria with autonomous HPTH

---

morphologic evidence of defective mineralization and increased bone resorption may be observed in patients with renal hypercalciuria (Table 1).

In our opinion, all these abnormalities, which may also be observed in primary hyperparathyroidism, <u>strongly suggest primary renal calcium leak with secondary hyperparathyroidism when they are present in patients with normal or low total and ionized serum calcium.</u>

The role of inhibitors of crystal formation and aggregation in the pathogenesis of calcium stones is not clear. A diminished excretion of inhibitors was observed by Baumann et al. (8) in 44% of their stone-forming patients, and it was attributed to a reduced excretion of inorganic pyrophosphate. Other inhibitors have been identified in some mucopolysaccharides and glycoproteins (15).

REFERENCES

1. Adams, N.D., Gray, R.W. and Lemann, J.: The calciuria of increasing dietary protein intake. Clin. Res. 26: 719 A, (1978).
2. Allen, L.H., Oddoye, E.A. and Margen, S.: Protein-induced hypercalciuria: a longer term study. Am. J. Clin. Nutrit. 32: 741-749, (1979).

3. Backman, U.,Danielson, B.G., Johansson, G., Ljunghall, S. and Wikstrom, B.: Incidence and clinical importance of renal tubular defects in recurrent renal stone formers. Nephron 25: 96–101, (1980).

4. Baker, L.R. and Mallinson, W.J.: Dietary treatment of idiopathic hypercalciuria. Brit. J. Urol. 51: 181–183, (1979).

5. Barilla, D.E., Townsend, J. and Pak, C.Y.C.: An exaggerated augmentation of renal calcium excretion after oral glucose ingestion in patients with renal hypercalciuria. Invest. Urol. 15: 486–488, (1978).

6. Barilla, D.E. and Pak, C.Y.C.: Pitfalls in parathyroid evalu- ation in patients with calcium urolithiasis. Urol. Res. 7: 177–182, (1979).

7. Barilla, D.E., Zerwekh, J.E. and Pak, C.Y.C.: A critical evalu- ation of the role of phosphate in the pathogenesis of absorptive hypercalciuria. Min. Electr. 2: 302–309, (1979).

8. Baumann, J.M., Bisaz, S., Felix, R., Fleisch, H., Ganz, U. and Russel, R.G.G.: The role of inhibitors and other factors in the pathogenesis of recurrent calcium-containing renal stones. Clin. Sci. Mol. Med. 53: 141–148, (1977).

9. Bordier, P., Rickewart, A., Gueris, J. and Rasmussen, H.: On the pathogenesis of so-called idiopathic hypercalciuria. Am. J. Med. 63: 398–409, (1977).

10. Brannan, P.G., Morawski, S., Pak, C.Y.C. and Fordtran, J.: Selective jejunal hyperabsorption of calcium in absorptive hypercalciuria. Am. J. Med. 66: 425–428, (1979).

11. Broadus, A.E., Dominquez, M. and Bartter, F.C.: Pathophysio- logical studies in idiopathic hypercalciuria: use of an oral calcium tolerance test to characterize distinctive hypercalciuric subgroups. J. Clin. Endocr. Met. 47: 751–760, (1978).

12. Caldas, A., Gray, R.W. and Lemann, J.Jr.: The simultaneous measurement of vitamin D metabolites in plasma: studies in healthy adults and in patients with calcium nephrolithiasis. J. Lab. Clin. Med. 91: 840–849, (1978).

13. Coe, F.L., Canterbury, F.M., Firpo, J.J. and Reiss, E. Evidence for secondary hyperparathyroidism in idiopathic hypercalciuria. J. Clin. Invest. 52: 134–142, (1973).

14. Coe, F.L. and Kavalich, A.G.: Hypercalciuria and hyperuricosuria in patients with calcium nephrolithiasis. New Engl. J. Med. 291: 1344–1348, (1974).

15. Coe, F.L.: Nephrolithiasis. Pathogenesis and treatment. Year Book Publ., Chicago, 1978, p. 141.

16. Coe, F.L.: Forum commentary. Kidney Intern. 16: 646-647, (1979).
17. Coe, F.L.: Personal communication, 1980.
18. Gray, R.W., Wilz, D.R., Caldas, A. and Lemann, J. jr.: The importance of phosphate in regulating plasma 1,25(OH)2-vitamin D levels in humans: studies in healthy subjects, in calcium stone-formers, and in patients with primary hyperparathyroidism. J. Clin. Endocr. Met. 45: 299-306, (1977).
19. Hodgkinson, A.: Evidence of increased oxalate absorption in patients with calcium-containing renal stones. Clin. Sci. Mol. Med. 54: 291-294, (1978).
20. Lawoyin, S., Sismilich, S., Browne, R. and Pak, C.Y.C.: Bone mineral content in patients with calcium urolithiasis. Metabolism 28: 1250-1254, (1979).
21. Lemann, J. jr., Adams, N.D. and Gray, R.W.: Urinary calcium excretion in human beings. New Engl. J. Med. 301: 535-541, (1979).
22. Maschio, G., Vecchioni, R. and Tessitore, N.: Recurrence of autonomous hyperparathyroidism in calcium nephrolithiasis. Am. J. Med., in press.
23. Modlin, M.: The interrelation of urinary calcium and sodium in normal adults. Invest. Urol. 4: 180-189, (1966).
24. Muldowney, F.P.: Diagnostic approach to hypercalciuria. Kidney Intern. 16: 637-648, (1979).
25. Pak, C.Y.C., Ohata, M., Laurence, E.C. and Snyder, W.: The hypercalciurias: causes, parathyroid function and diagnostic criteria. J. Clin. Invest. 54: 387-399, (1974).
26. Pak, C.Y.C.: Physiological basis for absorptive and renal hypercalciurias. Am. J. Physiol. 237: F 415-423, (1979).
27. Pak, C.Y.C., Sakhaee, K., Crowther, C. and Brinkley, L.: An objective evidence for the beneficial effect of a high fluid intake in the management of nephrolithiasis. In press.
28. Paul, A.A. and Southgate, D.A.T.: The composition of Foods, IV ED., H.M.S.O., London, 1978, p. 280.
29. Peacock, M., Knowles, F. and Nordin, B.E.C.: Effect of calcium administration and deprivation on serum and urine calcium in stone-forming and control subjects. Brit. Med. J. II: 729-731, (1968).
30. Peacock, M. and Nordin, B.E.C.: Tubular reabsorption of calcium in normal and hypercalciuric subjects. J. Clin. Pathol. 21: 353-358, (1968).
31. Pizzarelli, F., Ciccarelli, C., Parlono, S. and Maggiore, Q.: Il ruolo della dieta a basso contenuto di calcio e di ossalato

nell'ipercalciuria da aumentato assorbimento intestinale di
calcio. Abstract XX Congr. Soc. Ital. Nefrol., 1979.

32. Robertson, W.G.: Risk factors in calcium stone formation. In:
    Proceedings of the VIIth International Congress of Nephrology.
    Karger, Basel, 1978, p. 363.

33. Robertson, W.G., Heyburn, P.J., Peacock, M., Hanes, F.A. and
    Swaminathan, R.: The effect of high animal protein intake on
    the risk of calcium stone-formation in the urinary tract. Clin.
    Sci. 57: 285-288, (1979).

34. Shen, F.H., Baylink, D.J., Nielsen, R.L., Sherrard, D.J., Ivey,
    J.L. and Haussler, M.R.: Increased serum 1,25 - dihydroxy-
    vitamin D in idiopathic hypercalciuria. J. Lab. Clin. Med. 90,
    955-962, (1977).

35. Smith, D.A. and Mackenzie, J.C.: In Fleisch, H., Blackwood,
    H.J.J., Owen, M.: Calcified Tissues, Springer, Berlin, 1965,
    p. 211.

36. Thompson, W.G.: Practical dietetics. D. Appleton Co., New York,
    1895, p. 476.

37. Zerwekh, J.E. and Pak, C.Y.C.: Selective effects of thiazide
    therapy on serum 1,25 - dihydroxy vitamin D and intestinal cal-
    cium absorption in renal and absorptive hypercalciurias.
    Metabolism 29: 13-17, (1980).

DEFINITION, TYPES AND MANAGEMENT OF NORMOCALCAEMIC HYPERCALCIURIA IN RENAL LITHIASIS

B. Pinto, F. J. Ruiz-Marcellan and J. Bernshtam

Laboratorio de Exploraciónes Metabolicas
and Servicio de Urologia
Ciudad Sanitaria Barcelona
and Servicio de Análisis Clínicos
Aragón 420 - Barcelona - Catalunya (Spain)
and Residencia Juan XXIII, Tarragona - Catalunya (Spain)

DEFINITION OF HYPERCALCIURIA

In Catalunya normocalciuria is considered as a 24 hour urinary calcium excretion of 171 ± 48 mg when the subjects ingested 1 g of calcium (1). A level higher than 250 mg/24 h was considered as hypercalciuria (1).

TYPES OF HYPERCALCIURIA

Hypercalciuria may be associated with normo or hypercalcemia. Normocalcaemic hypercalciuria may arise from: a) increased intestinal absorption, b) bone resorption (2), and c) increased renal excretion (3).

STUDY OF HYPERCALCIURIA

The definition and classification of hypercalciuria involves a two phase study. Phase 1 attempts to detect hypercalciuria in recurrent stone-formers. Phase 2 aims to classify the different

337

types.

Phase 1. Detection of metabolic abnormalities involves the
determination in blood and 24 h urine samples of the concentrations
of calcium, magnesium, phosphate, uric acid, and creatinine. Oxalate
is determined in the 24 h urine samples. Ammonia, pH and titrable
acidity are also measured. Phosphate, uric acid and creatinine
clearances are then calculated.

All the subjects are kept on their standard diet (in Catalunya
1 g of calcium/24 h). The presence of hypercalciuria is confirmed
by three separate 24 h determinations.

Phase 2. In this phase the subjects are mantained on a diet
containing 400 mg of calcium/24 h for 12 days. On the 5th day,
after fasting for 10 hr, 4-hr urine samples are collected and calcium, magnesium and creatinine are determined, followed by the calculation of calcium: magnesium and calcium: creatinine ratios.
Following this collection 250 mg calcium chloride containing 25
μ Ci of ^{45}Ca is administered orally to each subject. At 1, 2, and 3
hr after receiving the radioactive dose, the participants drink 200
ml of distilled water. Urine is collected during the 4 hr period
after the ^{45}Ca administration. Blood plasma is also obtained on the
4th hr. In these 4 hr blood and urine samples, creatinine is determined as well as the calcium and creatinine clearances. Twenty-
four hour urine collections are then continued for 7 additional days.
Radioactivity is counted in all blood and urine samples.

CLASSIFICATION

Absorptive hypercalciuria is defined as the urinary excretion
during the 7 day collection of more than 5% of the radioactivity
from the oral dose of ^{45}Ca.

Resorptive hypercalciuria is defined as an increase of the
urinary calcium: creatinine ratio greater than 0.12 when the
subjects were maintained on a diet containing 400 mg calcium/day for
at least 5 days.

An increase in the radiocalcium clearance greater than
1 ml/min/1.73 m^2 is considered as excretory, due to renal leak.

Table 1. Mineral alterations in recurrent stone formers.
          The study was performed on 450 patients.

Type	N°	Percentage
Normocalcemic hypercalciuria	220	48.8
Crystallization defects	54	12.0
Hyperoxaluria	41	9.1
Uric acid lithiasis	32	7.1
Hyperoxalo – calciuria	28	6.2
Hyperparathyroidism	10	2.2
Tubular complex syndrome	9	2.0
Cystinuria	3	0.6
No abnormal findings	53	11.7

INCIDENCE

Of 450 recurrent renal stone formers, hypercalciuria was detected in 48.8% (Table 1). It was the most frequently found abnormality. However, the patients we study are usually referred to
us from other urologists and possibly these percentages are not
reliable, as they represent a highly selected group of patients.
This is confirmed by the fact that uric acid lithiasis was compara-
tively less frequent in our material than in other series from the
same region.

Resorptive hypercalciuria alone was not detected in any of these
450 patients. The most frequent finding was mixed or complex hypercalciuria (72.7%) (Table 2) followed by the absorptive (16.9%) and
excretory (10.1%) varieties. Mixed hypercalciuria was found in
various combinations: a) absorptive plus renal (16.9%);b) resorptive
plus absorptive (16.9%); c) resorptive plus renal (13.5%) and
d) resorptive plus absorptive and renal (25.4%).

TYPE OF STONES

Of the 450 stone formers 220 had hypercalciuria. The hyper-

Table 2. Characteristics of the different types of hypercalciuria (H.) (mean values).

Type of subjects	Percentage	Urinary calcium	Urinary calcium: creatinine ratio	7 day excretion of urinary 45Ca	Radiocalcium clearance
		mg/24 h		%	ml/min/1.73 m^2
Controls	–	171	0.085	3.4	0.75
Absorptive H.	16.9	310	0.083	15.5	0.67
Renal or excretive H.	10.1	317	0.093	2.2	1.88
Complex H.	72.7				
a) Absorptive plus renal	16.9	307	0.10	11.7	1.44
b) Resorptive plus absorptive	16.9	302	0.20	18.2	0.78
c) Resorptive plus renal	13.5	305	0.26	2.9	1.99
d) Resorptive plus absorptive and renal	25.4	303	0.24	17.5	1.54

Table 3. Type of stones in hypercalciuric patients.

Type of stones	N°	Percentage
Oxalate (weddelite – whewellite)	1028	61.0
Carboapatite	223	13.2
Struvite	63	3.7
Mixed (uric acid plus carboapatite)	107	6.3
Mucoprotein	1	0.05
Unknown	263	15.6

calciuric patients formed 1,685 stones. Oxalate stones were the most frequent (61.0%) followed by calcium phosphocarbonate (13.2%) as shown in table 3. The hypercalciuric patients were infected in 34.5% of cases.

MANAGEMENT

Absorptive hypercalciuric patients were treated with cellulose phosphate, at the usual dose of 5 g three times daily. Higher amounts were administered if the urinary calcium excretion was greatly increased. Excretory hypercalciuria was treated with 100 mg of chlorothiazide on alternate days. Mixed hypercalciuria was treated depending on the absorptive or excretory components. The resorptive component was not treated. Management was correctly followed for four years in 63.6% of the hypercalciuric patients (Table 4). About ten per cent of the patients took the treatment irregularly and twenty five per cent defaulted. Calciuria decreased in all groups. Of the patients that correctly followed the treatment two subgroups could be detected. Subgroup A occasionally had hypercalciuria in spite of the treatment. Subgroup B never had hypercalciuria throughout the management period.

The risk of stone formation decreased in all the treated patients by a factor of 22.2 to 1 (mena number of stones per patient before and after treatment). This ratio appears to be lowest in

Table 4. Effect of treatment on the calciuria.

Treatment modalities	Number of patients	Percentage	Calciuria (mean) pre treatment	post
		%	mg/24 h	
Treatment not followed	57	25.9	329	--
Irregular treatment	23	10.4	363	227
Treatment regularly followed	140	63.6	330	240
a) Occasional hypercalciuria	23	10.4	330	240
b) Hypercalciuria never observed	117	53.1	333	164
TOTAL	220	100	338	--

patients submitted to irregular treatment (8.8 : 1), highest in patients who regularly followed the treatment especially in group B (30.3 : 1). This ratio was relatively high also in subgroup A (21 : 1).

COMMENT

Hypercalciuria is the most frequent finding as a cause of renal stone formation.

Absorptive and renal or excretory hypercalciuria appear to be two separate causal factors in the hypercalciuria associated with renal lithiasis. Resorptive or bony hypercalciuria has not been identified as the sole factor responsible for this disorder in our series. Bone involvement seems however to occur in primary hypercalciuria in patients with renal lithiasis in association with the other types. Mixed hypercalciuria appears to be a transformation or complication of the two simple types (absorptive or excretory) (1).

Treatments that decrease the urinary calcium excretion below 200 mg/24 h always reduce the risk of stone formation.

REFERENCES

1. Pinto,B. and Ruiz-Marcellan, F.J.: Different types of hyper-
    calciuria in patients with renal lithiasis and evidence of the
    calcium renal waste. Rev. Esp. Fisiol. 35: 311, (1979).
2. Nordin, B.E.C. and Peacock, M.: Urinary Calculi (Cifuentes, L.,
    Rapado, A., Hodgkinson, A., editors), S. Karger. Basel 1973,
    p. 119.
3. Pak, C.Y.C.: Lithiasis Renal (Pinto, B. editor) Salvat, Bar-
    celona, 1976, p. 257.

OXALIC ACID LITHIASIS AND THE HYPEROXALURIC SYNDROMES

M. Marangella, M. Bruno, B. Fruttero and F. Linari

Divisione Nefrologia e Dialisi
Ospedale Mauriziano Umberto I
Corso Turati, 46
10128 Turin (Italy)

ABSTRACT

The Authors present their results on the intestinal absorption and renal handling of oxalate in normal subjects.

The major hyperoxaluric syndromes – primary hyperoxaluria (hOx), enteric hOx and mild hOx in idiopathic stone disease – are discussed. The results of studies in 20 enteric hyperoxaluric patients and 161 idiopathic calcium oxalate stone formers are reported, with special reference to the pathogenesis and treatment of these syndromes.

————

Oxalic acid (Ox) is a strong organic acid ($pK_{a1}= 1.2$; $pK_{a2}= 4.27$) whose clinical importance derives from the very low solubility in biological fluids of its calcium salts, so that Ox is present in about 80% of the human urinary calculi. At the physiological urinary pH, Ox is virtually all in the dissociated form. Calcium oxalate (CaOx) solubility does not vary significantly from pH 5 through 7 (9):

$K_{sp, th}$ CaOx H_2O = 2.5 x 10^{-9} mol^2 (T = 37° C) (8)

$K_{sp, th}$ CaOx $2H_2O$ = 5.0 x 10^{-9} mol^2 (T = 37° C) (3).

Table 1.   Oxalate kinetics in normal subjects: data from our
           laboratory.

Intestinal Absorption (^{14}C-Ox 10 μCi)	15.2% (7.9 - 23.7)
Renal Clearance (^{14}C-Ox infusion)	182 ml/min (142 - 220)
Clear. Ox / Clear. Creat.	1.42 (1.15 - 1.67)
Plasma Oxalate Concentration	2.0 μmol/l (1.3 - 2.36)
Exchangeable Pool	62 μmol (43 - 81)
24 hr Ox Excretion	0.33 μmol (0.09 - 0.58)

Table 2.   Oxalate urinary excretion in 20 Crohn's disease
           patients vs. 50 healthy controls.

	CONTROLS (50)	CROHN'S DISEASE (20)	P
OXALURIA mg/day	29.53 $\pm$ 11.29	38.29 $\pm$ 20.04	<0.025
OXALURIA mM/l	0.26 $\pm$ 0.09	0.43 $\pm$ 0.27	<0.001
R. S. CaOx	0.89 $\pm$ 0.18	1.07 $\pm$ 0.19	<0.001

R. S.= Relative Supersaturation.

The average dietary content (100–900 mg/day) contributes only 5–15% to the daily Ox excretion; 30% derives from ascorbic acid metabolism and 60% from other endogenous sources.

In our laboratory Ox intestinal absorption has been determined by ^{14}C-Ox oral administration; the activity recovered in a 48 hour urine collection was assumed as the percentage of absorption of the oral administered dose.

Our results in healthy volunteers (table 1) show a somewhat higher percentage of absorption as compared to previous findings; but it must be pointed out that all the subjects were studied while on a low calcium diet.

Ox pharmacokinetic studies were performed by ^{14}C-Ox infusion: our results are in agreement with previous reports (5, 14).

Daily Ox excretion was determined by an improved colorimetric method (10): oxaluria varied in normals within a wide range. We found a significant correlation between net oxalate excretion and urine volume, the relationship being linear (Ox = 0.0132 Vu + 11.453; r = 0.6804, p $<$ 0.001).

Hyperoxaluria can occur in many pathological conditions, but, from the clinical point of view, the most important are primary hyperoxaluria (hOx), enteric hOx, and mild hOx in idiopathic renal lithiasis.

PRIMARY HYPEROXALURIA

This is an inherited disease in which two distinct enzymic defects (type I and II) lead to increased endogenous synthesis and urinary excretion of oxalic acid. The two types of hOx are clinically indistinguishable: almost all the patients die in the first two decades of life from renal failure and oxalosis. The disease is poorly influenced by therapy, no drug having so far shown any significant effect (6).

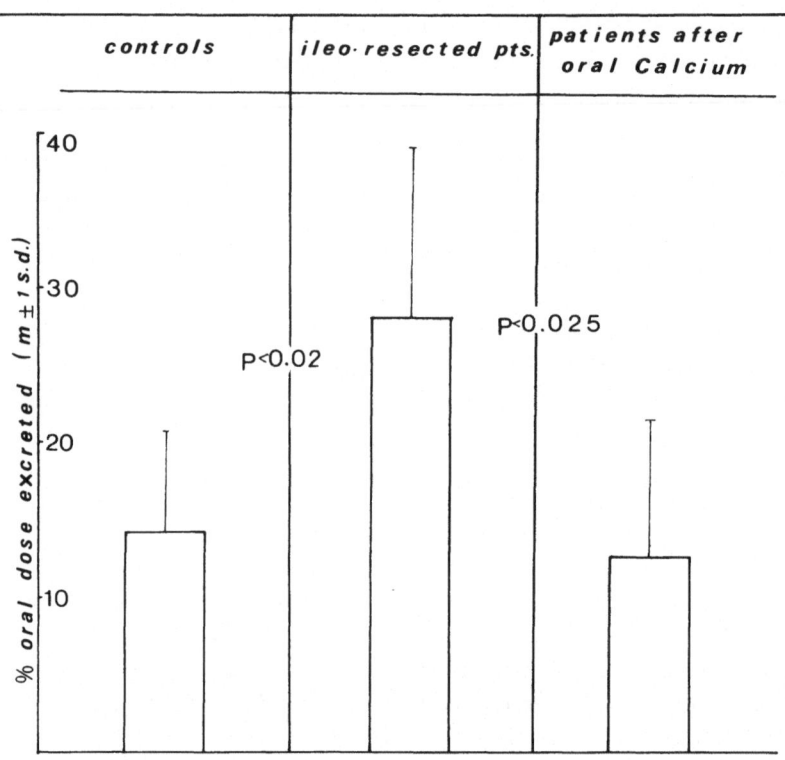

$^{14}C \cdot OX$ INTESTINAL ABSORPTION

Fig. 1.   14C-Ox intestinal absorption in healthy volunteers
          and in patients with ileal resection before and after
          oral supplements of calcium (19 mM/day).

ENTERIC HYPEROXALURIA

     This is a relatively more common type in which high Ox excretion
occurs in patients with ileal disease and resection or in patients
with other chronic intestinal disorders, characterized by fatty and
bile acid malabsorption.

     We studied 20 patients affected by Crohn's disease, of whom
11 had ileal resection and 1 ileostomy; in 13 there was a history
of renal lithiasis. The results (Table 2) show a significant increase
of Ox excretion and consequently a high degree of urinary supersaturation as calculated by means of the nomograms of Marshall and

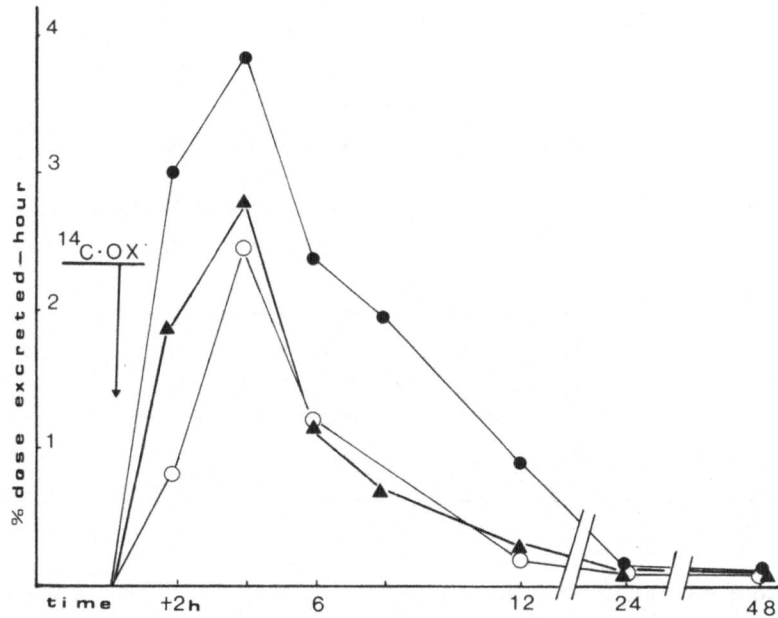

Fig. 2. 14C-Ox absorption kinetics in patients with ileal
resection on a low calcium diet before (full circles)
and after calcium supplements (open circles) as
compared to normal subjects on the same low calcium
diet (triangles).

Robertson (7). When 6 of these patients were placed on a vegetable-
rich diet, they showed an increase of net Ox daily excretion up to
average values of 0.51 ± 0.14 (S.D.) mmol/24 hs.

In 6 other patients with Crohn's disease and ileal resection
exogenous Ox absorption was studied by means of 14C-Ox oral administration (10 μCi oxalic acid, 20 mCi/mmol). To obtain reliable results, all the subjects studied were maintained on low calcium
intake for three days before and during the test. The amount of
14C-activity recovered in a 2 days urine collection was significantly higher in these 6 patients as compared to 5 controls. Furthermore the percentage of the oral dose absorbed returned to control
values when oral calcium supplements were given to the patients
(Fig. 1).

In the healthy controls only 18.6% of the overall absorbed

Table 3. Ox urinary excretion in 161 idiopathic stone-formers
          vs. 50 healthy controls.

	CONTROLS (50)	STONE FORMERS (161)	P
OXALURIA mg/day	29.53 ± 11.29	39.72 ± 16.95	< 0.001
OXALURIA mM/l	0.26 ± 0.09	0.26 ± 0.11	n.s.
R. S. CaOx	0.89 ± 0.18	0.96 ± 0.18	< 0.025

^{14}C-Ox, was excreted within the first 6 hours of urine collection;
in the same period the mean value for the patients was 34.4% (Fig. 2).
This could mean that Ox hyperabsorption probably takes place in
the lower tracts of the intestine. It is believed that fatty and
bile acid malabsorption cause hOx both by the decrease of free
calcium availability in the intestinal lumen, because of the formation of calcium-fatty acid complexes, and by the increase of
mucosal permeability of the colon induced by fatty and bile acids
(1, 2). Our results support this hypothesis.

Table 4. Calcium and Oxalate daily excretions in patients
          after 4 day calcium-oxalate controlled diet.

	NORMOCALCIURICS (37)	HYPERCALCIURICS (22)	P
CALCIURIA (mg/ 24 h)	190.0 ± 55.8	351.7 ± 107.8	< 0.001
OXALURIA (mg/24 h)	33.39 ± 12.87	43.21 ± 11.41	< 0.005

MILD HYPEROXALURIA IN IDIOPATHIC STONE DISEASE

The role of Ox as a stone risk factor is well recognised, however uncertainties still remain about the prevalence and the pathogenesis of the mild hOx which is often a feature of idiopathic stone disease.

Therefore we determined Ox excretion in 161 idiopathic stone formers: the results (table 3) show a significant increase of net daily Ox excretion in the stone-formers as compared to 50 controls. Our findings are in agreement with other reported series (11, 13).

Three mechanisms have been invoked to explain this hOx:

1) renal hyperoxaluria, in those patients whose net oxalate excretion is increased because of high urine volume. In the stone-formers we found a significant direct relationship between oxaluria and urine volume (Ox = 0.0154, vol. = 11.917, r = 0.5840, p 0.001).

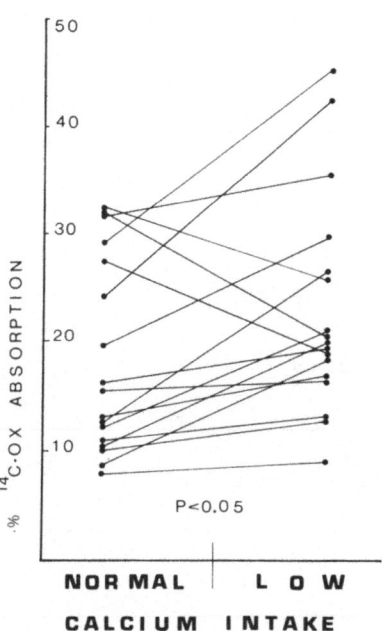

Fig. 3.   Intestinal absorption of oxalate in hypercalciuric
          patients in relation to calcium intake.

Table 5. Data after normal and low calcium intake in 40
          stone-formers.

	NORMAL Ca INTAKE	LOW Ca INTAKE	P
OXALURIA mg/die	33.03 ± 11.75	40.98 ± 15.14	< 0.01
OXALURIA mM/l	0.204 ± 0.060	0.262 ± 0.097	< 0.001
CALCIURIA mg/die	289.5 ± 140.4	216.3 ± 119.9	< 0.001
CALCIURIA mM/l	4.36 ± 2.38	3.13 ± 1.95	< 0.001
R. S. CaOx	0.890 ± 0.152	0.908 ± 0.193	n. s.

Table 6. Data before and after thiazide
          therapy in 45 stone-formers.

	BEFORE	AFTER	P
CALCIURIA mg/die	325.9 ± 119.2	221.6 ± 96.5	< 0.001
OXALURIA mg/die	37.04 ± 11.13	30.34 ± 9.15	< 0.02
R. S. CaOx	0.981 ± 0.123	0.889 ± 0.189	< 0.02

2) Endogenous hyperoxaluria, secondary to an inherited metabolic defect (4) or to an abnormally high intake of animal proteins (12).

3) Absorptive hyperoxaluria: oxalate intestinal hyperabsorption secondary to calcium hyperabsorption by a mechanism somewhat similar to that described for enteric hyperoxaluria. Ox excretion was in fact somewhat higher in our hypercalciuric patients as compared to the normocalciuric. The differences became striking when the stone-formers were studied after 4 days of calcium-oxalate controlled diet (Ca = 850 mg; Ox = 60-100 mg): the results (table 4) show a significant increase of Ox excretion in the hypercalciuric as compared to normocalciuric patients.

Therefore calcium and oxalate absorption and excretion seem to be interdependent and this must be taken into account when therapeutic approaches are tried. In the dietary management of hypercalciuria, Ox balance should be carefully monitored because too low a calcium intake with no change in Ox dietary content causes an intestinal Ox hyperabsorption (Fig. 3) and, therefore, a significant increase of oxaluria. The results listed in table 5 show no change of urinary saturation with CaOx and, presumably, no change in stone-risk, after calcium dietary restriction, in spite of a significant decrease of calciuria.

Better results have been obtained by the association of thiazide and allopurinol, which we have used in 45 hypercalciuric stone-formers: as shown in table 6, after 3-6 months of therapy we observed a significant decrease both of calcium and of Ox urinary excretion and, consequently, a striking decrease of CaOx urinary supersaturation.

The clinical results of one year follow-up are in agreement with the physico-chemical findings, more then 90% of the treated patients being free of stone recurrences.

REFERENCES

1. Dobbins, J.W. and Binder, H.L.: Effects of bile salts and fatty acids on the colonic absorption of oxalate. Gastro-enterology 70: 1096, (1976).

2. Earnest, D.L., Johnson G., Williams, H.E. and Admirand, W.H.: Hyperoxaluria in patients with ileal resection: an abnormality in dietary absorption. Gastroenterology 66: 1114, (1974).

3. Gardner, G.L.: Nucleation and Crystal Growth of Calcium Oxalate Trihydrate. J. Crystal Growth 30: 158, (1975).

4. Hautmann, R. and Lutzeyer, W.: Calcium oxalate stone disease: congenital defect of metabolism? J. Urol. 116: 687, (1976).

5. Hodgkinson, A., Wilkinson, R. and Nordin, B.E.C.: The concentration of oxalic acid in human blood. In: Urinary Calculi. Int. Symp. Renal Stone Res., Karger, Basel, pp. 18-23.

6. Linari, F., Malfi, B., Fruttero, B., Giorcelli, G. and Vacha G.: L'Ossalosi. In: Attualità Nefrologiche e Dialitiche (D'Amico G., Petrella E., Sorgato G. and Vendemia F., Milano, 1977), pp. 279-289.

7. Marshall, R.W. and Robertson, W.G.: Nomograms for the estimation of the saturation of urine with respect to calcium oxalate. Clin. Chim. Acta. 72: 253, (1976).

8. Nancollas, G.H. and Gardner, G.L.: Kinetics of Crystal Growth of Calcium Oxalate Monohydrate. J. Crystal Growth. 21: 267, (1974).

9. Pak C.Y.C., Ohata, M. and Holt, K.: Effect of diphosphonate on crystallization of calcium oxalate in vitro. Kidney Int. 7: 154, (1975).

10. Pellegrino, S., Zaffino, C., Tondolo, M., Ronzani, M., Ariano, M. and Marangella, M.: Valutazione di un metodo colorimetrico per il dosaggio dell'ossalato nelle orine. Giorn. It. Chim. Clin. 5: 171, (1980).

11. Robertson, W.G., Peacock, M. and Nordin, B.E.C.: Activity product in stone-forming and non-stone forming urine. Clin. Sci. 34: 579, (1968).

12. Robertson, W.G., Heyburn, P.J., Peacock, M., Hanes, F.A. and Swaminathan, R.: The effect of high protein intake on the risk of calcium stone formation in the urinary tract. Clin. Sci. 57: 285, (1979).

13. Thomas, J., Melon, J.M., Thomas, E., Steg, A. and Albouker, P.: The role of oxalic acid in oxalic nephrolithiasis. In: Urinary Calculi. Int. Symp. Renal Stone Res., Karger, Basel, 1973, pp. 57-66.

14. Williams, H.E., Johnson, G.A. and Smith, L.R. Jr.: The renal clearance of oxalate in normal subjects and patients with primary hyperoxaluria. Clin. Sci. 41: 213, (1971).

EDITORIAL NOTE

The role of oxalate excretion in calcium oxalate lithiasis is often neglected. The difficulty of obtaining reliable quantitative measurement of serum concentration and urinary excretion of oxalate is responsible, at least to some extent, for the tendency to think about calcium and to forget about oxalate. It has been clearly shown, however, that hyperoxaluria is a more critical risk factor than is hypercalciuria. Hyperoxaluria appears to be an important factor in promoting endemic primary bladder stone disease in children, which is still often found in Thailand and in other countries in Asia and northern Africa.

This interesting paper by Marangella, Linari and their co-workers points out a stimulating observation. Hypercalciuria is often accompanied by hyperoxaluria. No matter whether hypercalciuria is absorptive or renal. Even if calcium depletion is caused by a "renal leak" mechanism, intestinal absorption of calcium will be enhanced, perhaps through the intervention of the active derivatives of vitamin D3. If intestinal hyperabsorption of calcium occurs, the amount of free calcium in the intestinal content will be decreased. The concentration of free oxalate not bound to calcium will increase in the intestinal lumen and hyperabsorption of oxalate will ensue.

The therapeutic implications are also of interest. Correction of hypercalciuria by the use of thiazides is followed by a reduction of urinary oxalate excretion. This phenomenon may be at least as important in the prevention of recurrences of calcium oxalate stone formation, as is the decrease of calciuria, upon which the attention is usually focused.

M. Pavone-Macaluso

The abbreviation Ksp,th indicates: Thermodynamic solubility product.

URIC ACID NEPHROLITHIASIS - PREVALENCE, PATHOGENESIS AND RESULTS

OF CONSERVATIVE TREATMENT

L. Miano, S. Petta, M. Gallucci and S. Goldoni

Department of Urology
University of Rome (Italy)
National Centre of Research on Urolithiasis
Fiuggi (Italy)

ABSTRACT

In Italy about 25% of all renal calculi are composed of uric
acid and/or urates. This percentage includes also the cases with
mixed stones (uric acid + calcium oxalate and/or apatite). In our
series uric acid nephrolithiasis was idiopathic in 60% of the cases,
while in 40% it was associated with abnormal purine metabolism.
Our extended observation confirms our previous therapeutic experience
and is similar to that of other investigators in this field. Simple
dietary measures, adequate hydration and control of urine hyper-
acidity were considered satisfactory therapy. This was proved by
the decrease in frequency of new stone formation, and by complete
or partial stone dissolution, thereby decreasing the number of
surgical procedures. Allopurinol is the drug of choice in recurrent
uric acid lithiasis associated with hyperuricosuric states.

In Italy uric acid stones account for about 25% of all renal
stones. They become more frequent with increasing age (Fig. 1).
The incidence varies from country to country. In the United States
the overall rate is 5-10%, similar to that in the United Kingdom
(11, 15). On the other hand percentages similar to that found in
Italy have been reported in Germany, Czechoslovakia and France,

357

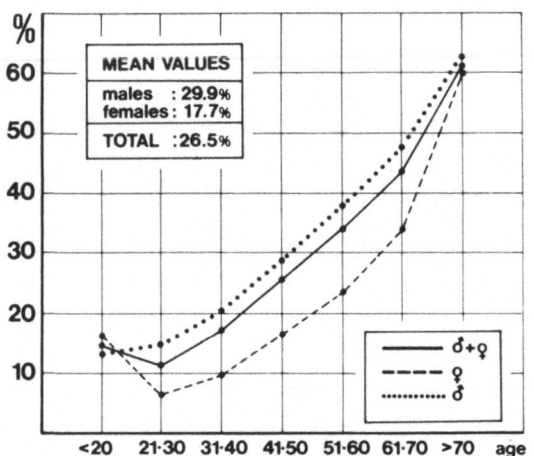

Fig. 1. Percentage of occurrence of uric acid stones in renal
         lithiasis by sex and age.

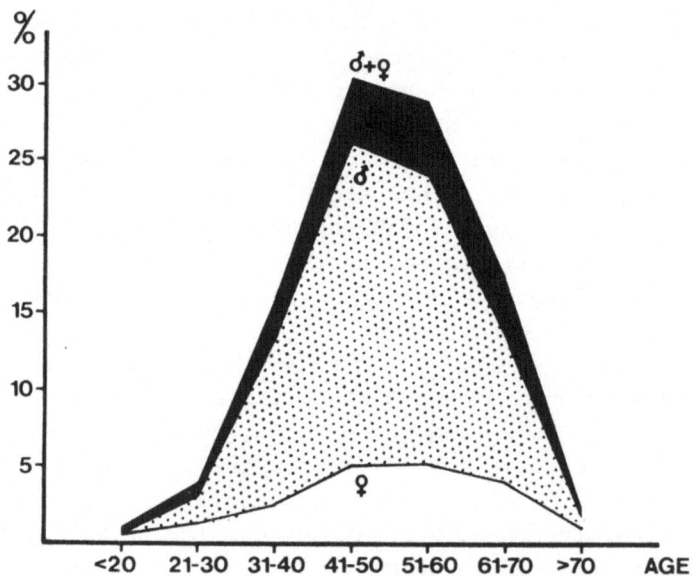

Fig. 2. Distribution of uric acid nephrolithiasis by age and
         sex.

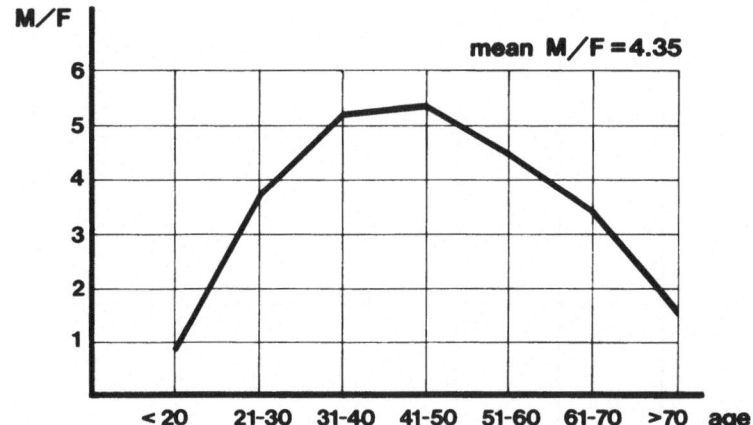

Fig. 3. Male/female ratio by age.

while the highest incidence is encountered in Israel where Atsmon et al. (1) have reported that 40% of stones were composed of uric acid.

Our data are based on a study of 1,002 patients with uric acid nephrolithiasis observed from 1970 to 1980 at the Urological Department at Rome University in connection with the National Centre for Research on Urolithiasis sponsored by "Ente Fiuggi S.p.A.". The prevalence of uric acid nephrolithiasis in population of Italy at large is estimated to be of the order of 0.07%. The maximum incidence of the disease falls between the ages of 40 and 60, no age group being exempt (Fig. 2). The overall male/female ratio is approximately 4:1 (Fig. 3), while in our calcium stone patients the same ratio is 2.4:1.

We encountered many uric acid stone patients who gave a family history of urolithiasis. Family occurrence of uric acid stones in gouty families would not be surprising in view of the fact that the gouty subjects often have uric acid stones. Our findings show that about 35% of cases presented a family history of urolithiasis in their first or second generation relatives (table 1).

What we call the "uric acid stone" often contains not only uric acid but also urates and calcium salts. The following substances

Table 1. Familial incidence of uric acid lithiasis.

FAMILIAL RELATION	MALES %	FEMALES %	TOTAL %
PARENTS	8.34	6.42	7.98
BROTHERS	11.90	12.30	11.98
GRAND-PARENTS	0.98	0.53	0.90
OTHER RELATIVES	5.64	8.02	6.09
MORE THAN ONE RELATIVE	6.87	10.16	7.49
T O T A L	33.74	37.43	34.44

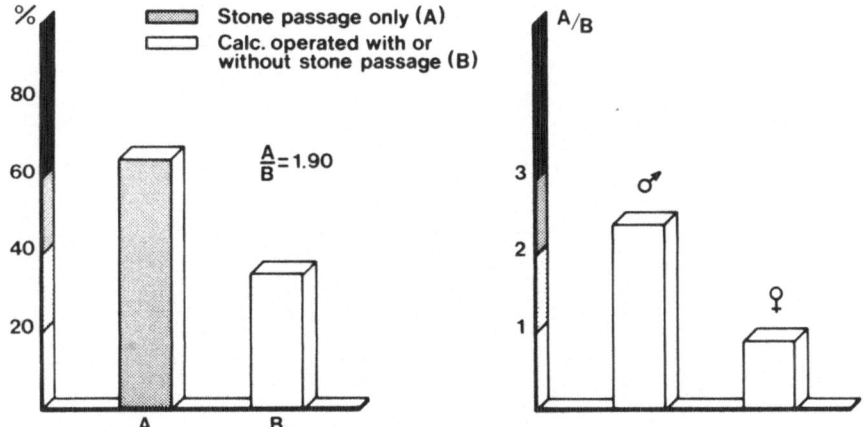

Fig. 4.   Distribution of the main clinical forms of uric acid
          nephrolithiasis by sex.

were found in our material: pure uric acid in either the anhydrous
form or the dihydrate; uric acid with either calcium oxalate, calcium
phosphate or magnesium ammonium phosphate or all together; uric
acid with either ammonium hydrogen urate or sodium hydrogen urate
monohydrate or both, and also with or without calcium salts. Ammonium
hydrogen urate has been found only as an occasional constituent of
renal uric acid calculi. In our own series of 500 urinary calculi
(mostly renal) submitted to crystallographic studies ammonium
hydrogen urate was found only in 3 cases. Sodium hydrogen urate
monohydrate was found only in 2 cases, in one of them as the only
component and in the other in combination with uric acid. Mixed
stones are frequently observed (7% of the cases) and are mainly
composed of uric acid (or urates) and apatite and/or calcium oxalate.
In several instances a sodium urate nucleus of variable size is
surrounded by a thick layer of calcium oxalate monohydrate.

    Uric acid nephrolithiasis is clinically characterized in a
high percentage of cases only by the recurrent passage of small,
smooth, reddish-brown stones and reddish "sand" (65% of our cases,
Fig. 4). As a matter of fact, the typical history for uric acid
stones, especially in male patients, is the voiding of large number
of small stones within short periods of time, alternating with free
intervals ranging from a few months to several years. On the other

Table 2. Retrospective study of uric acid stone recurrence in a group of patients not submitted to preventive medical treatment.

1,002 PATIENTS TOTAL RECURRENCE 74% (Follow-up 2-15 years)		
Episodic stone passage only	45%	M/F = 6.5
New urinary stone(s) treated with med. or surg. therapy	4%	M/F = 1.6
New urinary stone(s) with episodic spontaneous stone passage	25%	M/F = 2.9

hand the retrospective study of our series showed that in 35% of the
cases the stones were larger, with or without small ones, and it
was deemed necessary to proceed to surgical treatment. Today the
good results obtained with in vivo dissolution of stones has considerably decreased the number of patients who need surgery (see
treatment).

In a group of patients who presented radiographically documented
uric acid stones at the time of initial evaluation, the stone was
present in the kidney in 85% of the cases (45% as a single stone,
33% as multiple stones and 12% as a staghorn) and in the ureter in
15% (mainly as a single stone). Kidney calculi were present bilaterally in 10% of the cases.

The tendency to form new stones was studied retrospectively
in our patients before beginning the preventive medical treatment.
As shown in table 2, the recurrence rate is very high (74% of the
cases) and 1/3 of patients may require surgery. These facts emphasize
the importance of early prophylactic treatment to prevent severe
kidney damage.

PATHOGENESIS

Uric acid lithiasis is frequently associated with hyperuricosuric states as seen in primary gout, secondary gout (blood dyscra-
sias, including polycythemia vera, myeloid metaplasia, leukemia and
lymphoma) or enzyme deficiencies (Lesch-Nyhan syndrome, glycogen
storage disease) (8, 9, 10). In our experience abnormal purine
metabolism was present in 40% of the cases. Twenty-seven percent
of these cases presented primary gout, 8% secondary gout, while in
5% of the cases we found only hyperuricosuria without hyperuricemia.
The latter is related to excessive dietary purine intake or to uric
acid overproduction or to a renal tubular defect of urate reabsorption. In 60% of cases the lithiasis was classified as idiopathic,
no underlying disease having been detected. Moreover in some of
these cases, in accordance with other authors (1, 5), we observed
that urolithiasis preceded gouty arthritis or abnormal values of
serum uric acid by many years.

An increase in the urinary content of crystalloid is not the
only mechanism for the development of uric acid stones. As is well

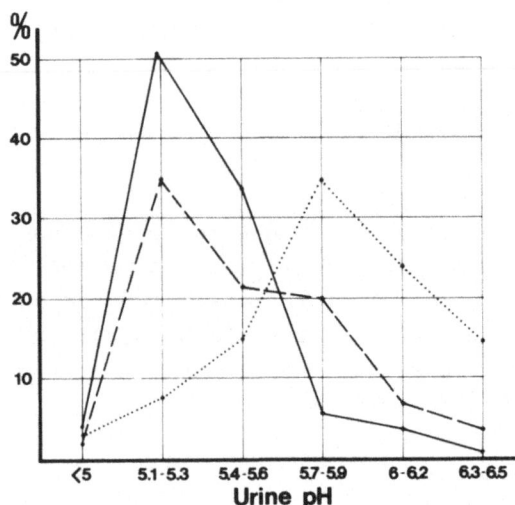

Fig. 5.   Early-morning urine pH levels in patients with idio-
          pathic uric acid nephrolithiasis (—), with primary
          gout and renal lithiasis (- - -) and in control
          subjects (...).

known, the pH of the urine strongly influences the solubility of
uric acid. Uric acid solubility increases greatly when urine pH
exceeds 6. Under normal circumstances the urine pH is subject to
wide diurnal fluctuation, being more acid at night and in the early
morning and less after meals. In patients with uric acid nephrolithiasis this circadian rhythm is lacking and the urine is ex-
cessively acid (12). Urine collected in the morning after overnight
fasting has a pH ranging from 5 to 5.5 in almost 80% of the patients
with idiopathic stones and in 60% of cases with gouty stones. The
same values are present only in 25% of the control subjects (Fig. 5).

TREATMENT

     Unlike the majority of calcium stones, uric acid calculi can
frequently be reduced in size by medical treatment. Therefore the
aims of therapy include not only the prevention but also the dissolution of the stones. From the therapeutic point of view, uric
acid lithiasis is today considered to be almost exclusively a medical
problem (13). The aspects of this therapy, such as efficacy of

adequate hydration, use of alkalinizing agents and/or allopurinol,
regulation of the diet and control of urinary tract infection, have
been well established for at least 15 years (1, 3, 6, 7, 8, 14,
17). Nevertheless the experience accumulated in time has allowed
us to obtain consistently better results.

The therapeutic approaches vary according to the presence of
abnormal purine metabolism and uric acid overproduction. The well-
known basic concept consists in the modification of urine pH towards
the alkaline side by the administration of alkalis such as sodium
bicarbonate (10-15 g/day), sodium citrate (3-6 g/day) possibly in
the form of lemon juice, Eisenberg's solution (citric acid, 40 g;
sodium citrate, 60 g; potassium citrate, 66 g; extractum aurantii,
6 g; with syrup to 600 ml; 15 ml or one spoonful three times daily)
or Uralyt-U [R]. The latter is more palatable to most people and may
have a longer and more even absorption and excretion phase (4).
Acetazolamide (250 mg x 2 daily) is also a useful alkalinizing
agent and may be employed when high sodium intake is contraindicated
or in cases of intolerance to the common alkalis. For preventive
purposes the urine pH should be around 6.5, whereas for litholytic
purposes the pH should be higher (ca. 7.5). In some patients in
whom surgical removal of stones had not been completely achieved,
we obtained good results by administering a solution of Tham 3.6 g%
or sodium bicarbonate 1.6 g% through the nephrostomy tube. When
there is evidence of abnormal purine metabolism it is advisable to
follow a diet (19) low in purine-rich foods; this may provoke
transiently excessive uric acid excretion. It is also useful to
reduce high caloric foods (mainly animal fats), refined sugars,
fructose and alcohol. In obese patients the caloric intake should
allow a gradual weight loss.

For prophylactic treatment we found it useful to employ allopurinol (100-300 mg/day) especially when hyperuricemia and hyperuricosuria occur simultaneously. When the treatment is for lytic
purposes, we still advise it in all cases because we believe that
it accelerates the process of dissolution of the stone. Many authors
consider that uricosuric drugs are contraindicated in these patients
(10, 16). However, we believe that these drugs, among which we
prefer benziodarone (50-100 mg/day), together with adequate oral
fluid and alkalis may be useful in cases of hyperuricemia and normal
or hypouricosuria. Piperazine (12-24 g/day) is also indicated and is
administered in form of citrate, which has a double alkalinizing

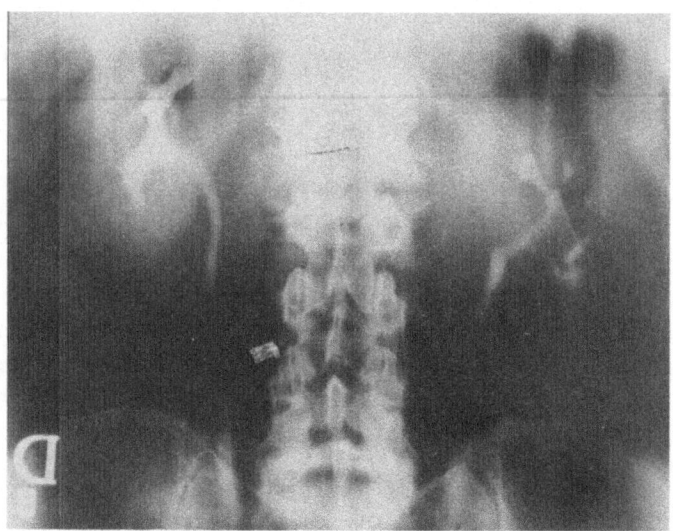

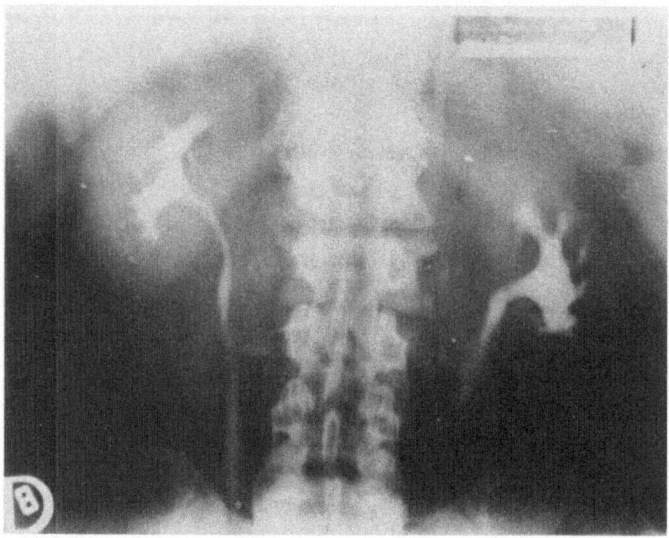

Fig. 6. Branched uric acid lithiasis of the left kidney.
         a) IVP at the time of diagnosis, b) IVP after 4 months
         of medical treatment: complete dissolution of the
         stone.

and uricosolubilizing action.

        The preventive treatment must be continued for a long time.
According to the gravity of the symptoms it may be necessary to

administer all of the previously mentioned remedies (high intake
of fluids, alkalis, allopurinol etc.); more often a cyclical or
discontinuous treatment with one or more of the already mentioned
drugs may be sufficient. When instituting treatment it is of extreme
importance to obtain the full cooperation of the patient by giving
as full an explanation as possible. It is important to see the patient at least once every 3 months, since we have noted that patients, when unsupervised, tend to stop treatment after having felt
well for a few months.

Proceeding along these lines we have obtained good results in
a high percentage of cases both in the litholytic and in the preventive therapy. Usually the symptoms disappeared, the urine analysis
became normal and a progressive shrinkage of the stones was observed.
The stones eventually disappeared within a period of time of 15 to
120 days (Fig. 6). We obtained a complete dissolution in 85% of the
cases and partial dissolution in 6%. In 17 patients conservative
treatment was of no avail. Of these patients 4 did not entirely
adhere to the prescribed treatment. Of the remaining 13 patients,
10 had recurrent or chronic urinary infection with high urinary pH
and in 3 cases the kidney was poorly functioning (table 3).

From the preventive point of view, the percentage of recurrences was 9% during an average follow-up period of 8 years (Table
4).

There are particular situations in which the type of therapy
has to be modified. If the patient has hypertension, or is on a
low sodium diet, sodium salts must be reduced or avoided in favour
of potassium salts, piperazine or acetazolamide as long as the
kidney function is normal. When hypercalciuria is present, the use
of the thiazides is advisable. In forms with recurrent or chronic
urinary infection the risks of secondary phosphate crystallization
are high. If the infection is difficult to control with antibiotics,
surgery is to be preferred. The same considerations apply in patients with medium-sized or large, partially calcified uric acid
stones, who have chronic urinary infections, especially if due to
urea-splitting bacteria. On the other hand, in cases of mixed but
sterile lithiasis, we believe medical therapy should be attempted.
This is because the weak radiopacity of the stone is related to
the size of the stone itself rather than to the presence of calcium
salts; with this procedure we have successfully treated 10 cases

Table 3. Results of litholytic medical treatment.

DISSOLUTION	Nº of patients	STONES < 1 cm ∅	STONES > 1 cm ∅	URIC ACID STONES with INFECTION
COMPLETE	157 (85%)	88	69	10
PARTIAL	11 (6%)	3	8	4
FAILURE	17 (9%)	7	10	10
T O T A L	185	98 (53%)	87 (47%)	24 (13%)

Table 4. Results of preventive medical treatment.

		N° of Patients	Recurrence
IDIOPATHIC URIC ACID LITHIASIS	Pure	230	15 (6.5%)
	Mixed	13	4 (30%)
		243 (62%)	19 (7.8%)
URIC ACID LITHIASIS WITH HYPERURICEMIA and/or GOUT	Pure	137	14 (10%)
	Mixed	12	3 (25%)
		149 (38%)	17 (11%)
T O T A L		392	36 (9.2%)

Follow-up ranging from 1 to 8 years.

(Fig. 7). We also believe that medical treatment is indicated in patients with a non-functioning or poorly-functioning kidney, when that condition has been present for a few weeks or at most a few months (Fig. 8). On the other hand surgery should be advised if a kidney has not been functioning for a longer time. Lastly, conservative treatment in a patient with a single kidney may be dangerous as anuria occasionally occurs by the plugging of the urinary tract with debris resulting from the rapid disintegration of the stone. Good results may also be obtained but the patient must be under strict control (Fig. 9). If that is impossible, surgery should be undertaken.

In conclusion, early institution of therapy may be successful in preventing stone recurrence in a large majority of uric acid stone patients. A correct diagnosis in a patient with large pure or mixed uric acid stones may prevent surgery since it is often

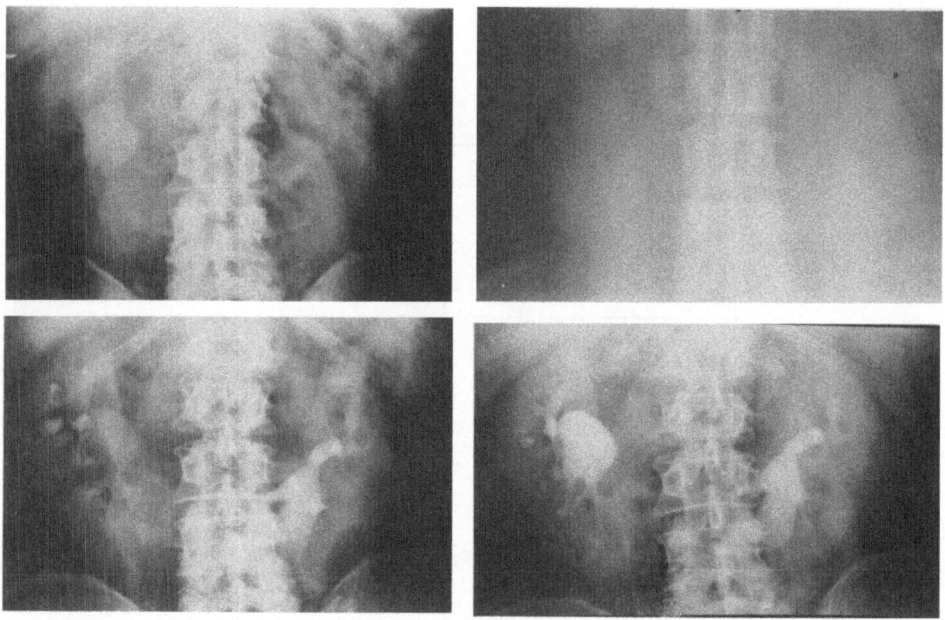

Fig. 7. Recurrent uric acid staghorn in a horseshoe kidney.
a) Plain X-ray showing a large weakly radiopaque
stone; b) IVP; c) Plain X-ray (tomography) and d) IVP
after 6 months of medical treatment: total dissolution
of the stone.

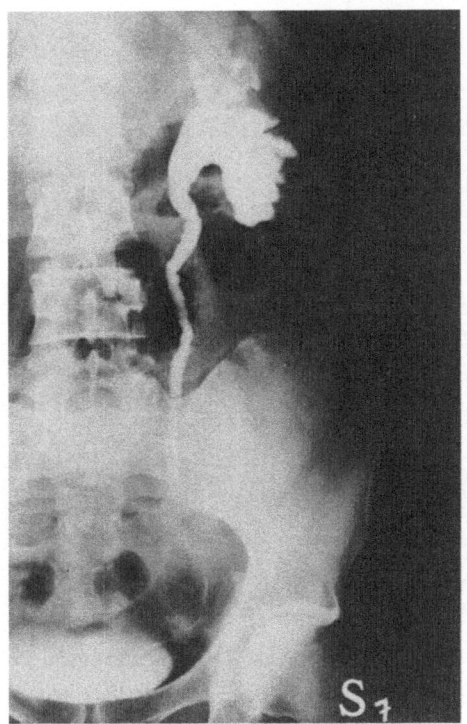
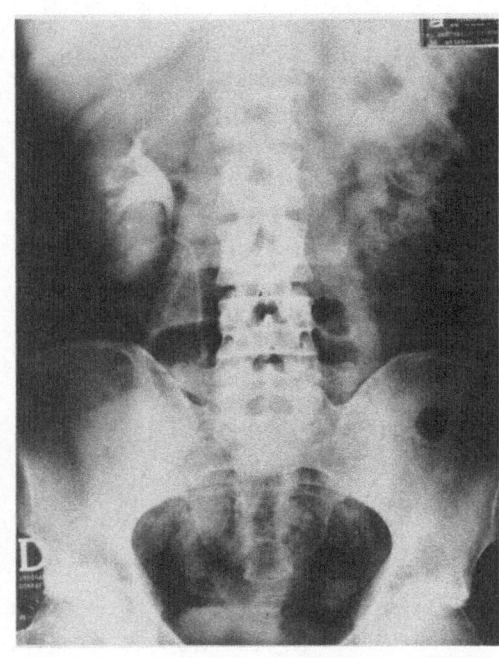
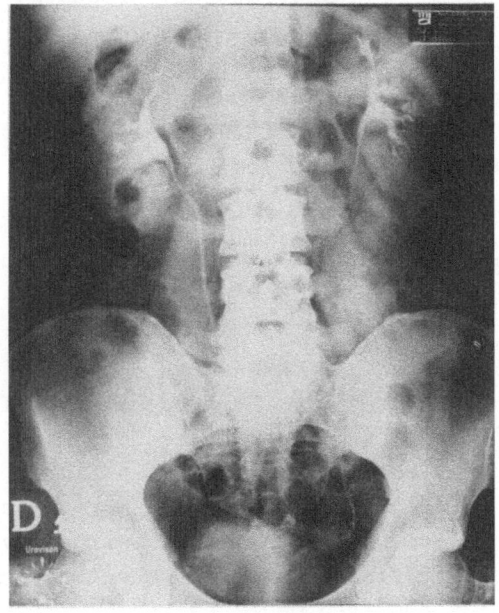

Fig. 8.

Male, 58 years old, with primary gout and recurrent stone
passage. a)IVP; non-functioning left kidney; b) Left
retrograde pyelogram; the
ureter is plugged by masses
of uric acid crystals
causing complete obstruc-
tion and severe hydroephro-
sis. Filling defects are
present in the calyces; c)
IVP after 3 months of medical treatment; the kidney
function is normal and
no stones are visible in
pelvis and calyces.

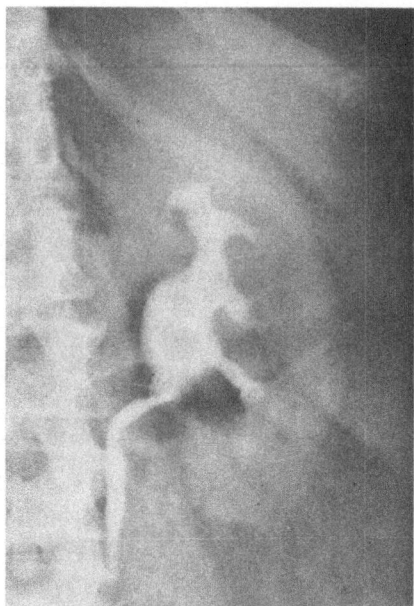 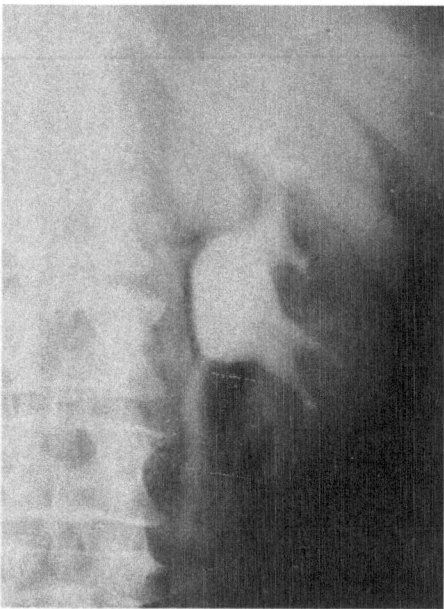

Fig. 9. Multiple uric acid stones in a single left kidney.
         a) IVP at the time of diagnosis, b) IVP after 3 months
         of medical treatment.

possible to dissolve the stones in vivo. We therefore agree with
Cottet (2) who stated that uric acid nephrolithiasis, if correctly
treated, is the most benign of all kidney stone disease.

REFERENCES

1. Atsmon, A., De Vries, A. and Frank, M.: Uric acid lithiasis.
    Elsevier, Amsterdam, 1963.
2. Cottet, J.: Le syndrome biochimique des lithiases urinaires,
    in "Handbuch der Urologie.Die Steinerkrankungen" Springer
    Verlag., Berlin, vol. X, 1961.
3. De Vries, A., Frank, M., Weinberger, A. and Oliver; Treatment
    of uric acid lithiasis. Observation on 658 patients. In:
    "Urinary calculi. Recent advances in aetiology, stone structure
    and treatment" edited by Cifuentes Delatte L., Rapado A.,
    Hodgkinson A., pp. 105-111 Karger, Basel, 1973.
4. Eisenberg, H., Connor, T.B. and Howard, J.E.: A useful agent for

oral alkali therapy. J. Clin. Endocr. 15: 503-505, (1955).

5. Frank, M., Lazebnik, J. and De Vries, A.: Uric acid lithiasis:
    a study of six hundred and twenty-two patients. Urol. Int. 25:
    32-46, (1970).

6. Freiha, F.S. and Hemady, K.: Dissolution of uric acid stones,
    alternative to surgery. Urology 8: 334-337, (1976).

7. Godfrey, R.G. and Rankin, T.J.: Uric acid renal lithiasis.
    Management by allopurinol. J. Urol. 101: 643-647, (1969).

8. Gordon, M.R., Carrion, H.M. and Politano, V.A.: Dissolution of
    uric acid calculi with Tham irrigation. Urology 12: 393-397,
    (1978).

9. Gutman, A.B. and Yu, T.F.: Uric acid nephrolithiasis. Amer. J.
    Med. 45: 756-779, (1968).

10. Gutman, A.B.: Views on the pathogenesis and management of
    primary gout. J. Bone J. Surg. 54A: 357, (1972).

11. Levinson, D.J. and Sorensen, L.B.: Uric acid stone. In: F.L.
    Coe "Nephrolithiasis. Pathogenesis and treatment" (Year Book
    Med. Publ., Chicago, 1978).

12. Micali, F. and Miano, C.: Considerazioni sul ruolo del pH uri-
    nario nella genesi della urolitiasi. Chir. Urol. 8: 293-305,
    (1966).

13. Miano, L., Arachi, N. and Cammarano, R.: La litiasi urica oggi:
    problemi di diagnosi e terapia. Chir. Urol. 9: 14-20, (1967).

14. Petritsch, P.H.: Uric acid calculi. Results of conservative
    treatment. Urology 10: 536-538, (1977).

15. Pyrah, L.N.: Renal calculus. Springer Verlag, Berlin, 1979.

16. Terhorst, B. and Melchior, H.: Rezidivprophylaxe bei Urolithi-
    asis. Z. Urol. 62: 761, (1969).

17. Uhlir, K. The peroral dissolution of renal calculi. J. Urol.
    104: 239-247 (1970).

18. Yu, T.F.: Nephrolithiasis in patients with gout. Postgr. Med.
    63: 164-170, (1978).

19. Zollner, N. and Griebsch, A.: Diet and gout. Adv. Exp. Med.
    Biol. 41B: 433-442, (1977).

CLINICAL FEATURES AND MANAGEMENT OF CYSTINE NEPHROLITHIASIS

L. Miano*, I. Antonozzi**, M. Gallucci*, S. Morano**
and S. Petta

*Department of Urology
 University of Rome (Italy)
**Laboratory of dysmetabolic oligophreni s
 University of Rome (Italy)

ABSTRACT

The Authors report the clinical features of cystine nephrolithiasis in 57 patients with cystinuria observed from 1972 to
1980. The disease affects both sexes with equal frequency and
severity. Symptom usually begin early, mostly between 15 and 20
years, but may develop at any age. Treatment consists of the administration of adequate oral fluids to obtain a copious urine
volume, of oral alkali and of α-mercaptopropionylglycine. Successful
results, from litholytic and from preventive medical treatment,
were obtained in a high percentage of cases.

Of the various forms of lithiasis, cystine stone formation is
not one of the most common (1-2% of the cases) (10). Nevertheless
conditions arising from this disorder are often complex, due to the
frequency and severity of clinical manifestations which, with time,
lead to irreversible impairment of renal function. Cystinuria is a
complex inherited disorder characterized by excretion in the urine
of large quantities of the aminoacids cystine, lysine, arginine,
ornithine and of the mixed disulphide of cysteine and homocysteine.
If it were not for the relative insolubility of cystine in urine,

375

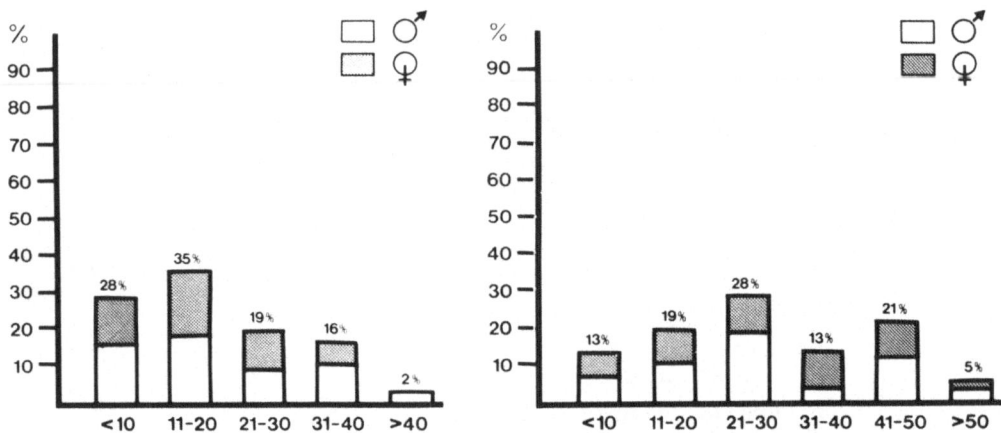

Fig. 1. Distribution of cystine stone disease by age and sex:
         a) at onset and b) at the time of diagnosis.

resulting in the formation of calculi in the urinary tract, this
disorder would be only a metabolic curiosity. The stones are composed
usually almost entirely of cystine, but occasionally they are mixed
with apatite or calcium oxalate.

Regarding the pathogenesis there are two distinct and apparently separate mechanisms responsible for cystinuria: firstly an
abnormality in the transport of certain aminoacids across the renal
tubules and secondly a failure of transport of aminoacids across
the mucosa of the jejunal epithelium. While the intestinal defect
allows one to distinguish three different types of cystinuria (13),
the nature of the renal tubular abnormality is still uncertain.
Findings of reduced fasting plasma cysteine, high renal arterio-
venous difference for cystine and mixed disulphide cysteine-homo-
cysteine in the urine suggest a defect in the renal tubular cell
transport of cysteine rather than of cystine (1, 5, 16).

CLINICAL FEATURES

Between the years 1972 and 1980, 57 patients with cystinuria
and cystine stone disease were examined at the Urological Department
of the University of Rome in connection with the National Centre of
Research in Urolithiasis sponsored by the "Ente Fiuggi". Of the pa-

Tab. 1. Familiarity of cystine stone disease.

FAMILIAL RELATION	MALES %	FEMALES %	TOTAL %
Parents	2.4	2.4	4.8
Brothers	24.4	14.6	39
Grand-parents	7.3	2.4	9.7
Other relatives	4.9	2.4	7.3
More than one relative	7.3	7.3	14.6
T O T A L	46.3	29.1	75.4

tients 30 were male and 27 female (M/F = 1.1). The age of onset, defined by the age at appearance of the first urinary tract symptom caused by urolithiasis, ranged from 2 months to 54 years with a mean of 18.6 years; 63% of patients had experienced their first symptoms before the age of 20 years and 28% before the age of 10 years (Fig. 1 a). On the other hand the age at the time of diagnosis ranged from 2 to 67 years with an average of 36.8 years (Fig. 1 b).

The disease is inherited as an autosomal recessive trait but there is compelling evidence for the existence of genetic hetero-geneity (2). About 70% of our patients recalled a family hystory of renal colic or urolithiasis in their first or second generation relatives; more than half of these were brothers (table 1).

Regrettably we found that 85% of cases had previously been wrongly diagnosed. In some of these cases the necessary analyses had not all been performed whereas others had been mistakenly identified as having different chemical types of lithiasis, such as calcium oxalate or mixed apatite and calcium oxalate (Fig. 2). Our own diagnosis was at first established by means of the analysis of the previously passed stones (68% of the cases), by urinary cystine determinations (28%) and by the presence of cystine crystalluria (4%).

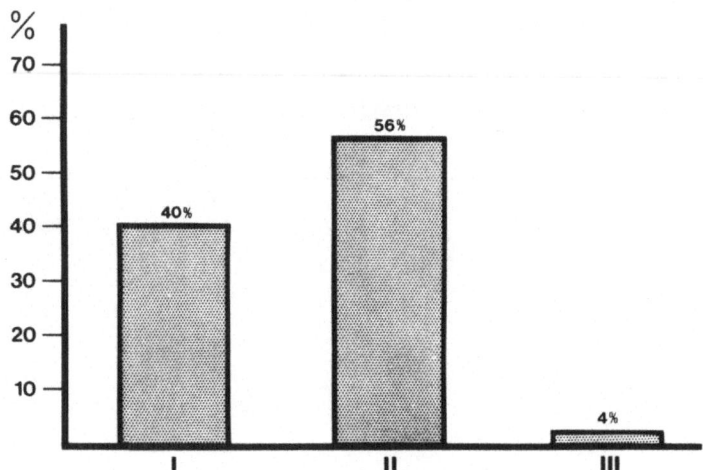

Fig. 2.   Diagnostic mistakes (47 cases, 85%).
          I   : Lithiasis chemically not identified.
          II : Ca oxalate (pure or mixed) lithiasis.
          III : Uric acid lithiasis.

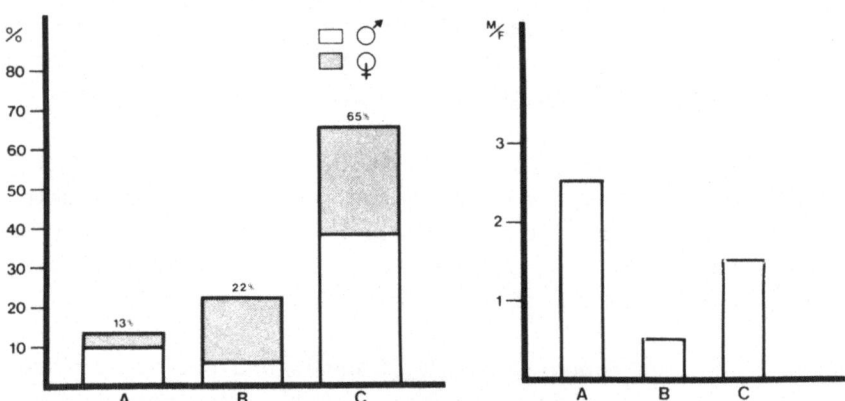

Fig. 3.   Clinical forms of cystine disease.
          A : History of stone passage.
          B : Lithiasis present at the time of diagnosis
              and/or previously removed.
          C : Lithiasis present at the time of diagnosis
              and/or previously removed with recurrent stone
              passage.

As shown in Fig. 3, the most frequent clinical expression of the disease is a renal stone, often of "staghorn" configuration with multiple small stones, preceded or followed by recurrent episodes in which stones are passed (average 4 stones/year in our patients). Passage of stones is more frequent in females than in males. A plain radiograph usually enable us to observe the presence of the stones; sometimes it is possible to suspect the disease from the variable radio-density of the calculi. These may be radiolucent, weakly radio-opaque or intensely radio-opaque, depending on the volume and the thickness of the stones. This peculiarity is strictly related to the sulphur content of the stone.

The incidence of urinary infection is very low at the onset of the disease. Almost all our patients with cystine stone disease and culture-proven urinary tract infection (28% of the cases) had been submitted to one or more surgical interventions or endoscopic procedures. The urinary infection had always arisen subsequently. Bacteria cultured from the urine were E. Coli (2 cases), Proteus species (6), Pseudomonas aeruginosa (3), Klebsiella (1); multiple organisms were isolated in 4 cases.

Over two thirds of the patients (41 out of 57) presented roentgenographically documented urolithiasis at the time of initial evaluation, 88% of these in kidney, 5% in ureter, 5% in both kidney and ureter and 2% in the bladder. Kidney calculi were present bilaterally in 10 patients (25%), unilaterally in 29 binephric patients (70%) and in a solitary kidney in 2 patients (5%). Branched or staghorn calculi were seen in 21 (51%); only 4 of these (9,7%) were bilateral.

Excluding cystoscopic procedures 42 patients has been submitted to a total of 88 urological operations (27 of these operations on 16 of the patients were done at the Urological Department of the University of Rome). Fifteen patients had undergone no surgical intervention either before or during their period of observation; some of them either passed stones repeatedly or their lithiasis had been treated medically. Overall, a total of 56 conservative operations (64%) had been performed on the kidney (pyelolithotomy, nephrolithotomy, partial nephrectomy), in addition to 6 nephrectomies, 22 ureterolithotomies and 4 cystolithotomies.

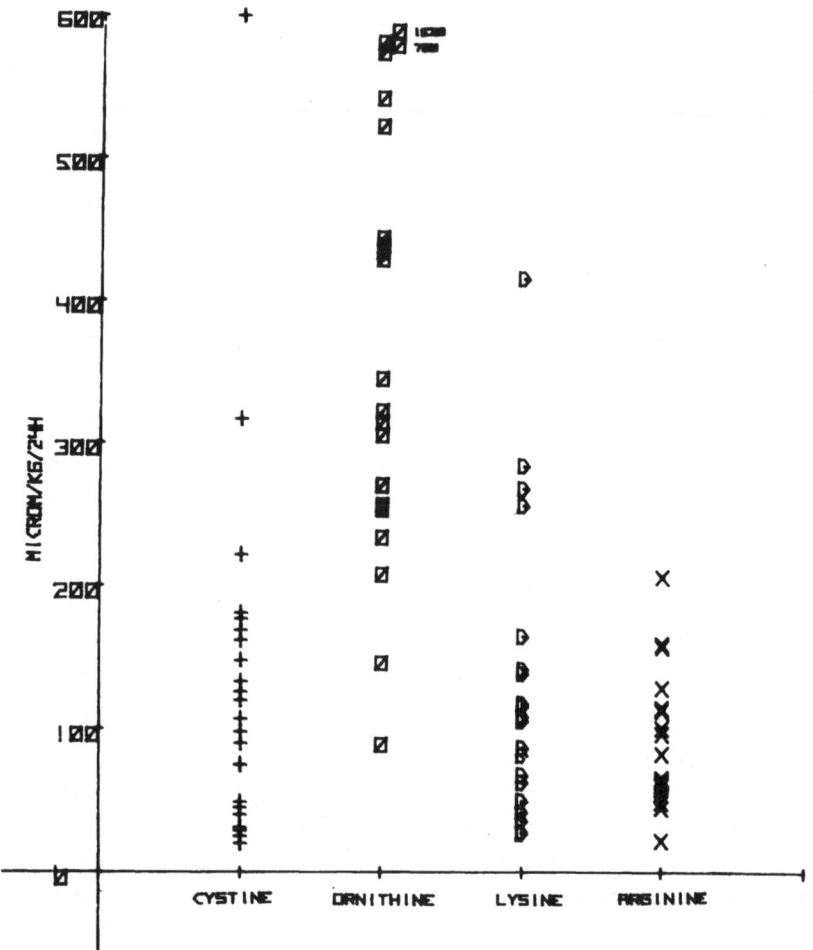

Fig. 4.   Urinary excretion of cystine, lysine, arginine and
          ornithine in 25 patients studied.

LABORATORY FEATURES

    Twenty-five patients had one or more 24-hour determinations of
daily urinary cystine excretion while they were not receiving any
drugs. For cases in which more than one 24-hour value was available,
the mean is reported. The results of determinations of aminoaciduria
indicated that half the patients (48%) excreted less than 50 μmol/
kg/24 h of cystine, in 32% the cystinuria ranged between 50 and
150 μmol/kg/24 h and in 20% was more than 150 μmol/kg/24 h (Fig. 4).

The 24-hour excretion of uric acid, calcium and phosphate was normal
in the great majority of the patients with a few exceptions (2
cases with mild idiopathic hypercalciuria and 2 cases with mild
hyperuricuria). The serum concentrations of uric acid, calcium and
phosphate was normal in all patients.

TREATMENT

The medical treatment of cystine stone disease has three main
objectives: a) to attempt to reduce cystinuria with a controlled
diet; b) to increase the solubility of cystine in the urine and
c) to attempt to transform cystine into more soluble compounds
(3-11).

As far as the first point is concerned, a low protein diet is
not, in our opinion, fundamental, particularly in infancy. A low
methionine diet, (primarily excluding eggs) without decreasing protein
intake, might be sufficient. In order to enhance solubility of
cystine it is, on the other hand, necessary to increase diuresis
($\sim$ 2.5 l/day) and to obtain alkalinization of the urine. The solubility of cystine is in fact closely related to the pH of the medium
and an increase in solubility is related to an increase in pH (4).
Alkalinizing treatment was performed preferably employing Eisenberg's
syrup or similar preparations (URALYT-U [R]); to prevent stone formation the urine should be kept at about pH 7, whereas for dissolution
of stone a pH of 8 or higher is recommended. In order to avoid risks
of secondary phosphate crystallization these drugs have not been
employed in cases presenting a concomitant chronic urinary infection,
particularly if due to ureolytic bacteria.

To convert cystine into more soluble compounds we have, since
1974, used α-mercaptopropionylglycine (MPG). This substance, with
cystine, forms a cysteine-MPG disulphur complex, both in the plasma
and in the urine, which is far more soluble than cystine. This drug,
which is non toxic and has a higher oxide-reducing capacity than
d-penicillamine, seems to be more efficient as a therapeutic agent
(5, 6, 7, 8, 12, 15, 17). For preventive purposes in adults 1 g/day
MPG (Thiola [R]) was given in 2 doses before meals, whilst for stone
dissolution 1.5 g/day was given in 3 doses; in children the dosage
was 30-40 mg/kg/day. Ideally the dosage should be established in
each case in relation to the concentration of cystine in the urine;

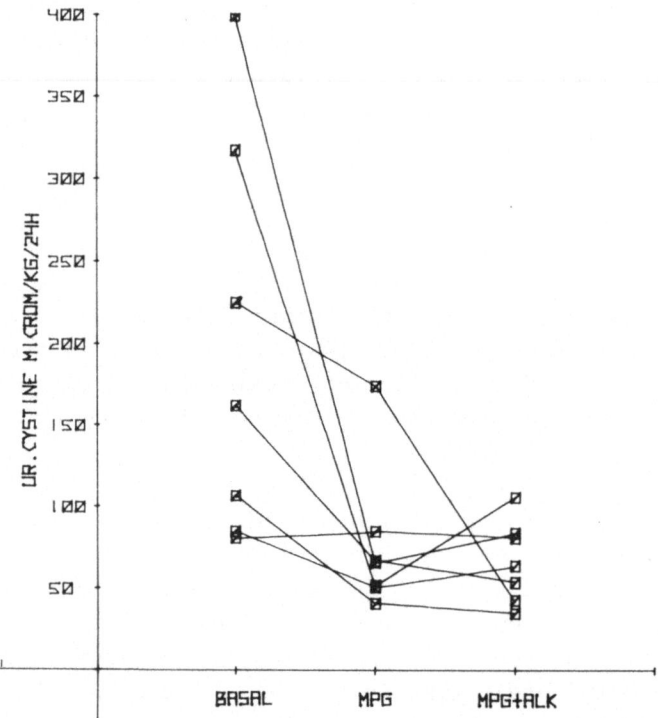

Fig. 5.   Changes in cystinuria following administration of
          MPG alone (1 g/day) and MPG (1 g/day) + alkali in a
          group of patients without stones at the time of study.

the optimal dose is that which reduces the concentration of cystine
in the urine below 200 mg/l.

The reduction of the urinary cystine levels was more marked in
those cases with higher basal values. Fig. 5 illustrates these
variations following administration of 1 g/day MPG alone and 1
g/day MPG + alkalis in a group of patients with cystinuria, but
without stones in situ at the time of the study. In our experience
(11), contrary to the observation of others, the combined administration of MPG and alkaline agents does not decrease the cystinuria
more than the administration of MPG alone. After administration of
MPG (1 g/day) we also observed a decrease of plasma cystine concentration (Fig. 6), the basal values of which are already below normal
in the cystinurics, ranging from 0.5-4 μmol per 100 ml (normal values:
5 μmol per 100 ml).

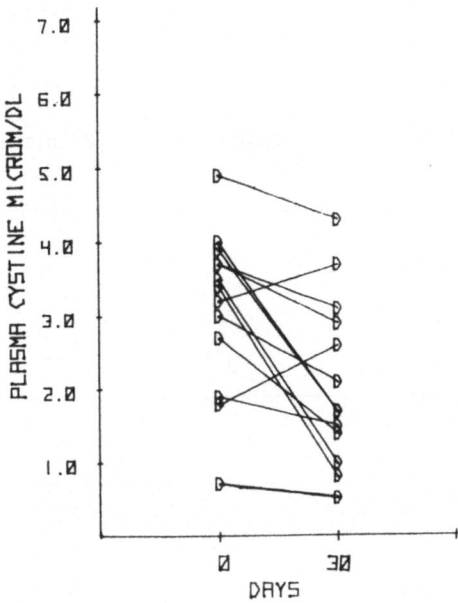

Fig. 6.   Changes of plasma cystine levels following MPG
          administration (1 g/day).

    The results of medical treatment are given in table 2. Complete
dissolution of the stones occurred in almost 50% of the cases, and
partial dissolution in 30%. Treatment was unsuccessful in 20%. Radio-
graphic evidence of one case of total dissolution of cystine stones
is presented in Fig. 7. Lack of success was partly due in 2 cases
to concomitant urinary infection and thus to precipitation of calcium and magnesium salts. Local litholytic treatment was performed
in 2 patients after the operation: in one case we employed a solution
of MPG 5% at pH 7.5 (1 l/day) and in the other a 3.6% tham solution
or 1,4% Na bicarbonate, always through the nephrostomy tube. Complete
dissolution of small residual stones occurred over a period of
about 20 days.

    The results of preventive medical treatment are given in table
3. In 96% of patients no recurrence was observed during the follow-
up (1 to 7 years) even in patients with concomitant urinary infection.
In the 5 patients (10%) with recurrences (3 without and 2 with
recurrent or chronic urinary infection), two presented struvite
stones secondary to the infection and to excessive alkalinization.

Table 2. Results of litholytic treatment.

DISSOLUTION	No. of patients	Stones < 1 cm. Ø	Stones > 1 cm. Ø	Urinary infection
Complete	11 (46%)	5	6	0
Partial	8 (33%)	2	6	2
Failure	5 (21%)	0	5	2
TOTAL	24	7	17	4

Table 3. Results of preventive medical treatment.

		No. of patients	Recurrence
CYSTINE LITHIASIS without urinary infection (ALKALI + MPG)	PURE	32	3
	MIXED	2	0
		34 (69%)	3 (8.8%)
CYSTINE LITHIASIS with recurrent urinary infection (ALKALI + MPG)	PURE	10	1
	MIXED	1	0
		11 (23%)	1 (9.1%)
CYSTINE LITHIASIS with chronic urinary infection (MPG)	PURE	3	1
	MIXED	1	0
		4 (8%)	1 (25%)
T O T A L		49	5 (10.2)

Follow-up ranging from 1 to 7 years.

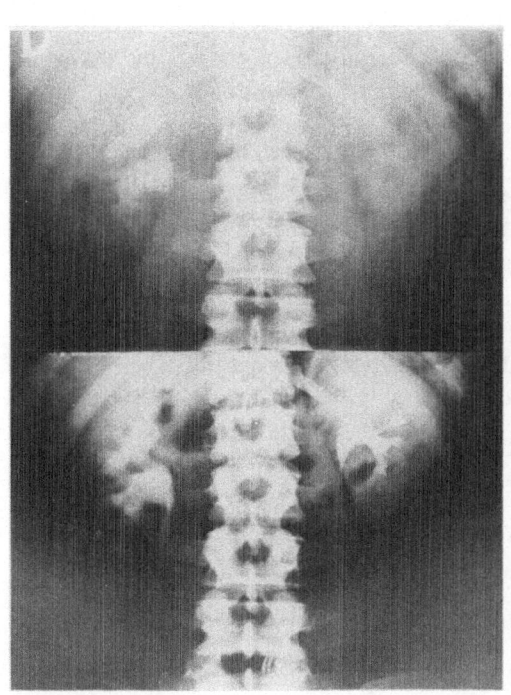
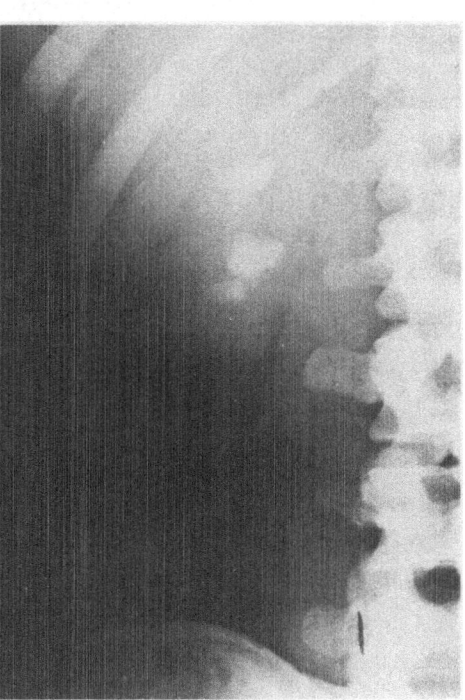
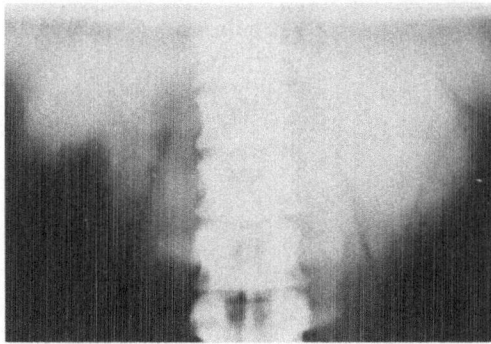

Fig. 7.

R.M. 20 year-old male. Recurrent staghorn calculi. a) plain x-ray; b) IVP; c) plain x-ray after 4 months medical treatment; d) plain x-ray after 7 months medical treatment: total dissolution of the stone.

The other three subsequently received d-penicillamine (1-1,5 g/day
or 30 mg/kg/day associated with vitamin B6) without improvement.
The last drug is still used by many authors (1-3-9) and it acts
similarly to MPG, but we no longer use it on account of the frequent
side effects (measles-like rash, fever, arthralgia, leukopenia,
proteinuria or even nephrotic syndrome). MPG on the other hand
usually has no toxic effects and is commonly employed as a "liver
protector". With penicillamine we experienced side effects in 7 cases,
i.e. 14% (skin reaction, in one case, mild transient proteinuria
in one, nephrotic syndrome in one, reduced sensitivity to taste in
two and slight gastric intolerance in two). In 2 of these cases
the drug was discontinued. The effectiveness of treatment and
detection of complications are best monitored by frequent re-evalu-
ation of patients by means of appropriate laboratory studies.
Considering the high incidence of recurrence (90-100% of the cases),
which is typical of this type of lithiasis, these results appear
to be particularly encouraging and lead us to the conclusion that
medical treatment, if correctly performed, considerably modifies
the prognosis in these patients.

REFERENCES

1. Bartter, F.C., Lotz, M. and Thier, S.O.: Cystinuria. Ann. Int.
    Med. 62: 796-822, (1965).
2. Caldwell, R.J., Townsend, J.I. and Smith, M.J.V.: Genetics of
    cystinuria in an inbred population. J. Urol. 119: 531-533,
    (1978).
3. Dahlberg, P.J., Van der Berg, C.J., Kurtz, S.B., Wilson, D.M.
    and Smith, L.H.: Clinical features and management of cystinuria.
    Mayo Clinic. Proc. 52: 533-542, (1977).
4. Dent, C.E. and Senior, B.: Studies in the treatment of cysti-
    nuria. Brit. J. Urol. 27: 317-332, (1955).
5. Kallistratos, G., Burchardt, P. and Freerksent, P.: Recent
    advances in the diagnosis and treatment of L-cystine stone
    disease. Folia Bioch. Biol. Graeca 15: 21-23, (1978).
6. King, J.S. Jr.: Treatment of cystinuria with α-mercaptopro-
    pionylglycine. Proc. Soc. Exp. Biol. Med. 199: 927-932, (1969).
7. Kinoschita, K., Yachiku, S., Takeuki, N., Kotake, T., and

Sonoda, T.: Treatment of cystinuria with α-mercaptopropio-nylglycine. Proc. 2nd Int. Symp. on Thiola, pp. 50-54 Santen, Osaka, (1972).

8. Hautmann, R., Terhost, B., Stuhlsatz, H.W. and Lutzeyer, W.: Mercaptopropionylglycine: a progress in cystine stone therapy. J. Urol. 117: 628-630, (1977).

9. Lotz, M., Potts, J.T. Jr. and Holland, J.M.: D-penicillamine therapy in cystinuria. J. Urol. 95: 257-263, (1966).

10. Miano, L.: Cistinuria e litiasi cistinica. Progr. Med. 28: 285-287, (1972).

11. Miano, L., Gallucci, M. and Petta, S.: Results of medical treatment of cystine lithiasis. Eur. Urol. 5: 265-272, (1979).

12. Remieen, A., Kallistratos, G. and Burchardt, P.: Treatment of cystinuria with Thiola (α-mercaptopropionylglycine). Eur. Urol. 1: 227-230, (1975).

13. Rosenberg, L.E., Downing, S. and Durant, J.L.: Cystinuria: biochemical evidence for three genetically distinct diseases. J. Clin. Invest. 45: 365-371, (1966).

14. Silbernagl, S.: Renal handling of aminoacids: recent results of tubular microperfusion. Clin. Nephrol. 5: 1-8, (1976).

15. Terhorst, B. and Stuhlsatz, H.W.: Cysteine Therapie mit Mercaptopropionylglycine (MPG). Urologe 14: 190-193, (1975).

16. Thier, S.O. and Segal, S.: Cystinuria in "The metabolic basis of inherited disease" edited by Stanbury J.B., Wyngaarden J.B., Fredrickson D.S., pp. 1504-1519 Mc Graw-Hill Book Comp., New York, (1972).

17. Zechner, O., Latal, D., Grundig, E. and Muller, M.: Klinische Aspekte der Cystinuria und ihre Behandlung mit Thiola. Wien. Klin. Wschr. 88: 101-105, (1976).

EDITORIAL NOTE

A few important points emerge from this careful study: 1. In the vast majority of cases (85% in this series) the diagnosis of cystine stones is not made when the patients are first seen and operated; 2. Most patients will form recurrent bilateral stones, requiring multiple operations, and many will eventually develop chronic renal failure; 3. Preventive medical treatment is effective in about 70% of cases; 4. Early identification of cystinuria and of the chemical composition of the cystine stones is essential and potentially life-saving, especially considering the very young age

of most patients; 5. MPG (thiopronine) is less toxic and at least
as effective as d-penicillamine. It certainly appears to be the
drug of choice at the present time.

M. Pavone-Macaluso

URINARY LITHIASIS IN CHILDREN - A REVIEW OF 100 CASES

M. Broyer*, M.F. Gagnadoux* and D. Beurton**

* Service de Néphrologie pédiatrique
** Clinique Urologique
   Hôpital Necker - Enfants Malades
   Paris (France)

ABSTRACT

From 1970 to 1979, 100 cases of urinary lithiasis have been referred to our service in the Hôpital des Enfants Malades. Sixty two were boys and 38 were girls, the usual sex ratio for stone disease in children. Age at diagnosis was below 2 years in 33 cases and below 5 years in 60 cases. The presenting symptoms were the following: hematuria (20 cases), pyuria (19 cases), abdominal pain or colic (18 cases), passage of a calculus (11 cases), proteinuria (10 cases), incidental finding on abdominal x-ray (4 cases), urinary retention (3 cases), renal mass (3 cases) and anuria (3 cases). The frequency of these symptoms varied according to age: below 2 years pyuria was the most frequent, between 2 and 5 years hematuria and above 5 years, pain and colic.

Various etiologies were found in this series. In 34 cases a metabolic cause was found, and in 12 cases urinary tract abnormalities existed prior to lithiasis. 54 cases remained idiopathic despite full investigations.

Idiopathic lithiasis seems to have become less frequent in the last five years since we have seen only 15 cases as compared with 28 cases in the five years before. Stones were generally located

in the renal pelvis or the ureter. Only 3 cases had bladder stones,
2 of whom came from tropical areas. Thirty five of these children
were successfully operated upon but recurrence of stones was observed in 2 cases who subsequently underwent a second operation.
The other 19 were treated medically and a few small stones were
passed spontaneously. In the latter group only one recurrence was
observed.

The frequency of metabolic causes is probably over-estimated
in this series due to the specialised nature of our service.

Primary hyperoxaluria (11 cases) was the most frequent, followed by familial cystinuria-lysinuria (7 cases) and by oxalic
lithiasis with a mild hyperoxaluria (6 cases). Two cases had hyperuricemia (related to Lesch-Nyhan disease) and two others developed
xanthine stones after cytolytic treatment for malignant disease.
Finally, six cases had hypercalciuria. One had a distal tubular
acidosis of Albright type, one had primary hyperparathyroidism, one
developed stones after long term steroid and vit D therapy, one was
immobilised in a cast and two had idiopathic hypercalciuria of renal
type.

The last group of patients (12 cases) developed stones in
association with a preexisting urinary tract abnormality of whom 2
had pyelo-ureteral stenosis and 2 mega ureters with reflux. Four
patients who had unilateral stones in a non-functioning kidney were
classified after histological study as having xanthogranulomatous
pyelonephritis.

––––––––

The incidence of urinary lithiasis in children varies according
to the geographical area. It is well known that in the rural areas of
some developing countries, bladder calculi are endemic and affect
especially young male children. In contrast urinary lithiasis in
western Europe and America is less frequent in childhood and affects
essentially the upper urinary tract. The purpose of this paper is
to describe a series of 100 consecutive cases gathered in a specialised section for pediatric nephrology at the Hôpital des Enfants
Malades in Paris.

METHODS

Using a manual chart-retrieval system we found that from Jan.
1970 to Dec 1979, 100 children with urinary lithiasis had been
hospitalised or followed in out-patient clinics in the service of
pediatric nephrology. Most of these patients underwent a complete
metabolic evaluation including calcium, phosphorus and uric acid
determination in plasma and urine, Brandt's nitroprusside test for
detection of cystinuria and oxalate urinary excretion determination.
When available the stones were chemically analysed and submitted to
X ray spectrography analysis.

PRESENTING SYMPTOMS

The presenting symptoms and the circumstances of diagnosis were
varied and quite different in infants and in older children (Table
1). Overt pyuria was the most frequent presenting symptom in the
youngest children but was much more rare above 5 years. Spontaneous
passing of stones was also very frequent in infants and rare above
2 years of age. Hematuria and abdominal pain or colic are on the
contrary, much more frequent in older children than in those under
2 years. Quite exceptional are the stones revealed by anuria or
retention of urine. Finally a certain number of cases were detected incidentally by discovery of a palpable mass in the abdomen,
following a plain x-ray of the abdomen and by discovery of proteinuria by systematic examination of urine.

PATIENT POPULATION

There were 62 boys and 38 girls aged 3 months to 17 years, 33
being below 2 years. Of these children 92 usually lived in France,
6 in North Africa and 2 in tropical countries. A complete survey
was performed in the last 24 cases of idiopathic lithiasis and
familial antecedents were detected in 6 of them.

AETIOLOGY

Aetiology was sometimes quite obvious but it remained unknown
in a few cases. Patients may be classified schematically in three

Tab. 1.   Presenting symptoms.

Age of patients	< 2 years	2-5 years	5-17 years
Pyuria	11	6	2
Stone passed	8	1	2
Hematuria	2	9	9
Abdominal pain	3	2	13
Retention of urine	1	1	1
Anuria	0	0	3
Abdominal mass	1	1	1
Incidental x-ray finding	1	0	3
Proteinuria	6	3	1

groups: 1) metabolic disorders, 2)urological abnormalities, 3)
idiopathic lithiasis.

Metabolic Causes

Metabolic causes are not frequently considered to be of great
importance in the aetiology of stone disease in children. Piel and
Roof reviewing 445 cases of lithiasis in children from U.K. and
U.S.A. found that only 17% of stones were of metabolic origin (4).
However in the present series a metabolic abnormality was found in
35% of cases but there was certainly a bias of reference accounting
for such a finding, as our service is specialised both in nephrology
and in metabolic diseases. Metabolic abnormalities leading to urinary
calculi may be divided in 4 principal groups: oxalate disturbances,
purine disorders, defects of tubular cystine reabsorption and hypercalciurias.

Oxalate lithiasis. 1) Primary hyperoxaluria is a rare disease
with a recessive mode of inheritance characterised by excessive
production of oxalate which accumulates in the form of calcium
oxalate crystals in many tissues and specially in the kidney, leading
to nephrocalcinosis and renal failure. High urinary excretion of

oxalate also causes urinary lithiasis.

Eleven cases in our series suffered from this disease.

The first symptoms were noted between 3 months and 12 years of age. These symptoms were always related to the stones, but some latent cases were only discovered in the terminal stage of renal failure. The diagnosis was ascertained by oxalate determination in 24 hour collections of urine; the results were always above 100 mg/day even when creatinine clearance was decreased to around 5 ml/min corrected for body surface area.

The stones in this disease are characteristic, being very opaque, with a variable shape and having a number of rounded mammillary processes. They are usually formed of pure calcium oxalate mono-hydrate (whewellite). This disease is always associated with pro-gressive destruction of the kidneys due to nephrocalcinosis and to interstitial nephritis rather than to the stone formation.

It is important to emphasize that sudden worsening of renal function may follow surgery for renal lithiasis. This occurred in 2 patients from our series, in whom short nephrotomies were performed in order to remove the stones.

In another case terminal renal insufficiency occurred after ureterotomy without any nephrotomy. It is possible that modifications of hydration and/or of renal haemodynamics had provoked massive oxalate deposition in the kidney. Therefore we think that the indi-cations for surgery must be very conservative in these patients.

Apart from severe cases rapidly developing renal failure, there were other patients who formed recurrent stones for many years, but whose renal function remained in the normal range. For example, one child in this series passed 3-5 calculi/year from the age of 2 years. He remained without any kidney deterioration and with a normal kidney biopsy at 11 years, in spite of a high urinary oxalate ex-cretion.

High doses of oral pyridoxine are sometimes able to decrease oxaluria and a high fluid intake may assist in limiting the re-currence of stones, but there is no effective treatment for this disease. The prognosis is therefore very poor, as these patients

do very badly on dialysis and have a constant recurrence of the
disease after kidney transplantation.

2) Secondary oxalosis with its high urinary excretion of oxalate
is favoured by excessive ingestion of oxalate or oxalic acid from
rhubarb, sorrel or tea and certain other foods, by pyridoxine deficiency and by short or inflammatory bowel syndrome.

These causes are exceedingly rare in children and not one such
cause has been found in our series. Oxalate lithiasis in the short
bowel syndrome was described in adult patients, whose diets contained
too much oxalate and lipid resulting in excessive absorption of
oxalic acid from the gut. Careful dietary prescription must avoid
this complication in children with such intestinal abnormalities.

3) Oxalate lithiasis with bordeline hyperoxaluria. One third
of cases of adult lithiasis belong to this group, which is however
seldom observed in children. Nevertheless 6 cases of oxalic lithiasis
with bordeline oxaluria were·noted in this series. Their principal
features are shown in table 2. The age at diagnosis of urinary
lithiasis was 3 months to 13 years. In 4 out of 5 children, for
whom familial antecedents were studied, at least one parent had a
history of stone formation and in one case recurrent stone formation
was found in four successive generations. Urinary oxalate excretion
was between 30 and 54 mg/24 h, which is just above the normal limits
for our laboratory (20 - 30 mg/24 h). The stones, when analysed,
were made of dihydrated calcium oxalate (wheddelite) or of a mixture
of mono and dihydrated calcium oxalate.

Treatment currently proposed for these patients is, as in the
adult, a high fluid intake, pyridoxine and magnesium salts.

Purine lithiasis. Primary gout is exceptional in children.
Uric acid stones, however, can be found in some situations where
uric acid synthesis is greatly increased as Lesch-Nyhan syndrome
or in leukaemia or other malignant diseases, usually after chemothe-
rapy.Uric acid stones may also occur in the absence of hyperuricaemia
or even hyperuricosuria, if constant acidity of the urine favours
uric acid precipitation.

Other urine compounds such as xanthine or adenine may also be
found in children's stones. Xanthine lithiasis is a possible compli-

Tab. 2. Oxalate lithiasis with bordeline hyperoxaluria.

Case	Age at diagnosis	Oxaluria mg/day	Operation	Recurrence	Familial antecedents
1	5 yrs	32 – 46	–	–	Mother
2	3 mos	38 – 54	+	(residual stone)	Mother
3	10 mos	36	–	–	Father, uncle
4	4 yrs	30 – 35	–	–	Mother, grand mother, grand grand mother
5	13 yrs	40	+	+	Nil
6	4 yrs	31	+	–	Unknown

cation of familial xanthinuria, a disease where the enzyme xanthine-
oxidase is lacking and uric acid is absent from plasma. Xanthine
or other oxypurines may accumulate and form gravel and calculi in
patients treated with xanthine-oxidase inhibitors. Recently a new
variant of purine metabolism disorder, the lack of adenine-phospho-
ribosyl-transferase was described. In this disease, the only symptoms are related to the occurrence of calculi. Plasma uric acid is
normal or low, and the diagnosis is made by the analysis of the
stones, which are made of 2,8 hydroxyadenine, and by the evidence
of the missing enzyme in red cells (2).

In this series we found 2 cases of Lesch-Nyhan syndrome with
typical encephalopathy including self-mutilation; uric acid lithiasis
was discovered very early in one case (age 13 months) and after 5
years in the other case. Recurrence of stones was prevented by
allopurinol. Two cases of xanthine-gravel obstruction of the ureters
also occurred in patients with lymphoblastosarcoma after treatment
by chemotherapy together with allopurinol. Oligo-anuria developed
in both patients and urine flow returned after ureteral catheterization, which liberated sand and mud made of oxypurines and xanthine.

Cystinuria. The reported frequency of 5% of cystine stones in
childhood lithiasis (4) is in agreement with the present series, as
we observed 7 cases of cystine lithiasis in our 100 patients. Cystine
stones occur as the only clinical symptom of a tubular defect affecting cystine and dibasic amino-acid reabsorption and which is
inherited as a recessive or an incompletely recessive gene. Pure
cystinuria without dibasic aminoaciduria exists also but is very
rare. The seven cases of this series are all typical cases of
cystinuria-lysinuria. Three were revealed by symptoms of stones at
6 months; the diagnosis was made between 4 and 12 years in the other
cases. Brandt reaction was positive in the urine in all of the cases
at the first examination. Cystinuria was between 260 and 600 mg/24 hr.
The range of the follow up of these patients is 2 to 23 years.
It is said that cystine stones are often non opaque; we do not
confirm this point and very regularly found visible stones on plain
x-rays in almost all the cases with symptoms, except when the stones
were too small and were easily passed in urine. Preventive therapy
of stone recurrence, including increased fluid intake and alkalini-
sation of urine, can be effective but this it difficult to maintain
permanently, so that recurrence is frequent in these patients. Some
of the patients of this series (to be truthful, not all of them were

very disciplined or well followed) underwent up to 10 operations in
23 years.

D-penicillamine seems to be able to dissolve cystine stones
in situ. We attempted to treat 5 patients by penicillamine in order
to avoid repeat surgery. This treatment was really effective in only
one patient in whom one calyceal stone was totally dissolved. In
another case two big stones in the renal pelvis decreased in size
and passed through the ureter and urethra after one month of penicillamine treatment, but in the same child, two years later, a big
bladder stone did not decrease at all under the same treatment. In
one case penicillamine was effective in preventing recurrence for
3 years, but the drug was discontinued and a new stone appeared one
year later. Finally penicillamine was of little value or was inactive in 2 other cases.

Hypercalciurias. Known causes of hypercalciuria are listed on
table III; most of them are represented in this series. Primary
hyperparathyroidism is a classical but very rare cause of lithiasis
in children. Parathormone assay and echosonography of the neck allow
one to confirm the diagnosis, which should be suspected in patients
with recurrent calcium lithiasis with hypercalciuria, when variable
hypercalcaemia and decreased phosphataemia and tubular reabsorption
of phosphate are found. Albright's distal tubular acidosis could
also be revealed by hypercalciuric lithiasis. The diagnosis can be
made from the constant association of lithiasis with the characteristic papillary nephrocalcinosis from which the stones are originated.
Acidosis, hypokalaemia and hypocitraturia are also constantly associated in this syndrome, in which the characteristic defect of
H^+ excretion can be found. Thus urinary lithiasis is only one
symptom among others in this disease. It is noteworthy that out of
15 cases of Albright tubular acidosis that we have collected in the
last 10 years, only one was complicated by lithiasis.
Immobilisation by paralysis or by fracture and setting in a plaster
cast, or by any other cause is followed by calcium mobilisation
from bones, hypercalcaemia and hypercalciuria. One patient in this
series had both legs in a plaster for one month before the first
symptoms of lithiasis developed. Excessive vitamin D intake and
long-term daily steroid administration were found to cause recurrent
lithiasis in a child who received both medications for 3 years because of dermatomyositis. Each of these drugs is able to induce
urinary lithiasis; steroids are frequently implicated in a recent

Table 3. Causes of hypercalciuria in children.

- PRIMARY HYPERPARATHYROIDISM

- DISTAL TUBULAR ACIDOSIS

- IMMOBILISATION OSTEOLYSIS

- EXCESS OF VITAMIN D OR SIMILAR COMPOUNDS

- CUSHING SYNDROME, STEROID THERAPY

- HYPOTHYROIDISM

- IDIOPATHIC HYPERCALCIURIA

---

review of adolescent lithiasis (5). Other definite causes of hypercalciuria, like sarcoidosis, hypothyroidism or Cushing syndrome,
are not represented in this series.

Two cases of idiophatic hypercalciuria can be included in this
chapter. From a clinical point of view two forms of idiopathic hypercalciuria had been described in children, i.e. the form described
by Zetterström (10) where lithiasis and hypercalciuria are the only
symptoms and the form described by Royer et al. (6) where hypercalciuria is associated with nanism, osteoporosis and other renal
symptoms such as hyposthenuria and proteinuria. Lithiasis seems
however to be rare in this latter form. The two cases in this series
were of the Zetterström type.

The physiopathology of hypercalciuria in children remains obscure. Disturbance of phosphate metabolism and/or of 1,25 (OH)$_2$
cholecalciferol synthesis has been advocated in the renal forms of
adult hypercalciuria. Hyperabsorptive (intestinal) hypercalciuria
has also been described in adults. Very little information is yet
available on this matter in childhood.

Lithiasis Associated with Obstructive Uropathy and Infection.

In some series of children's lithiasis, obstructive uropathy represents more than 60% of cases (8). It is sometimes difficult to distinguish between the anatomical derangements which existed before lithiasis and which predisposed to stone formation, and the anatomical derangements which are the consequence of the obstructive and infected stones. In this series, we have deliberately selected, for inclusion in the former group, eleven cases with significant malformations. Four cases were known to suffer from urinary tract abnormalities before the discovery of lithiasis. Complex congenital bilateral malformations were present in three (one with bilateral ureterostomy) and one had a neurological bladder. In seven cases lithiasis and urinary tract abnormalities were discovered at the same time. The abnormality consisted, in 2 cases, of pyelo-ureteral stenosis and, in 5 cases, of megaureter with reflux (2 bilateral, 3 unilateral). All these cases were infected and all underwent surgical removal of the stones. When analysed, the stones were formed of magnesium and ammonium phosphate. No recurrence was observed in these patients.

Xanthogranulomatous pyelonephritis is not uncommonly found in this group of patients (3); we collected 3 such cases with a uniform symptomatology. Age at the diagnosis was 8 months, 3 and 6 years. All were infected with proteus and had massive pyuria; radiological examination disclosed numerous renal calculi disseminated in the area of a non functioning kidney which looked like a renal mass. In the youngest case the affected kidney was enlarged but still functioning. The diagnosis was confirmed after examination of the nephrectomy specimen, which showed a characteristic histological pattern with large zones of necrosis and with xanthomatous cells with foamy cytoplasm and pyknotic nuclei, as well as a marked inflammatory reaction. As is usual in this disease, definitive healing was obtained after nephrectomy on the affected side.

The physiopathology of xanthogranulomatous pyelonephritis remains unclear. It may be the consequence of obstruction by stone and infection, but the very peculiar histological findings suggest other factors, such as abnormal lipid metabolism, vascular or lymphatic obstruction.

Idiopathic Lithiasis

There is a high percentage of idiopathic lithiasis in this

series (52%). Half of these cases occurred below 2 years of age, 60%
were male and 40% female, but in young children the proportion of
males increased to 75%. Infection was frequently noted in these
patients and proteus was the most usual micro-organism isolated
from urine (64%), sometimes in association with other bacteria.

Lithiasis was more often found on the left side and was rarely
bilateral (three cases). Eleven stones were of the staghorn type.
With regard to the location, nine stones were located in one or
several calyces, 14 in the renal pelvis, 11 in the ureter, and only
3 in the bladder. Two of the vesical calculi were referred from
tropical areas (Guyana and Tahiti), corresponding perhaps to the
classical endemic bladder lithiasis.

In ten patients a minor urinary tract abnormality was found
after pyelography and cystography: 7 had vesico-ureteral reflux,
one had a duplex ureter and 2 had non obstructive urethral valves.
These cases are totally different from the group with lithiasis
complicating obstructive uropathy.

The evolution of these cases was variable: eleven patients
finally passed their stones through the urethra and 34 others
underwent stone removal. Only one patient underwent nephrectomy;
the remaining patients still have their stones. Only two patients
had recurrent stone formation after surgery.

Renal sequelae are rare: 4 or 5 cases presented a small area
of segmental atrophy of the kidney, and one had severe permanent
hypertension. The composition of the stones was determined in 29
cases. Very few were made of pure material: Ca oxalate (n = 3),
urate (n = 3), calcite (n = 1) and apatite (n = 1). The other 21
stones were made of a combination of carbonate, phosphate, urate
and/or oxalate on one hand and calcium, ammonium and/or magnesium
on the other.

Twelve combinations have been found in these 21 stones, the 2
most frequent being carbonate-apatite-Mg-NH_4 phosphate (n = 5) and
oxalate-apatite (n = 4). It is interesting to note that the two
cases of bladder stones coming from tropical areas were made of
urate, as is usual under such circumstances (9).

Looking for familial antecedents in the families of 24 cases

of idiopathic lithiasis, we found that 6 had at least one parent
having suffered from a urinary stone.

In this series we did not find any example of the phosphatic
lithiasis associated in young children with hypotonia and hip abnormalities described by Royer and David (7).

CONCLUSIONS

Almost half the cases of childhood lithiasis could be related
to a precise cause in our series. For this reason a complete
screening for metabolic causes is justified in such patients, as
special treatment may be indicated to prevent recurrence.

Surgical removal of calculi must be undertaken: 1) when the
size of stone is too large .to allow the expulsion through ureter
or urethra; 2) when the stone is obstructive; 3) if there is associated recurrent or permanent urinary infection. The results of
surgery are generally good (1). Nephrectomy is rarely necessary. In
idiopathic lithiasis recurrences are very rare. Nevertheless, many
problems have yet to be solved in the story of urinary lithiasis in
children, as in adults, since the physiopathology of origin and
development of a number of these calculi remain a mystery.

REFERENCES

1. Beurton, D. and Cukier, J.: La lithiase urinaire de l'enfant,
    à propos de 106 observations. J. Urol. Néphrol. 80: 219, (1975).
2. Debray, H., Cartier, P., Temstet, A. and Cendron, J.: Child's
    urinary lithiasis revealing a complete deficit in adenine
    phosphoribosyl transferase. Pediat. Res. 10: 762, (1976).
3. Habib, R., Levy, M. and Royer, P.: La pyélonéphrite chronique
    xanthogranulomateuse chez ·l'enfant. Arch. Franç. Péd. 25: 489,
    (1968).
4. Piel, C. and Roof, B.: Renal calculi in children. In Rubin and
    Baratt M.: Pediatric nephrology (Williams & Wilkins, Baltimore
    1975) p. 760.
5. Rambar, A. and Mac Kenzie, R.: Urolithiasis in adolescents.
    Am. J. Dis. Child. 132: 1117, (1978).
6. Royer, P., Mathieu, H., Gerbeaux, S., Frederich, A., Rodriguez-

Soriano, J., Dartois, A.M. and Cuisinier, P.: L'hypercalciurie idiopathique avec nanisme et atteinte rénale chez l'enfant. Ann. Pediat. 38: 147, (1962).

7. Royer, P. and David, L.: La lithiase urinaire phospho-calcique multiple et non récidivante du nourrisson avec retard psyco-moteur et anomalies des hanches. Arch. Franç. Péd. 26: 607, (1969).

8. Sinno, K., Boyce, W. and Resnick, M.: Childhood urolithiasis. J. Urol. 121: 662, (1979).

9. Valyasevi, A. and Van Reen, R.: Pediatric bladder stone disease; current status of research. J. Pediat. 72: 546, (1968).

10. Zetterström, R.: Modern problems in paediatrics.I vol. (Karger Basel/New York, 1958) p. 478.

SOFT RENAL CALCULI (MATRIX STONES)

M. Camey and A. Le Duc

Service d'Urologie
Centre Médico-Chirurgical Foch
92151 Suresnes – Paris (France)

ABSTRACT

Soft calculi or so called matrix calculi appear to be very
rare. Less than 50 cases have been published. The matrix substance
A (mucoprotein), which usually represents only 2,5% of the weight
of fully calcified stones can reach 65% in soft calculi. However,
there is no difference in the architectural arrangement of the matrix
skeleton between the two types of lithiasis.

Soft calculi belong to infective renal calculi in which neither
metabolic disturbances nor intratubular matrix precursor can usually
be found. Calculogenesis starts in the major collecting system. The
polymerisation of urine mucoprotein, induced by urea-splitting infection, produces a calculogenic matrix substance. Precipitation of
apatite also occurs. The immature matrix calculus, initially plastic
(soft), will eventually develop to form a phosphatic staghorn calculus.

Our experience, based on 25 cases of soft calculi, is rather
similar to previous reports, except for some particular points,
such as the participation of oxalic or uric crystals in the calculogenesis of soft calculi. Our renal biopsy findings are described.

Among these 25 cases, 9 can be considered as "malignant"

(leading to renal failure or life threatening). The surgical treatment requires complete removal of the stones, control of the urinary infection and the prevention of recurrences by fibrinolytic and mucolytic infusion in the major collecting system.

––––––––

"Soft calculi", generally called "matrix stones" by English-speaking authors, are a very particular and rather rare kind of lithiasis. Their most characteristic macroscopic feature is their putty-like substance, more or less white, generally smooth and shining. Sometimes they are represented by several, more or less ovoid, pieces. Sometimes they spread through the pelvis and calyces, as soft staghorn calculi. Usually little hard concretions (which are nuclei of mineralisation) can be felt within this substance.

The first description was made in 1909 by Gage and Beal (8). Later, the main publications came from Boyce and King (3), Innes Williams (13), Allen and Spence (1), Bruezière (5), Camey and Le Duc (6), Küss et al. (9), Malek et al. (10), and, more recently, from Wickham (12).

BIOCHEMICAL FINDINGS

The most common urinary calculi contain both a crystalline salt, usually the main component, and a small amount (2,5% on average) of non crystalline mucoprotein matrix. This matrix is an organic substance consisting of approximately one third mucopolysaccharide and two thirds mucoprotein as shown by Boyce and Garvey (2). Concerning the composition of soft calculi, the average percentage of matrix is over 65%. The explanation for the almost complete absence of crystalline material from matrix calculi is lacking. No real experimental study has been carried out to solve this problem, probably because of the scarceness of matrix calculi (less than 50 cases published among adults), so that the explanation for their formation is usually included in general concepts of routine calculus formation. The precise role of matrix in stone formation has been the source of considerable confusion and speculation. According to Boyce and coworkers (2, 3, 4), the formation of an organic matrix is the primary event in calculogenesis, which is then followed by deposition of crystalline material, whereas other workers suggested that matrix is simply a fortuitous inclusion in what is basically a mineralogic process.

In 1976, from a review of the literature and from a personal study of 100 cases of surgically treated calculi, Wickham (12), pointed out the following facts:

- The architectural arrangement of the matrix skeleton is entirely similar in matrix calculi and in fully calcified stones.

- Patients who suffered from non phosphatic stones had abnormalities of Ca metabolism (in 80% of the cases), and biopsy evidence of intranephronic calculosis (10).

- Patients who suffered from infected phosphatic stones had only 19% of metabolic abnormalities and entirely normal biopsies.

Thus, Wickham proposed two systems of calculogenesis:

- Firstly, the metabolically induced system with a sterile urine: disturbed renal metabolism induces proximal tubular damage with deposition of intracellular matrix precursor + apatite (Randall plaques or Carr's concretions) resulting in spheroliths in tubules, then in mature calculi such as small oxalate stones.

- Secondly, in the infectively induced system, which starts in the major collecting system,urea-splitting infection induces polymerisation of normal urinary mucoprotein (Tamm-Horsfall protein) and produces matrix substance which, together with apatite, leads to a mature matrix stone of staghorn type.

Since 1953, in our department of urology, we have operated upon 25 cases of soft calculi. From the study of these 25 cases many analogies concerning clinical and bacteriological findings were apparent, in comparison with previous reports, but some differences concerning biochemical and histological findings were observed.

CLINICAL FEATURES

- There were 20 female and 5 male patients, the usual ratio for matrix and infected calculi, their ages ranging from 21 - 70 years.

- The symptomatology was essentially that of any infected
urinary tract obstruction with lumbar pain and fever. In 5 cases,
acute renal failure was the main symptom.

- Urinary infection was constant. It was mainly due to urea-
splitting organisms, namely Proteus, which was found in 15 cases.
In 7 cases an association of several gram-negative rods was found.

- The roentgenographic appearance of these matrix calculi was
generally a radiolucent, filling defect. However, nuclei of pre-
existing calcified lithiasis and some areas containing faintly
stippled calcifications may be identifiable on the plain film. On
the I.V.P. soft calculi often appeared as a radiolucent staghorn
mass. Sometimes they were found above a calcified stone of the
ureter.

In other cases, retrograde pyelography was required to demonstrate the radiolucent masses within the collecting system. In the
radiological differential diagnosis blood clots, uric acid calculi,
and sloughed papillae should be taken into consideration, rather
than papillary tumours.

Metabolic disturbances were uncommon. Global hypocalciuria
often occurred in patients with renal failure (4 cases) and no study
of a possible unilateral hypocalciuria (masked by the healthy opposite kidney) has been carried out. Nevertheless, some of our findings
seem to be rather at variance with Wickham's concepts.

1) In 22 cases, matrix calculi seemed to have grown around a
nucleus of phosphate or, more rarely, oxalate (2 cases) or uric
acid (1 case). One case of cystinic soft calculus has been reported
by Suhler and Fedida (11). In one of our cases without previous
lithiasis a soft radiolucent staghorn calculus appeared grey and
its crystalline part was mainly represented by oxalates.

2) In the softest parts of the matrix calculi, which appeared
as the newest ones, a mixture of polynuclear leucocytes, urothelial
fragments and a little fibrin could be found.

3) Renal biopsy performed in 13 cases showed not only interstitial and tubular lesions, but also, in some cases, the presence
of necrotic casts, containing polynuclear leucocytes, necrotic

cells and a matrix-like substance which could be found within the
tubular lumina. These findings correspond very closely to the
microscopical appearance of soft calculi. In our view, they may
represent an immature phase of soft stone formation, which is
slightly different from the usual "adult" matrix stones. From our
findings, a possible intra-renal origin of matrix calculi can be
suggested. It is obvious that, in the case of obstruction, urease-
splitting infection may involve the distal tubules, as well as the
collecting system.

EVOLUTION

From our study it appears that soft lithiasis is a serious
disease.

- In 5 cases, febrile acute renal failure was the first symptom;
in 3 cases calculi were bilateral, in 2 cases in solitary kidneys.

- Only 6 kidneys involved by matrix calculi reverted to a normal
functional and morphological urographic appearance. The other 19
cases showed either atrophy or at least pyelo-caliceal and parenchymal scars. Eight such cases can be considered as "malignant
lithiasis", according to the definition proposed by Camey at the
1980 congress of the European Urological Association.

SURGICAL TREATMENT

Complete removal of the calculi and effective treatment of the
urease-releasing infection are absolutely necessary in order to
avoid recurrence.

1) Before the operation, the presurgical investigations must
include:

- an accurate bacteriological investigation, including in vitro
sensitivity to antibiotics and the exact identification of the
location of the calculi not only in the intra-renal cavities but
also in the ureter, where their assessment is sometimes difficult
(retrograde uretero-pyelography may be essential);

- A search for abnormalities of the urinary tract (pyelo-
ureteral junction, uretero-vesical junction) is also vital.

2) During the surgical procedure, location by x-rays or by
needle cannot be used because of the radiolucency and soft consistency of the stones. The intrasinus approach, the visual exploration
of the calyces with our cold-light sucking device and pulsatile
flushing of the cavities can lead to a complete clearance of calculi from the kidney.

3) At the end of the procedure, a nephrostomy is performed,
so that irrigations can be employed postoperatively in order to
prevent recurrences. We used this technique in 1967 for the first
time, to treat a malignant soft lithiasis:

A 61 year old female, after an operation for incontinence
suffered from bilateral pyelonephritis, bilateral lithiasis and
had been operated on by right rephrectomy, and removal of a poorly
calcified stone in the left kidney. She was admitted one year later,
in 1967, with febrile acute renal failure (creatinine clearance =
5 ml/min). Retrograde pyelography showed a radiolucent mass in the
pelvi-calyceal system. Staghorn soft calculi were removed as an
emergency. Renal function improved, but in the post-operative days
a white soft substance,.showing the same structure as soft calculi
and containing fibrin, constantly came out from the nephrostomy
catheter. A recurrence of a soft staghorn calculus was observed two
months later. This calculus was removed and a nephrostomic flushing
system was put into the pelvi-calyceal system. Ten days later, a new
radiolucent mass appeared in the pelvis, and for the first time
perfusion of the cavities was performed using a fibrinolytic, mucolytic and antibiotic solution. Fifteen days later, the kidney was
clear. The creatinine clearance rose to 40 ml/min in the following
months and has since remained stable. The I.V.P. is satisfactory.

Since this observation, we perform retrograde pyelography
through the nephrostomy catheter on 10th day after pyelostomy for
soft calculi. If no radiolucent mass is found, the flushing system
is removed. If radiolucent bodies are present, a continuous local
perfusion is done, using 500 ml physiologic saline + 1 vial of
Varidase (20,000 U streptokinase + 5,000 U streptodornase) + antibiotics for 8 hours. Every 8 hours, 3 cc N-acetylcysteine is
injected in the cavities and the perfusion is stopped for half an

hour. This treatment is maintained until pyelography shows disappearance of any radiolucent masses. This may take from 10 to 40 days.

RESULTS

Between 1967 and 1975, we have operated upon 22 soft calculi. 20 patients were regularly followed and 2 were lost to follow-up. Three recurrences were observed, two being radio-opaque, and one radiolucent. In the last case this patient had an intubated cutaneous ureterostomy. Two patients developed chronic renal failure leading to haemodialysis, but in one case polycystic disease of the kidneys and in the other multiple sloughed papillae favoured this poor evolution.

REFERENCES

1. Allen, T.D. and Spence, A.M.: Matrix stones. J. Urol. 95: 284-290, (1966).
2. Boyce, W.H. and Garvey, F.K.: The amount and nature of the organic matrix soft urinary calculi: a review. J. Urol. 76: 213-227 (1956).
3. Boyce, W.H. and King, J.S.: Present concepts concerning the origin of matrix stones. Annals of the New York Academy of Science, 104: 563-578, (1963).
4. Boyce, W.H.: Ultrastructure of human renal calculi. In: Proceedings of the International symposium on renal stone research. Karger,Basel 247-255, (1972).
5. Brueziere J., Lasfargues, G. and Gallet, J.P.: Les coagulums uroproteiques chez l'enfant. Journal d'Urol. et de Néphrol. 74: 91-103, (1968).
6. Camey, M.and Le Duc, A.: Les calculs mous du rein (à propos de 6 observations). J. d'Urol. et de Néphrol. 74: 258-266, (1968).
7. Finlayson, B., Vermeulen, C.W. and Stewart, E.J.: Stone matrix and mucoprotein from urine. J. Urol. 86: 355-363, (1961).
8. Gage, H. and Beal, H.W.: Fibrinous calculi in the kidney. Ann. Surg. 48: 378-387, (1908).
9. Küss, R., Denis, M. and Dimopoulos C.: Les calculs mous de l'appareil urinaire chez l'adulte. J. d'Urol. et de Néphrol. 75: 1-13, (1969).
10. Malek, R.S. and Boyce, W.H.: Intranephronic calculi: its sig-

nificance and relationship to matrix in nephrolithiasis. J.
Urol. 42: 511-518, (1973).

11. Suhler, A. and Fedida, G.: Calcul mou pyélique de nature
     cystinique. Ann. Urol. 8: 163-167, (1974).

12. Wickham, S.E.A.: Matrix and the infective renal calculi. Brit.
     J. Urol. 47: 727-732, (1976).

13. Williams, D.I.: Matrix calculi. Brit. J. Urol. 35: 411-415, (1963).

STRUVITE STAGHORN LITHIASIS

A. Rousaud

Fundación Puigvert
c/ Cartagena 340-350
Barcelona - 13 (Spain)

ABSTRACT

Struvite staghorn calculi grow rapidly, easily recur and quickly impair renal function. In this report we consider the incidence and the clinical and radiologic aspects of this condition, based on our experience with one hundred patients. We also discuss the formation, evolution and medical and surgical treatment with particular reference to renal cooling, diagnostic procedures and the use of acetohydroxamic acid.

––––––––

Struvite stones show rapid growth, frequent recurrence and a tendency to lead to renal destruction. These factors make this type of lithiasis worthy of the most careful treatment.

Until recently no really effective treatment has been available. However our therapeutic results have recently improved, thanks to the development of new surgical techniques and, above all, to the introduction in the therapeutic field of the urease inhibitor acetohydroxamic acid, which offers a new hope for all patients having a struvite staghorn calculus.

CLINICAL, RADIOLOGIC AND ANALYTICAL CHARACTERISTICS OF THE STRUVITE
STAGHORN CALCULI

Normally, the struvite staghorn calculus is formed rapidly and
without clinical signs. Occasionally, some frequency of micturition
accompanied by pyuria may warn of its presence.

It is predominantly a disease of adulthood. In our statistics
the maximum incidence was found between 20 and 42 years of age. The
preponderance of the female sex and of the right kidney are evident.

The latter observations may be related to the frequency of
cystitis in the female and the presence of a pregnancy in many of
our cases.

The radiologic characteristics of struvite staghorn calculi
are very typical. The rapid growth of this type of lithiasis in the
presence of infection provokes the inclusion of a large volume of
organic tissue in its structure, which makes it barely opaque to
x-ray explorations.

The calculus is formed in the renal cavities, so there is no
need to look for papillary lithogenic foci. If there are deformities
of the renal papillae, they are due to the destructive action of the
ureolithic bacteria upon the renal epithelium.

LITHOGENIC FACTORS

The fundamental factor in the genesis of struvite staghorn
calculi is urinary infection due to urea-splitting bacteria, present
in 100% of our cases (83% identified as Proteus Mirabilis), which
cause alkalinity of the urine.

Urease, the enzyme produced by these bacteria, is responsible
for struvite stone formation. It is also called amidohydrolase and
easily hydrolyses C-N non peptidic chains of lineal amides such as
carbamide (urea). Urease has a nephrotoxic action, favouring the
intracellular infection of the renal epithelium and necrosis due to
alkali formation.

Mechanism of the Formation of Struvite Staghorn Calculi

The presence of stone-forming bacteria in the urine provokes
the massive production of urease. This enzyme splits urea into
carbonic anhydride and ammonia, in watery solution. The enormous
increase of bicarbonate in the urine is followed by a massive mo-
bilization of the buffer systems of the urine trying to counteract
this progressive alkalinization.

This buffer system is formed of urinary phosphate, which com-
bines with ammonia and with free magnesium in the urine, to form
magnesium ammonium phosphate (struvite),which rapidly precipitates
and grows.

Other stone-forming factors listed in table 1 may contribute
to their development.

Evolution of the Struvite Staghorn Calculus

Tubular acidosis, a recurrent urinary infection, or both combined, can provoke the conditions needed to develop a struvite calculus. Once the calculus is formed, the circle is closed; the infection is maintained and this favours stone growth. This process –
if not interrupted – causes severe lesions of the renal epithelium
and later involvement of the interstitial spaces which inevitably
leads to the loss of the kidney.

Classically this vicious circle is interrupted by conservative
surgery and antibiotic therapy.

After conservative surgery, three different circumstances may
impair the results, as observed in many of our cases:

1. Persistence of infection by urea-splitting organisms leading
to a rapid recurrence.

2. Persistence of a residual stone nucleus, due to the difficulty of detecting these poorly radio-opaque stones, is frequently
present in this type of surgery. In addition, the brittle consistency
of struvite often leads to multifragmentation in the course of stone
removal.

Table 1. Struvite staghorn calculi.

ORGANIC AND FUNCTIONAL LITHOGENIC FACTORS.

- CONGENITAL MALFORMATIONS: Horseshoe kidney, megaureter,
  hydronephrosis, etc.
- STENOSIS OF URETERO-PELVIC JUNCTION
- VESICO-RENAL REFLUX
- NEUROGENIC BLADDER
- URINARY DIVERSION: Ileal bladder, ureterosigmoidostomy,
  nephrostomy, etc.

METABOLIC LITHOGENIC FACTORS.

- They are usually of little importance in this type of
  lithiasis.
- In rare cases: associated hypercalciuria.

The residual stone grows rapidly and closes the circle again
leading to renal destruction.

3. Recurrences can occur despite sterilization of the urine if
the tubular acidosis is not corrected.

In our experience (see table 2), only 39% of the patients who
underwent surgical intervention had no recurrences.

Following the use of a medical and surgical treatment, the
percentage of success was raised to 57%. This figure is expected to
improve further when we obtain the first results of patients treated
with the urease inhibitor acetohydroxamic acid.

COMBINED THERAPEUTIC PROTOCOL OF STRUVITE STAGHORN CALCULUS

This protocol is based on two essential points:

a) Surgical intervention.
b) Later medical treatment.

Table 2.   Results of treatment.

A. SURGERY: 36 cases

    Recurrent stone formation   48%
    Residual lithiasis   13%
    Good results   39%

B. SURGICAL AND MEDICAL TREATMENT: 64 cases

    Recurrent stone formation   31%
    Renal functional impairment
    due to residual stones   12%
    Good results   57%

The Surgical Intervention

There is insufficient time for a description of the different types of surgical approach used in our series for the treatment of struvite staghorn calculi.

However, a few observations related to this topic deserve to be emphasized.

The surgical treatment of struvite staghorn calculi presents some difficulties independently of their size. Usually, a staghorn calculus involves all the upper collecting system.

Its development provokes serious lesions in the epithelium and in the renal parenchyma.

Its soft consistency makes its removal difficult, with the inherent risk of leaving a residual stone.

As a result, we must employ a surgical technique with a wide exposure, which offers the possibility of thoroughly exploring all the calyces and with the least possible renal damage.

These circumstances justify three basic points, included in

Table 3. The use of radio-isotopes in staghorn lithiasis.

PREOPERATIVE EVALUATION

It will influence the surgeon's decision towards:

a) Abstention from surgery.
b) Nephrectomy.
c) Conservative surgery.

POSTOPERATIVE EVALUATION

It will allow a comparison with the preoperative condition, indicating improvement, impairment or stabilization of renal function.

METHODS EMPLOYED

Computer-assisted gamma-camera.

a) Scintigraphy: evaluation of renal function in the whole kidney and in its various parts.

b) Sequential scintigraphy and renograms.
   Will evaluate the degree of obstruction and renal functional involvement.

c) Britton's index (Renal Plasma Flow) can be obtained.
   If above 20%, conservative renal surgery is advised.

---

our surgical protocol:

A. Pre and post operative isotopic study. Since the adaptation of the gamma-camera to the computer, this exploration gives extremely exact data on the functional condition of the renal parenchyma. The study of renal function can be performed on the whole organ or, separately, on parts of it (see table 3). A computerized figure is obtained, which indicates the relative renal plasma flow (ERPF). It is known as Britton's index and is given as a percentage of normal values.

We think that conservative renal surgery is advisable only

if Britton's index is above 20%.

The preoperative study is very useful for the evaluation of
the zones of the parenchyma which have good function and which
must be preserved in the course of the surgical operation.

B. Intraoperative renal refrigeration.
In some cases the size of the staghorn calculus is such that a rather
complex surgical approach is needed. In 8% of our cases, we have
used bivalve nephrotomy, not involving the poles (Boyce), to obtain
a satisfactory surgical result.

It is necessary to perform renal refrigeration to preserve the
parenchyma from possible ischemic damage. We have used transcavitary
renal hypothermia (Rousaud) occasionally combined with contact
hypothermia (see fig. 1) obtaining good results in all instances
as confirmed in the post-operative isotopic control. Intrapelvic
refrigeration is performed as follows: a preliminary clamping of the
lumbar ureter, followed by air injection into the pelvic cavities leads
to a better detachment of the stone from the calyceal walls. A first
intraoperative film is taken. After arterial clamping, an intravenous plastic catheter, connected to a refrigerated ringer lactate
solution, is inserted into the upper calyx. The use of an indwelling
bladder catheter is advocated throughout the perfusion.

C. Postoperative renal irrigation. Litholysis.
In all operated
patients, we have placed a small nephrostomy catheter. Through this
catheter, a continuous perfusion has been performed after the 5th
day (when the suture in the pelvis is consolidated) with an acidic
solution, as shown in table 4.

D. POSTOPERATIVE MEDICAL TREATMENT.
After the removal of the
staghorn calculus and after the immediate post-operative period,
the patient is transferred to the metabolic unit under medical
management to achieve three main aims:

1. Treatment of persistent infection. In most of our treated
cases the causal infection has not been completely eradicated after
the surgical intervention.

Antibiotic treatment, guided by urine culture and sensitivity
studies, cultures from the calculus and serum antibody titration,

RINGER LACTATO

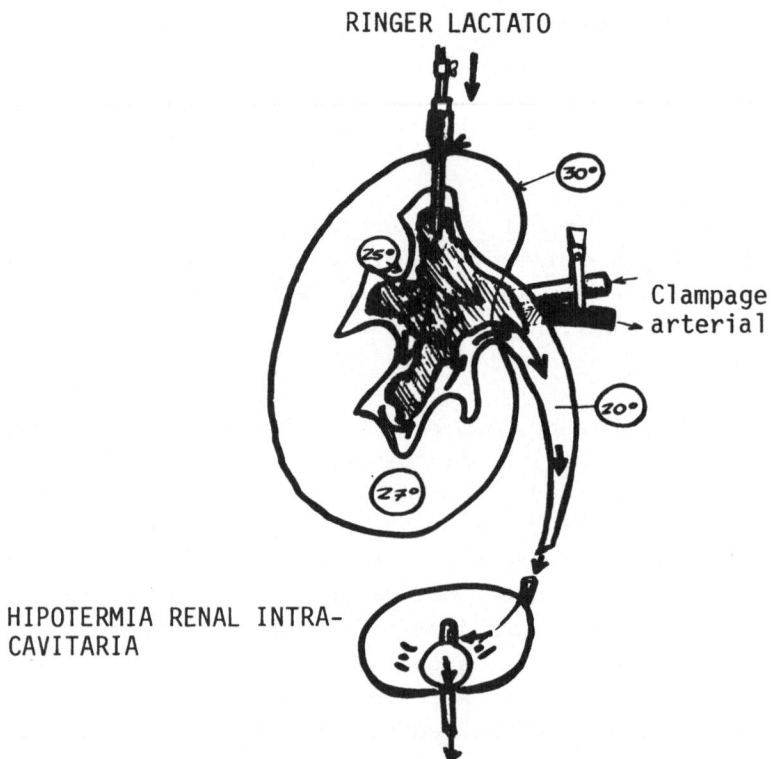

Fig. 1.   Transcavitary renal hypothermia.

is instituted. Follow-up, on out-patient basis, was carried out
every 15 days for at least three months until permanent sterilization
of the urine had been obtained.

This was achieved in 16% of our patients. The other patients
have maintained a urinary alkalinity that has facilitated reinfec-
tion. This explains 31% of the recurrences.

2. Treatment of the concomitant metabolic abnormalities. A
metabolic study is undertaken in all patients who have been operated
upon.

The trouble most commonly seen is tubular acidosis. This can
usually be corrected with small doses of sodium bicarbonate.

Hypercalciuria of renal type has very seldom been detected,

Table 4.   Composition of the various solutions employed for
           litholysis.

A. LITOLYS SCHARPER   (Dormia's solution)

   Na EDTA (Sodium ethylen-diamino-tetraacetate).
   Na NTA (Sodium nitryl-triacetate).
   In 450 ml bottles.

B. D.R. SOLUTION

   20% Chlorhexidine gluconate   5 ml
   N-acetyl-cysteine   900 mg
   Acetic acid   2 ml
   1,8% Methylene blue   2-3 drops
   Sterile distilled water to total volume of   1000 ml

   In case of intolerance to n-acetyl-cysteine, streptokinase
   (400.000 u) and streptodornase (100.000) are employed instead.

C. F.R. SOLUTION

   α - chymotrypsine   100 mg
      Normal saline   1000 mg

---

when found the patient has been treated with oral hydrochlorothiazide.

   The treatment of the metabolic abnormalities is absolutely
essential if the urine is to be sterilized.

   3. Treatment with the urease inhibitor, acetohydroxamic acid.
This substance is the basic drug for the treatment of this type of
lithiasis.

   Acetohydroxamic acid has a strong structural similarity to
carbamide (urea).

$$H_2N-C = O-NH_2 \qquad\qquad H_3C-C = NOH_2$$

UREA   ACETOHYDROXAMIC ACID

Table 5.   Protocol of treatment with acetohydroxamic acid.

There is only one fundamental indication: struvite lithiasis associated with urea-splitting organism infection.

1. SELECTION of patients entered in this study depends on the above mentioned indication. A complete metabolic study is also requested.

2. TREATMENT is given only after the clearest explanation of the aims, duration and side effects including:

   a. The aim of the treatment:

      - definitive sterilization of urine;
      - urinary acidification;
      - prevention of lithiasis recurrence;

   b. Duration of treatment: 10 - 15 weeks.

   c. Dosage: one 125 mg capsule twice daily.

   d. Possible side effects: headache, gastric troubles, haemolytic anaemia, phlebitis.

   e. Daily control of urinary pH with Merck paper strips (Neutralit n. 9564).

3. CONTROLS are performed at weekly intervals:

   a. Complete blood count.

   b. Urinary ammonia and titratable acidity.

   c. Urinary sediment, culture and pH.

   A good result from the treatment will be manifested by reduction in urinary ammonia and pH, as well as by progressive sterilization.

It inhibits the action of urease by a competitive mechanism with carbamide.

The acid is hydrolyzed into substances easily excreted by the urine. A progressive lowering of pH and urinary ammonia excretion is obtained. All this, added to its synergic action with some antibiotics (trimethoprim - sulfamethoxazole), makes it a useful drug in the struggle against struvite stone recurrences.

This substance is administered in capsules of 250 mg synthesized with 99% of purity. It is absorbed through the alimentary tract and more than 50% is excreted in the urine. Four side-effects have been detected: reversible haemolytic anemia, gastric troubles, headaches in the first weeks of treatment and phlebitis of lower extremities.

In preparing our protocol we have followed the guidelines given by Griffith et alii in their outstanding work on this subject (see table 5).

The results of the combined and medical treatment have been really encouraging. They are summarized in table 2.

REFERENCES

1. Griffith, D.P., Gibson, J.R., Clinton, C. and Musher, D.M.:
    Acetohydroxamic acid: clinical studies of a urease inhibitor
    in patients with staghorn renal calculi. J. Urol. 119: 9,
    (1978).

INDICATIONS FOR SURGERY AND SURGICAL TECHNIQUES IN STAGHORN CALCULI

M. Camey and A. Le Duc

Service d'Urologie
Centre Médico-Chirurgical Foch
92151 Suresnes - Paris (France)

ABSTRACT

The indications for surgical intervention in staghorn calculi
have changed considerably during the past 25 years. Conservative
surgery, formerly rare, is now practically the rule. There is usually
no reason to postpone the procedure once the lithiasis is discovered.
Sometimes, the stone has to be removed as an emergency, in the
presence of acute infection or renal failure. In the majority of
cases a thorough preoperative check-up is required, including (in
addition to intravenous pyelogram) lateral x-ray of the kidney,
tomography, isotopic assessment of separated renal function and
possibly arteriography.
    Two main types of surgical procedures, the sinus and the
transparenchimal approaches, are usually carried out.
    The choice of the procedure depends upon various factors such
as the extent of the staghorn calculi, the intrarenal situation of
the pelvis, the thickness of the parenchyma, possible previous
intervention, renal insufficiency and also on the preference of the
surgeon and his team for a particular technique with which they
are more familiar.

    Partial nephrectomy may be resorted to, in order to clear the kidney of the stones or to remove lithogenic areas. When staghorn calculi are responsible for severe renal failure, they must be removed in the most atraumatic way (intrasinus approach and multiple small nephrotomies) avoiding clamping of the renal pedicle.

———————

Seven main questions must be answered when dealing with the indications for surgery in cases of staghorn calculi:

1 — Is conservative surgical treatment necessary?

2 — When is the procedure to be carried out?

3 — What kind of assessment has to be made before the operation?

4 — What kind of procedure must be chosen to fit different cases?

5 — Can partial nephrectomy be useful in certain cases?

6 — How can one solve the problem of associated stenosis of the pyelo-uerteral junction?

7 — How can one deal with staghorn calculi complicated by renal failure?

I — The indications for surgery for staghorn calculi have changed considerably during the past 25 years. Formerly conservative surgery was considered dangerous for the kidney and was very often followed by stone recurrence, whereas staghorn calculi were believed to be well tolerated by the patient.

Since that time, it has been shown (6) that staghorn lithiasis, sometimes apparently well tolerated by the patient, practically always resulted in kidney failure. When the stones are bilateral or in solitary kidneys this process may be very slow and take 20 years or even more. The process may also be acute, with lumbar pain, high fever, possibly septicemia, and acute renal failure; the stone has then to be removed as an emergency (and sometimes, the kidney itself must be removed, if it has been destroyed by infection).

Conservative surgery is nowadays the rule, except when the stone
has already resulted in pyonephrosis as, owing to improvements in
surgical techniques and antibiotic treatment, most kidneys can be
cleared of calculi, and infection and recurrence can be avoided.

II - There is usually no reason to delay once the diagnosis is
made except perhaps for poorly calcified stones when the surgical
procedure has to be carried out mainly by nephrotomies. Here it may
be preferable to wait for some months in order to operate upon more
calcified and harder stones.

III - In addition to a satisfactory intravenous pyelogram which
must show the thickness of the parenchyma (by tomography if necessary) other investigations can be useful. Plain lateral and oblique
x-rays of the kidney can show the location of the calyceal stones
in front of, or behind, the middle plane of the kidney. Tomography
can also be used for this purpose. Nevertheless, because of the
obliquity of the kidney, upper calyceal stones often appear posterior whereas the lower ones appear anterior, even when they are all
in the same middle plane of the kidney. Arteriography may be useful
in case of recurrent staghorn lithiasis or when an upper pole partial
nephrectomy is contemplated. Isotopic evaluation of separated renal
function, (using D.M.S.A. as a tracer) may help to decide between
conservative surgery or nephrectomy. It is the best way to assess
long term results of surgical treatment as far as renal function
is concerned. Angioscintigraphy with Tc gluconate shows the vascular
supply of the kidneys. Poor vascularisation is usually a sign of
severely and permanently reduced parenchyma. Accurate urinary bacteriological control is necessary to achieve post-operative sterilisation
by adequate antibiotic therapy. It is our practice to begin this
treatment 2 days before the surgical procedure in order to avoid
peroperative soread of the infection (systemic and local) as much
as possible.

IV - The aim of every surgical procedure for staghorn lithiasis
is complete removal of the calculi in order to achieve eventual
sterilization of the urine and to avoid recurrence, while preserving
as much as possible of the kidney function. Actually, clearing the
kidney of a large and complex staghorn calculus always involves a
certain risk for the renal parenchyma. The choice of the procedure
is therefore very important. The two main types of surgical procedures are the sinus and the parenchymal approaches.

A - the sinus approach (Gil Vernet) with or without pyelocalyceal endoscopy, flushing, coagulum or small nephrotomies. This
requires, at most, only brief clamping of the renal pedicle.

We use a lumbar incision leading as directly as possible to
the renal hilum and sinus in order to provide the surgeon with a
direct axial view into the calyces. The posterior vertical lumbar
approach is rather narrow, and we have been using a postero-external
lumbar incision (4) for the last 15 years with complete removal of
the 12th rib and section of the lower part of the diaphragm after
dissection of the pleural sinus. The patient in the lumbar position
leans 60° ventrally on the operating table. This incision usually
provides easy nephroscopy, control of the renal pedicle and performance of x-rays.

Small fragments can be glued into a pyelocalyceal coagulum made
of a mixture of thrombocyte-enriched plasma + human fibrinogen and
thrombin + calcium chloride, and removed together with the coagulum.

The calyces can be rinsed with saline using a forceful, pulsatile flushing device (such as the one used for teeth cleaning).

12 years ago we designed (5) a cold-light (fiberoptic) sucker
with a flushing system, in order to remove calyceal stones under
visual control. New flexible endoscopes are being tried in Germany
and will perhaps improve the endoscopic possibilities.

Nevertheless, it may be impossible to remove some mushroom-
headed projections of the staghorn calculi through a sinus approach
without splitting the calyceal neck, with a risk of massive bleeding. These fragments may be removed by small nephrotomies on the
top of the calyces without damaging the interlobar arteries. Clamping
the renal artery is usually either unecessary or brief or may involve
only one of its branches.

B - the transparechymal approach (extensive longitudinal
bivalve nephrotomy, intrasegmental approach, small paravascular
nephrotomies),which requires clamping of the renal pedicle and
usually some form of protection of the renal parenchyma against
ischemia by cooling or by inosine administration.

This approach is usually far more rapid than a long intrasinus dissection. Three main methods are available.

1 - Bivalve nephrotomy which endangers the vascularisation of
a significant part of the kidney.

2 - Intersegmental approach (Boyce and Elkins) (2) which allows
a very adequate access to the pelvic portion of the kidney but is
likely to tear the large intrarenal veins although it avoids the
segmental arterial supply. Even when the pelvis is exposed, peripheral calculi must be extracted through the calyceal neck with a
risk of bleeding.

3 - Paravascular nephrotomies (Wickham) (11) allow the extraction of the major portion of the calculi through a pyelotomy and
the removal of peripheral portions of the stone through multiple
radially-sited small nephrotomies avoiding the peripheral intrarenal
vessels, both arteries and veins, by gentle dissective displacement.

All these transparenchymal approaches require clamping of the
renal artery and some form of protection of the renal parenchyma
against ischemia. Inosine can be used locally or by systemic perfusion (9, 10) but generally affords no more than 60 minutes of
ischemia without risk of damage. Cooling the kidney is the most
commonly used means of protection. It can be achieved by external
methods with special devices including circulation of cold fluids
or by crushed ice. It can also be achieved by perfusing the kidney
with cold solutions (Collins, Ringer lactate) after clamping and
incising the renal artery and vein. The combination of intraluminal
balloon occlusion of the renal artery and hypothermic perfusion
(Marberger) (7) obviates the need for dissection of the renal artery,
specially on multiply operated kidneys. Nevertheless, all these
methods may be complicated by thrombosis of the renal artery which
may be immediate or delayed. We have met such a complications after
puncture of the artery for cooling.

In order to reduce vascular damage by nephrotomies, avascular
lines can be sought by clamping branches of the renal artery under
thermographic control.

The choice of the procedure depends on several factors:

1 - The size, shape and location of the renal pelvis and

calyces and consequently of the staghorn calculus when it completely
fills the cavities. An intrarenal pelvis associated with a very
narrow hilum and massive stone is best approached by longitudinal
nephrotomy (intrasegmental if possible), whilst staghorn calculi
with multiple mushroom-headed projections are best removed by multiple radiating nephrotomies. A wide extrarenal pelvis with wide
major calyces without mushroom-headed stones allows a sinus approach.

2 - the state of renal function: it has been proved by experimental studies on dogs (6, 10) and by isotopic assessment on
man (1) that the degree of functional depression following the
procedure is minimal with the sinus approach, intermediate with
the paravascular intersegmental one, and maximal following the
bivalve procedure.

3 - Fibrosis of the pedicle, hilum and sinus due to multiple
procedures for recurrent lithiasis may contraindicate the sinus
approach.

The preference of the surgeon and his team are also important.
Nevertheless, we think that the urologist and his team must be
able to carry out any procedure and must have at their disposal
all the necessary technical devices.

In our department, except when operating on central or multiple
thick mushroom-headed stones, we generally prefer the sinus approach
and remove most of the calculi under visual control and flushing.
The remaining stones are located by x-rays and according to their
size, number and location, and to the thickness of the parenchyma,
are removed by nephrotomies with or without clamping of the renal
artery, and with or without cooling.

Occasionally a fragment located in a small minor calyx per-
pendicular to a major calyx in thick parenchyma cannot be reached
either through the sinus or by a short nephrotomy. This situation
seldom warrants an extensive nephrotomy with its vascular risk,
unless it concerns a recurrence of stones remaining from previous
procedures. It must be remembered that, despite complete removal,
the recurrence rate is generally above 20% and mainly depends on
persistent urinary infection with urea-splitting organisms. If such
organisms persist, acetohydroxamic acid may be indicated (see chapter
by Martelli and coworkers).

Ex situ surgery with autotransplantation of the kidney may exceptionally be carried out for staghorn calculi in solitary kidneys when the stone is particularly complicated or associated with other kidney disease (e.g. tumour, aneurysm, extensive stenosis of the ureter).

V - Partial nephrectomy may be useful in some cases: for clearing the kidney of fragments remaining in a pole which cannot be reached either by a sinus approach or by nephrotomy, and for removing a pathologic area (e.g. localized lithogenic area, distended or infected wide calyx, tuberculosis). In these particular cases the polar resection may complete the sinus approach.

VI - The pyelo-ureteral junction sometimes appears to be stenosed below some staghorn calculi, usually because of local oedema and paradoxical dilatation of the lumbar ureter. Nevertheless, during the procedure a fibrous stenosis may be discovered and, if untreated, it may induce stasis, infection and later recurrence. On the other hand, surgical repair by resection and anastomosis involves a risk of leakage of urine because of the poor quality of the tissues. Uretero-calyceal anastomosis, and/or plasty with peritoneum and omentum flaps may possibly obviate this complication.

VII - When staghorn lithiasis has induced severe renal failure (3) with a creatinine clearance below 40 ml/mn, the aim of the procedure is to retrieve as much as possible of the renal function in order to postpone haemodialysis for as long as possible. Any procedure likely to decrease the blood supply of the remaining nephrons, e.g. clamping of renal artery and extensive nephrotomies, must be avoided. A sinus approach with small nephrotomies is the best procedures as the parenchyma is usually thin and the bleeding minimal. Drainage by nephrostomy is compulsory in such cases. Induced polyuria by furosemide is very useful in order to maintain both the permeability of the nephrostomy catheter and the maximum renal function in the post-operative period.

REFERENCES

1. Beurton, D., Gonties, G., Malloun, A., Pascal, B. and Cukier, J.: Lithiase rénale complexe, évaluation chiffrée de la fonction rénale à distance de l'intervention, après néphrotomie et clampage. 18th Congress of International Society of Urology, Paris, June 1979.

2. Boyce, W.H. and Elkins, I.B.: Reconstructive renal surgery
    following anatrophic nephrolithotomy, follow-up of 100 consec-
    utive cases. J. Urol. 111: 307-312, (1974).

3. Camey, M.: L'insuffisance rénale en Urologie. Report to the
    65th Congress of the French Urological Association, Masson,
    Paris, 1971.

4. Camey, M.: La voie d'abord lombaire postero-externe du rein.
    J. Urol. Néphrol. 80: 920-924, (1974).

5. Camey, M.: L'explorateur rénal à lumière froide. J. Urol.
    Néphrol. 80: 918-920, (1974).

6. Iguchi, H., Matsura, T., Akiyama, T. and Kwita, T.: Experimen-
    tal studies on renal circulation and function after the nephro-
    tomy. 18th Congress of International Society of Urology, Paris,
    June 1979.

7. Marberger, M. Guenther, R., Alken, P., Rumpf, W. and Ranc, M.:
    Inosine alternative or adjunct to regional hypothermia in the
    prevention of post ischemic renal failure. Europ. Urol. 6:
    95-102, (1980).

8. Petkovic, S. and Ostojie, B.: Le destin du rein et du malade
    avec le calcul coralliforme du rein. 18th Congress of Interna-
    tional Society of Urology, Paris, June, 1979.

9. Ventura, M., Acampora, F., Cagna, G., Genesi, D. and Guarda,
    F.: Clinical and experimental experience of hypothermia and
    inosine infusion versus ischemic renal surgery. 18th Congress
    of International Society of Urology, Paris, June 1979

10. Wickham, S.E.A., Fernando, A.R., Mendry, N.F., Watkinson, L.E.
    and Whitfield, H.N.: Inosine:clinical results of ischemic renal
    surgery. J. Urol. 50: 465-468, (1978).

11. Wickham, S.E.A.: 250 cases of nephrolithotomy under hypothermia.
    18th Congress of International Society of Urology, Paris, June
    1979.

UREASE INHIBITOR THERAPY IN INFECTED RENAL STONES

A. Martelli, P. Buli and V. Cortecchia

Department of Urology
University of Bologna
Bologna (Italy)

ABSTRACT

Struvite (ammonia and magnesium triplophosphate) and carbonate-apatite lithiasis is the effect of urinary infection due to urease-producing bacteria, such as Proteus, Providencia and Klebsiella. The splitting of urea induced by bacterial urease brings about the formation of NH_3, the increase of bicarbonates and alkalinization of the urine, which leads in turn to supersaturation with regard to struvite and carbonate-apatite. It can be stated that formation of such calculi can only occur in the presence of ureolysis.

Treatment of these types of lithiasis, which were almost constantly prone to recurrence until recently, consists of: a) surgical removal of all stones; b) antibiotic therapy; c) administration of urease-inhibitors.

Acetohydroxamic acid (AHA) and hydroxyurea (UH) are the only clinically available drugs that can block urease activity.

In a group of 41 patients with infectious lithiasis a comparison between the two drugs was performed and the results accurately recorded.

AHA, at the daily dose of 500 mg is, in our view, the drug of choice. Its side-effects are less than following HU and its activity is greater, as far as normalization of ammoniuria and synergism or even potentiation of antibiotic therapy is concerned. The latter finding can be explained on the ground that bacterial urease and urinary ammonia formation represent important factors interfering with the sensitivity of urease producing bacteria to antibiotics.

———

Struvite and carbonate-apatite lithiasis (4), though posing no difficult aetiopathogenetic problems of interpretation is, due to its urease-dependent nature (7, 8, 11), the most difficult to treat from the preventive-medical viewpoint since there is:

1) marked antibiotic resistance on the part of the urease-producing bacteria (2);

2) limited experience, as yet, with urease-inhibitor drugs;

3) difficulty in acidification of the urine in many patients and the possibility of undesirable side effects;

4) possible co-existence of other metabolic disorders.

The aim of this paper is to study the anti-urease effects of two drugs, hydroxyurea and acetohydroxamic acid evaluating their true effectiveness both alone (3, 5, 6, 9, 16), and combined with antibiotic therapy (1).

MATERIAL & METHOD

A study was carried out on 41 patients (16 men and 25 women) between the ages of 16 and 58 who were carriers of either unilateral or bilateral struvite and carbonate-apatite stones.

One had only one kidney, and six presented recurrent stones.

All patients were treated with urease inhibitors after surgical removal of their stones.

All urine cultures were positive for urease-producing bacteria (table 1). Bacterial counts were greater than 10^5 bacteria/ml.

Table 1.   Infective stones. Number of cases treated with
           "urease inhibitors": 41 patients.

SEX:	16 males	35 patients
	25 females	TREATED AFTER SURGERY
AGE RANGE:	Males 26-58 yrs.	6 patients
	females 16-50 yrs.	STILL HAVING RENAL STONES

MICRORGANISM:		
	Proteus	65 %
	Providencia	20 %
	Staphylococcus	10 %
	Others	5%

(table 2) in 69% of cases.

Ammonia and urinary pH, evaluated on fresh urine samples with
ORION specific ion electrodes, were constantly above 140 mM/1 and
7.5 respectively (figure 1).

Those patients who, after a first cycle of three months of

Table 2.   Infective stones. Initial urine culture (41 cases).

10^5 col/ml	69%
$10^4 - 10^5$ col/ml	20%
$10^3 - 10^4$ col/ml	11%
CULTURE OF STONES	100%

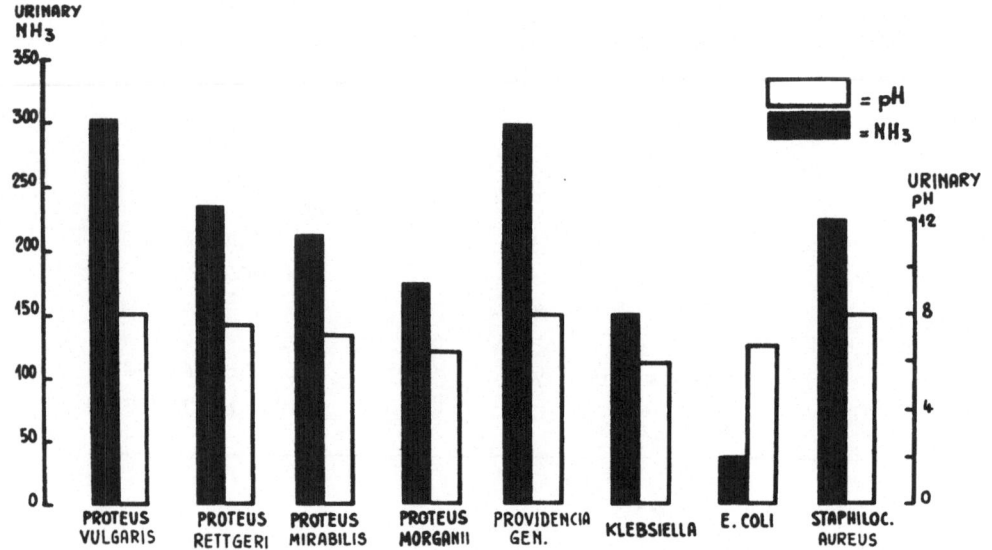

Fig. 1.   Urease splitting activity of some microrganisms by
          urinary NH3 and pH variations.

antibiotic treatment showed persistent urinary infection, were
subsequently treated with AHA or HU.

In such patients with persistent urinary infection AHA was
administered to 22 in doses of 500 mg/day; HU in 5 cases at 500
mg/day and in 6 at 1 g/day. The total number of treated patients
evaluable for side effects was 41, 28 for AHA and 13 for HU.

The control urine cultures, with specific bacterial counts,
were carried out monthly, the NH3 and urinary pH every two weeks.
The antibiotics advised on the basis of the antibiogram were Gentamicin, Tobramycin, Amikacin, Methenamine, Nitrofurantoin and
Cephalosporins. As general control parameters, both before and
during treatment, all patients were investigated and the following
tests were carried out: tests of renal and of liver function, serum
cholesterol, haematocrit, serum iron and red and white cell counts.

The treatment with AHA and HU lasted from a minimum of 6 to a
maximum of 14 months.

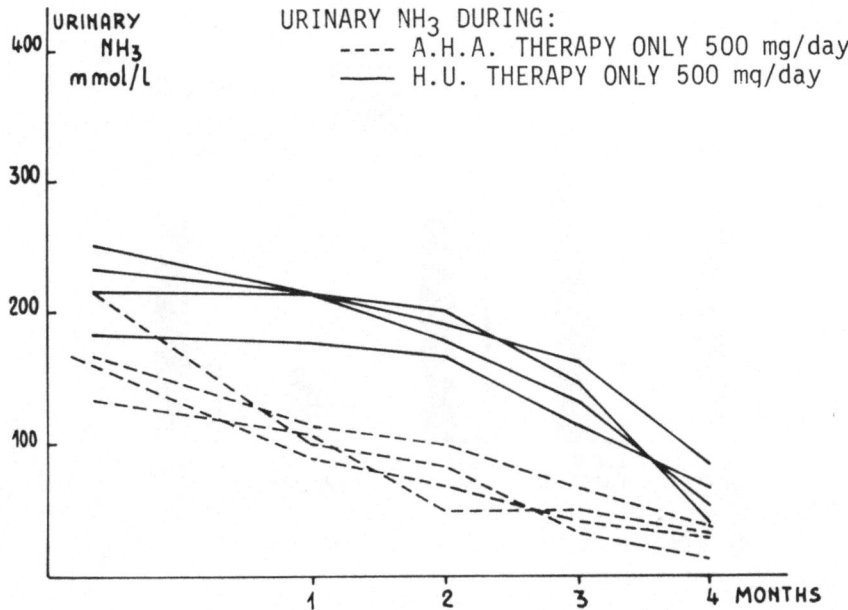

Fig. 2. Urinary ammonia excretion during treatment with
acetohydroxamic acid and hydroxyurea.

RESULTS

Of the 41 patients studied, persistent urinary infection from
urease-producing bacteria was found in 33 after 3 months of antibiotic treatment alone. Only 5 patients in the AHA-treated group
still showed positive urease urinary infection after 3 months of
treatment. In all these 5 patients the presence of stones was demonstrated, representing either an incomplete surgical removal or a
true stone recurrence at the beginning of treatment. Of the patients
receiving 500 mg/day HU, only 1 had sterile urine after 3 months
while, in the group receiving 1 g/day HU, all the patients treated
had sterile urine after four months' therapy.

The urinary ammonia and pH values in all the patients treated
with either AHA or HU, returned to normal. In the case of the latter,
however, the normalization times of these parameters coincided with
those for AHA only with the 1 g/day dose; with the 500 mg/day dose
the period was considerably longer and the results were not as good
(Fig. 2).

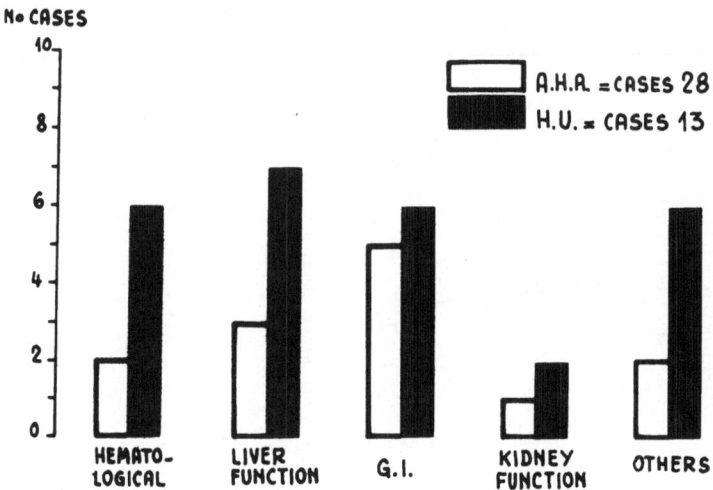

Fig. 3. Side effects on therapy with AHA and HU.

Moreover, AHA appeared to potentiate the activity of some antibiotics, such as Gentamicin, Methenamine and Nitrofurantoin, used
in the treatment of urinary infection.

No stone recurrence was met with in any patient during treatment.

The side effects appearing during administration of the two
drugs were clearly greater with HU and mainly consisted of gastrointestinal disturbances and/or haemolytic anaemia. In all cases they
regressed completely with reduction or suspension of the treatment
(figure 3).

CONCLUSIONS

The strict dependence of struvite and carbonate-apatite
lithiasis on the urease-producing bacteria (14-15), as well as the
results obtained on the incidence of recurrences (figure 4), and
on the treatment of urinary infection, confirm the need to combine
bacterial urease inhibitor drugs with antibiotic treatment (13).

Furthermore, it is clear (figure 5) that to eradicate the

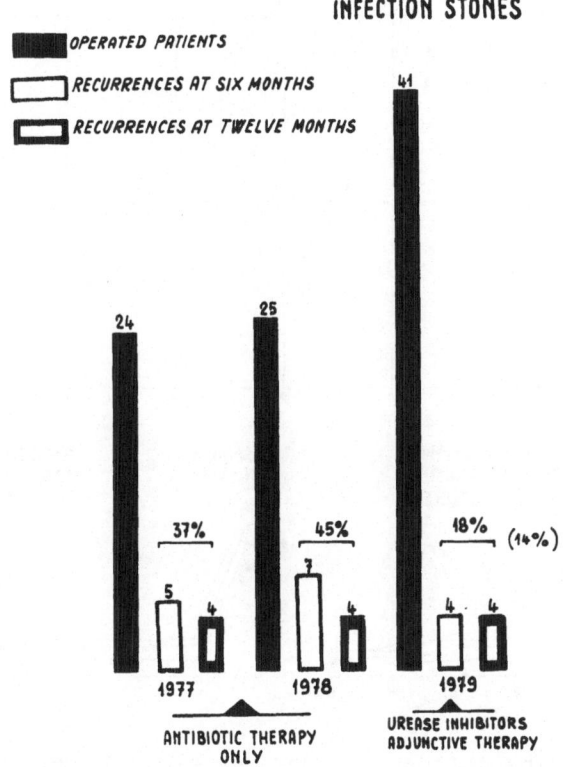

Fig. 4.   Incidence of recurrences with the association of
          urease inhibitor drugs and antibiotic therapy.

urinary infection, complete removal of the calculi is necessary;
otherwise their presence is a continuous and persistent source of
infection.

Of the two antiurease drugs used, the better results were
obtained with AHA, both for its minimal side effects and for its
more powerful urease inhibitor capacity.

Stone recurrence was avoided even in patients with urinary
infection not responding to specific antibiotic treatment.

Fig. 5 shows how the urinary ammonia values return to normal
during AHA treatment though a urinary infection with a bacterial
count over 10^5/ml persists.

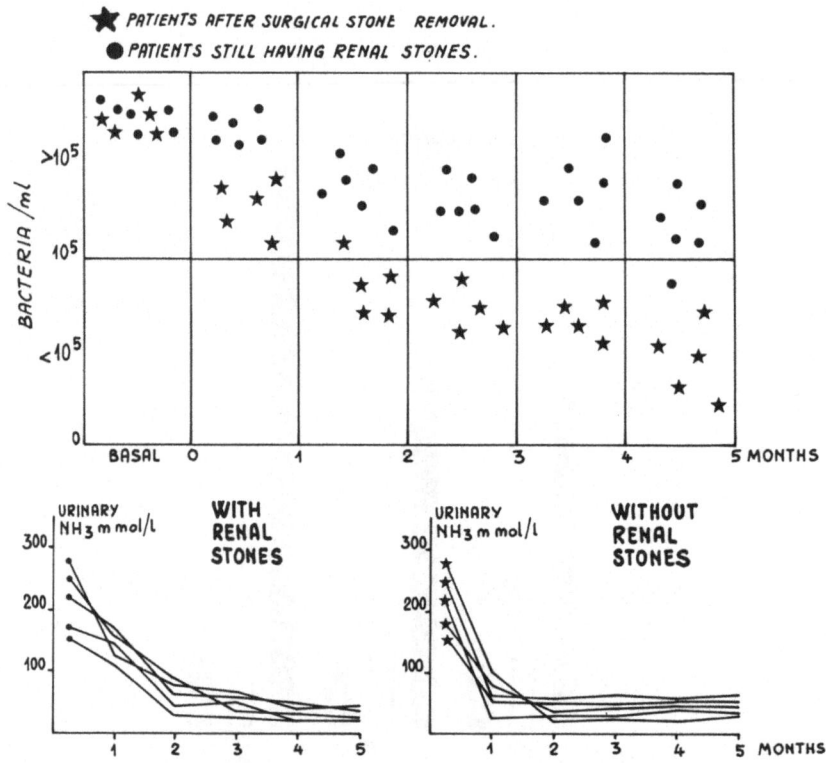

Fig. 5. Urine culture and urinary ammonia excretion in pa-
        tients treated with the association of acetohydrox-
        amic acid and antibiotic therapy.

It also appears that AHA strengthens the effects of the antibiotic on the bacteria through its antiurease action (12).

REFERENCES

1. Cooper, J.W., Jr.: Urinary tract infections: II. Optimal
    drug therapy. Hospital formulary 175-182, University of Rhode
    Island College of Pharmacy. Kingston, 1977.
2. Gale, G.R.: Urease activity and antibiotic sensitivity of
    bacteria. J. Bact. 91: 499, (1966).

3. Gale, G.R. and Atkins, L.M.: Inhibition of urease by hydroxamic acids. Arch. Int. Pharmacodyn. 180: 289, (1969).

4. Griffith, D.P.: Struvite stones. Kidney Int. 13: 372, (1978).

5. Griffith, D.P., Gibson, J.R., Clinton, C. and Musher, D.M.: Acetohydroxamic acid: clinical studies of a urease inhibitor in patients with staghorn renal calculi. J. Urol. 119: 9, (1978).

6. Griffith, D.P. and Musher, D.M.: Prevention of infected urinary stones by urease inhibition. Invest. Urol. 11: 228, (1973).

7. Griffith, D.P., Musher, D.M. and Itin, C.: Urease: the primary cause of infection-induced urinary stones. Invest. Urol. 13: 346, (1976).

8. Hamilton-Miller, J.M.T. and Gargan, R.A.: Rapid screening for urease inhibitors. Invest. Urol. 16: 327, (1979).

9. Kumaki, K., Tomioka, S., Kobashi, K. and Hase, J.: Structure-activity correlation between hydroxamic acids and their inhibitory powers on urease activity. I. A quantitative approach to the effect of hydrophobic character of acyl residue. Chem. Pharm. Bull. 20: 8, (1972).

10. MacLaren, D.M.: The influence of acetohydroxamic acid on experimental Proteus pyelonephritis. Invest. Urol. 12: 146, (1974).

11. Musher, D.M., Griffith, D.P., Yawn, D. and Rosen, R.D.: The role of urease in pyelonephritis due to Proteus urinary infection. J. Infect. Dis. 131: 171, (1975).

12. Musher, D.M., Griffith, D.P., Tyler, M. and Woelfel, A.: Potentiation of the antibacterial effect of methenamine by acetohydroxamic acid. Antimicrob. Agents Chemother. 4: 101, (1974).

13. Musher, D.M., Saenz, C. and Griffith, D.P.: Interaction between acetohydroxamic acid and twelve antibiotics against fourteen gram-negative bacterial pathogens. Antimicrob. Agents Chemother. 5: 106, (1974).

14. Smith, L.H.: Renal lithiasis associated with infection. Mod. Treat. 4: 505, (1967).

15. Smith, L.H.: Hydroxyurea and infected stones. Urology 11: 274, (1978).

PROPHYLAXIS AND METAPHYLAXIS OF UROLITHIASIS

W. Lutzeyer, F. Hering and J. Hannappel

Department of Urology
Medical Faculty
Rheinisch-Westfälische Technische Hochschule (RWTH)
Aachen (W. Germany)

INTRODUCTION.

The incidence of urolithiasis in adults ranges from two to three per cent, comparable with the occurrence of diabetes.

In the prevention of urolithiasis, analysis of the urinary concrement by infrared-spectroscopy or x-ray diffraction and knowledge of the patient's metabolic abnormalities are essential.

Different incidences of metabolic disorders or urinary tract infection (UTI), were found in various groups of patients, related to their stone composition. Hypercalciuria with an incidence of 20 to 30 per cent is the most frequent metabolic disorder found in patients with calcium containing stones (Fig. 1). Hyperoxaluria is found less often, i.e. in 12 per cent of calcium oxalate stone formers and in 7.5 per cent of patients forming calcium phosphate stones. Hyperuricosuria is the main pathogenic factor in urate and uric acid lithiasis. It was also present in about 20% of cystine calculi in our series.

Calcium oxalate and phosphate calculi occur in 69.2 and 9.9 per cent, respectively, and have a high rate of relapses.

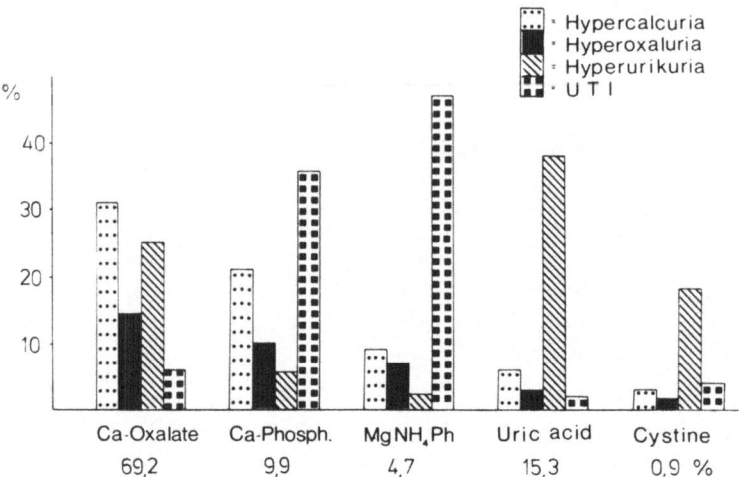

Fig. 1.   Metabolic disorder in 818 patients with calcium
          stones. RWTH, Aachen 1974-1978.

LABORATORY INVESTIGATIONS

    Hypercalciuria and primary hyperparathyroidism (HPT) were
investigated by a modification of Nordin's schedule: The calcium/
creatinine ratio in the urine was measured in morning sample after
overnight fasting and in a 24 hour urine sample following an oral
calcium load of 1 g. Patients were prepared adequately on a low
calcium diet prior to the test. Determinations of serum total and
ionized calcium immunoreactive PTH and urinary cyclic AMP were
routinely performed (table 1).

    A fasting calcium/creatinine ratio under 0.15 rising above 0.30
following an oral calcium load was consistent with absorptive hypercalciuria. A ratio above 0.15 in both tests and a minimal rise after
an oral calcium load revealed resorptive hypercalciuria.

    Mixed forms of hypercalciuria show a high ratio during the
fasting state and a maximal rise after an oral calcium load.

    To demonstrate an intermittent or permanent hyperoxaluria of
primary or secondary genesis, we used gas-chromatography to measure
oxalate elimination in 8 urine aliquots passed during 24 hours.

Table 1. Diagnostic work-up for the assessment of various
types of hypercalciuria.

Type	Diagnostic parameters			
	Ca/Creat.-ratio in urine		Urine	Serum
	4-hour morning sample after overnight fast	24-hour sample and oral calcium load (1000 mg/d)	c-AMP	Ca,Ca^{++} PTH
absorptive	< 0.15	> 0.30		normal
resorptive renal	> 0.15	> 0.15 minimal rise		normal or elevated
mixed	>0.15	>0.15 maximal rise		normal

Cystinuria was examined by amino-acid chromatography.

THERAPY

In consideration of the latest scientific findings about
saturation products and stone formation products as well as precipitation of lithogenous substances, the first priority is sufficient fluid intake by day and by night to maintain a dilute urine.

In a long term study of hypercalciuria treatment, 39 patients
with absorptive hypercalciuria were investigated. Of these, 27
without hypertension, edema, cardiac or renal insufficiency were
given sodium cellulose phosphate (SCP[R]) 5 g three times a day
orally. SCP[R] is an intestinal ion exchanger with a special affinity
for calcium, which is subsequently eliminated in the faeces.

The remaining 12 patients with mild hypertension or minimal
edema were given Campanyl[R], an intestinal ion (potassium versus
calcium) exchanger, 7.5 g two times a day orally.

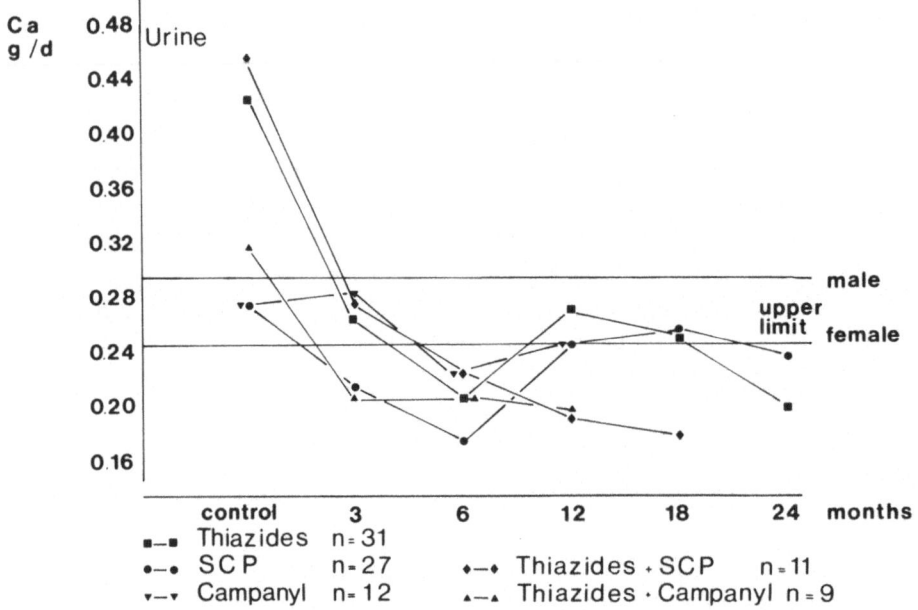

Fig. 2.   Urinary calcium excretion during treatment with
          Thiazides in combination with SCP or Campanyl.

        31 patients with resorptive or renal hypercalciuria were given
hydrochlorothiazide 25 mg three times a day orally. Patients with
primary HPT were excluded.

        20 patients with mixed forms of hypercalciuria were treated
with thiazide 25 mg two times a day orally in combination with
SCP[R] 5 g two times a day (11 patients) or Campanyl[R] 7.5 g a
day orally (9 patients).

        In all of these 90 patients treated with the above mentioned
drugs and their combinations urinary calcium decreased significantly
towards normal (Fig. 2). The best results were obtained with the
combination therapy.

        Based on calcium/creatinine ratios normal values were achieved
by thiazide alone or in combination with SCP[R] or Campanyl[R].
Treatment with Campanyl[R] alone resulted in a normalization of
total calcium excretion but not of the calcium/creatinine ratio,
which is the most reliable parameter (Fig. 3).

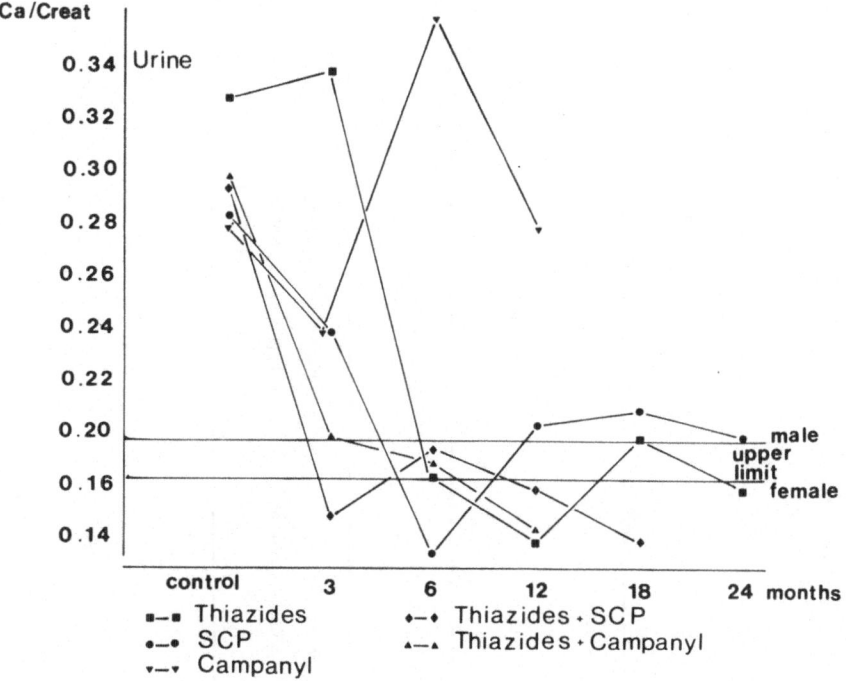

Fig. 3. Calcium/creatinine ratios in patients treated by
         Thiazides alone or in combination with SCP or Campanyl.

Our present views and some information concerning the treatment of hypercalciuric states are presented in table 2.

The most surprising side effect of the treatment of hypercalciuria was a significant increase of the urinary oxalate. The
highest value was observed following the use of intestinal cation
exchangers. Combination therapy with thiazide and cation exchangers
resulted only in a minimal increase of urinary oxalate excretion
(Fig. 4).

An increase in oxalate excretion is deemed to be much more
dangerous than hypercalciuria. After ingestion of calcium-binding
substances, substantial amounts of oxalate remain unbound and diffuse
freely across the intestinal wall. Normally, only approximately 5
per cent of ingested oxalate is absorbed.

Therapy of calcium oxalate calculi is the most difficult.
Therefore we use a FORTRAN IV computer program to calculate the

Table 2. Treatment of hypercalciuria

	precautions	dosage and administration	site of action	systemic side effects
absorptive	hypertension edema	S C P 3 x 5 g/d orally	intestine	edema hypertension weight gain
	cardial or renal insufficiency			
	none	CAMPANYL 2 x 7.5 g/d orally	intestine	none
resorptive	gout, diabetes, hypotension	THIAZIDES 3 x 25 mg/d	renal distal tubule	hyperglycemia, hyperuricemia, hypokalemia, hypotension
mixed	above mentioned precautions	SCP 2 x 5 g/d or CAMPANYL 7.5 g/d with THIAZIDES 2 x 25 mg/d orally		minor side effects

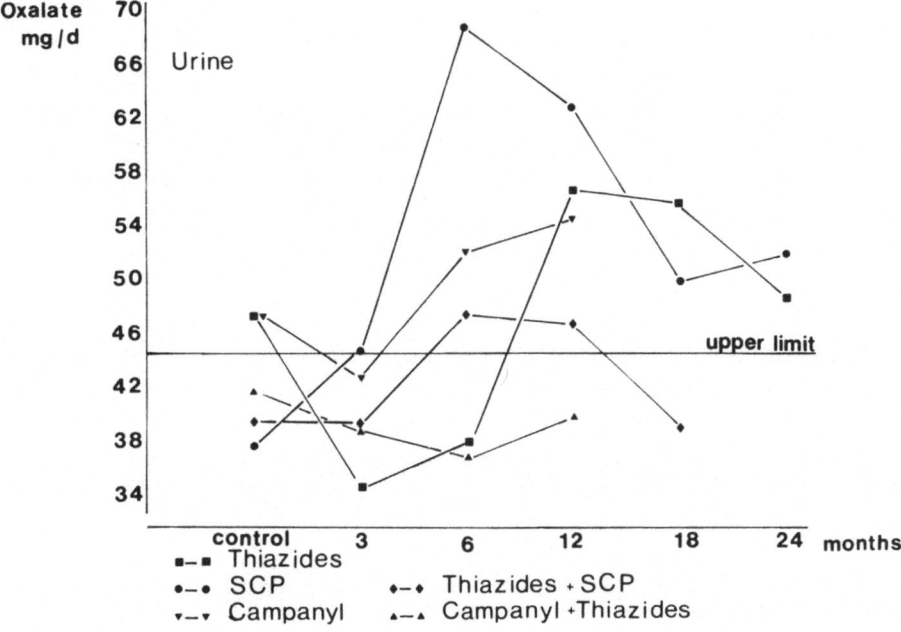

Fig. 4. Urinary oxalate excretion in patients treated by
         Thiazides alone or in combination with SCP or Cam-
         panyl.

relative supersaturation of the urine. Only in cases of a metastable
urine, with respect to calcium oxalate, is drug therapy initiated.

Treatment of calculi caused by infection (Ca-phosphate, Ca-
carbonate and Mg-ammonium phosphate) is also rather difficult. In
addition to the exclusion of a distal renal tubular acidosis which
was found in 10 per cent of the cases, four other cardinal principles
of treatment should be observed: control of infection, an ample
fluid intake, acidification of urine and treatment of any metabolic
disease.

Only in pure uric acid calculi and, with reservation, in cystine
calculi can oral treatment for stone dissolution be given. The
principle of such an attempt is based on neutralizing, or even
alkalizing the urine, using simultaneous treatment with xanthine-
oxidase blocking agents (allopurinol) in uric acid calculi or of
SH-group binding agents, D-penicillamine or mercaptopurinol (e.g.

Thiola) in cystine calculi.

A follow-up from 1974 to 1978 of 369 patients with recurrent urolithiasis of different stone composition – corresponding to 935 patients/years of treatment – shows that the renal stone formation rate decreased significantly from 1.21 stones per patient before onset of treatment to 0.09 stones per patient under conservative treatment.

CONCLUSION

In patients with stones in the urinary tract the combination of a high fluid intake, dietetic control, correct medical treatment, regular laboratory controls and cooperation with the patient, makes it possible to lower the recurrence rate to under 10 per cent.

LONG-TERM TREATMENT WITH THIAZIDE IN THE PREVENTION OF CALCIUM NEPHROLITHIASIS

N. Tessitore, A. D'Angelo, F. Pagano, C. Lo Schiavo,
A. Corgnati, A. Fabris, A. Lupo, E. Valvo, A. Tasca,
L. Oldrizzi, G. Previato and G. Maschio

- Cattedra e Divisione di Nefrologia,
  Istituti Ospedalieri, 37100 Verona (Italy)
- I Clinica Medica, Università di Padova (Italy)
- Clinica Urologica, Universita di Padova (Italy)
- Cattedra di Chimica e Microscopia Clinica,
  Universita di Padova, sede di Verona (Italy)

ABSTRACT

During the last 4 years, we have examined 292 patients with recurrent calcium nephrolithiasis. During 1845 years of lithiasis, they had formed 2340 stones, with a mean interval between two stone events of about 9 months. Hypercalciuria was observed in 56%, and hyperuricosuria in 22% of the patients.

A reliable follow-up after thiazide administration (hydrochlorothiazide 50 mg + amiloride 5 mg per day) was obtained in 161 patients, treated for 4-48 months (mean 18 months). During the treatment, 15 new stones were formed, in contrast with the 305 predicted. In our experience, thiazide (associated with allopurinol in patients with hyperuricosuria) is very effective in preventing the recurrences of calcium nephrolithiasis.

During the last four years, we have examined 292 patients with recurrent calcium nephrolithiasis. During 1845 years of lithiasis,

they had formed 2340 stones   (approximately 8 stones per patient)
with a mean interval between two stone events of about 9 months
(0.788 years). In all patients, an investigation of calcium, magnesium and uric acid metabolism and an x-ray control were performed
prior to the treatment. Absolute hypercalciuria, defined as a 24
hour urinary calcium excretion of more than 250 mg in women and
300 mg in men (4), was observed in 24% of our stone-formers. Marginal
hypercalciuria, defined as a 24-hour urinary calcium excretion of
more than 150 mg per g. of creatinine (3), was observed in 32% of
our patients, while 44% of them had a normal calcium excretion. In
all patients, urinary calcium excretion was measured again after
at least 7 days of calcium-restricted diet (about 400 mg per day).
In this second assessment hypercalciuria, defined as a 24-hour
calcium excretion of more than 200 mg (5), was observed in 43% of
our stone-formers, while 57% had normal urinary calcium excretion,
that is less than 200 mg per day. In order to evaluate the pathophysiological mechanisms underlying stone formation, we attempted
to study the relationship between urinary calcium and creatinine
excretions in a fasting 2-hour sample and in 4-hour sample after
oral ingestion of 1 g. of calcium (6). A group of patients had no
metabolic alterations, that is a fasting and after-load urinary
calcium to creatinine excretion ratio in the normal range. A second
group of patients had absorptive hypercalciuria (normal to high
24-hour calcium excretion, normal fasting and high after load excretion). A third group of patients had renal hypercalciuria (high
24-hour and fasting calcium excretion, with a further increase after
calcium load). The urinary uric acid excretion was evaluated in all
patients. Hyperuricosuria (defined as a 24-hour excretion of more
than 750 mg per day) was observed in 22% of the patients. Only in
10% of our patients was hyperuricosuria associated with hypercalciuria. Long-term thiazide administration was prescribed for 200
patients with recurrent calcium nephrolithiasis (50 mg/24 hours of
hydrochlorothiazide associated with 5 mg of amiloride); 28 patients
were lost to follow-up and 11 (about 6%) had to stop the treatment
because of side effects such as hypotension, dizziness and skin
rashes. In about 30% of the patients, additional treatment with
allopurinol (100 to 200 mg per day) was prescribed. A reliable
follow-up was available in 161 patients, treated for periods of
time ranging from 5 to 48 months, with a mean period of about 18
months. During the treatment, 15 new stones were formed, in contrast with the 305 predicted; i.e. less than 5% of the predicted
stones were actually formed (Table 1). We have attempted to evaluate

Table 1. Effects of thiazide treatment in 161 patients
with recurrent calcium nephrolithiasis.

	P R E	P O S T
Years per patient	6.316	1.534
Number of stones	1257	15
Stones per patient	7.807	0.093
Stones per year	1,235	0.060
Mean interval between stones (yrs)	0.809	16.466
Stones predicted		305
New stones / predicted (%)		4.918

the clinical effects of the drugs in the various subgroups of patients. The patients with no metabolic alterations had a shorter
history of nephrolithiasis and formed fewer stones per patient before
starting any treatment than did those with renal or absorptive
hypercalciuria (7 stones per patient against 8.08 and 8.19, respectively); on the other hand, during treatment they formed fewer
new stones than did the patients with hypercalciuria both in
absolute (3 stones against 6, respectively) and in relative terms
(3.84% of the predicted stones against 5.77% - renal - and 4.87%
- absorptive -, respectively) (Table 2). Our experience adds confirmation to the well known effectiveness of thiazide in controlling
the recurrences of nephrolithiasis in patients with renal hypercalciuria (1) as well as in stone-formers with normocalciuria (2).
These data are in agreement with a multifactorial mode of action
of thiazide, not restricted to the well known hypocalciuric effect
(3, 7) and including the well documented increase in urinary excretion of some inhibitors of stone formation during thiazide therapy
(7). During thiazide therapy, an increase in urinary magnesium excretion was observed in patients with hypercalciuria and with normocalciuria; in the first group, the effect was statistically more
significant, so that the increase of urinary magnesium to calcium
ratio was nearly the same in the two groups of patients. Thiazide

Table 2. Effects of thiazide treatment in 161 patients with recurrent calcium
          nephrolithiasis.

	RENAL HYPERCALCIURIA	ABSORPTIVE HYPERCALCIURIA	NO METABOLIC ALTERATIONS
Number of patients	53	62	46
Pretreatment years per patient	7.45	6.87	4.26
Pretreatment stones	428	507	322
Stones per patient	8.08	8.19	7.01
Mean interval between stones (yrs)	0.92	0.84	0.61
Treatment years per patient	1.57	1.76	1.24
New stones formed	6	6	3
Stones predicted	104	123	78
New stones / predicted (%)	5.77	4.87	3.84

administration did not induce statistically significant changes in urinary uric acid excretion, while the association with allopurinol was obviously able to reduce uricosuria significantly. In our experience, the association of thiazide and allopurinol is very effective in the treatment of recurrent calcium nephrolithiasis. Allopurinol is helpful in reducing one of the most important risk factors in its pathogenesis, the urinary excess of uric acid, and it also controls one of the side effects of thiazide therapy, namely the increase in serum uric acid concentration.

REFERENCES

1. Aroldi, A., Graziani, G., Mioni, G., Cecchettin, M., Brancaccio, D., Galmozzi, C., Surian, M., Cantaluppi, A. and Ponticelli, C.: Thiazide diuretics in renal hypercalciuria. Proc. EDTA 16: 571-576, (1979).
2. Coe, F.L.: Treated and untreated recurrent calcium nephrolithiasis in patients with idiopathic hypercalciuria, hyperuricosuria and no metabolic disorders. Ann. Int. Med. 87: 404-410, (1977).
3. Coe, F.L.: Nephrolithiasis. Pathogenesis and Treatment. Year Book Publ. Chicago, (1978).
4. Hodgkinson, A. and Pyrah, L.N.: The urinary excretion of calcium and inorganic phosphate in 334 patients with calcium stones of renal origin. Brit. J. Surg. 46: 10-16, (1958).
5. Pak, C.Y.C., Ohata, M., Lawrence, E.C. and Snyder, W.: The hypercalciurias. Causes, parathyroid function and diagnostic criteria. J. Clin. Inv. 54: 387-399, (1974).
6. Pak, C.Y.C., Kaplan, R., Bone, H., Townsend, J. and Waters, O.: A simple test for the diagnosis of absorptive, resorptive and renal hypercalciurias. N. Engl. J. Med. 292: 497-500, (1975).
7. Yendt, E.R. and Cohamin, M.: Prevention of calcium stones with thiazide. Kidney Internat. 13: 397-409, (1978).

EDITORIAL NOTE

Throughout the text the term "thiazide" is employed, in spite of the fact that the patients were treated with a combination of hydrochlorothiazide with a non-thiazide diuretic, amiloride. Amiloride, a potassium-sparing diuretic, was given concomitantly

with thiazide in order to counteract the potassium-losing effect
of the latter.

The results presented by Tessitore and his coworkers with the
combination of hydrochlorothiazide and amiloride are comparable
with those obtained by Authors who employed thiazide alone. The
effect of amiloride on calcium excretion and prevention of stone
recurrence deserves to be clarified.

In addition, it should be noted that about 30% of patients also
received allopurinol. Thus, "long-term treatment with hydrochlorothiazide and amiloride - with or without allopurinol - in the
prevention of calcium nephrolithiasis" might be have been a more
descriptive title.

LITHOLYSIS: IN SITU DISSOLUTION OF KIDNEY STONES

E. Dormia

Department of Urology
Ospedale Provinciale
Lecco (Italy)

ABSTRACT

Litholysis can be obtained by oral treatment only in the case of uric acid lithiasis and in other rare situations. It can also be obtained by rinsing the stones with lytic solutions which are brought in close contact with the calculi by instrumental means.

The techniques, the solutions and the indications for instrumental lytholisis are described.

The results and the long-term experience are presented, with a follow-up of 11 years in some cases.

––––––––

Oral litholysis can be consistently performed only in uric acid lithiasis.

It can occasionally be successful in the treatment of cystine and, rarely, of phosphate stones. In all other cases, a dissolution of renal stones can be performed by endorenal perfusion, using a technique which involves the use of particular solutions, instruments and maneuvres.

In situ dissolution of various stones can be achieved if the

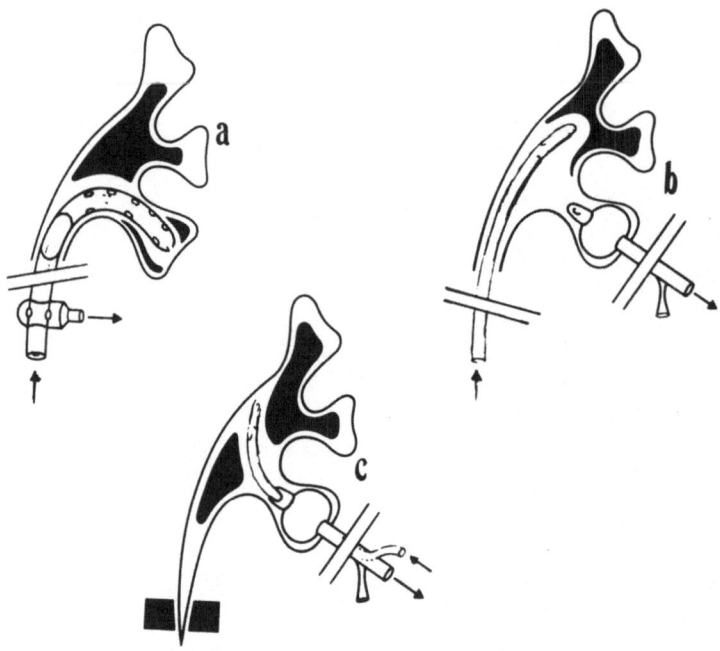

Fig. 1. Continuous perfusion.

calculi are rinsed with lytic solutions of different composition,
varying according to the various stone types. We can affirm today
that, at least theoretically, the dissolution of all stones is
possible.

The perfusion, obtained by instrumental endoscopy, can be either
continuous or intermittent, depending on whether the solvent flows
through the renal cavities continuously or by alternate instillation
and drainage. These two types of perfusion are performed in the
following ways:

Continuous perfusion (Fig. 1)

a) Inlet and outlet through a two-way ureteral catheter.

b) Inflow through a simple ureteral catheter and outflow from
   a nephrostomy tube or vice versa.

c) Outflow through a nephrostomy tube, the lumen of which

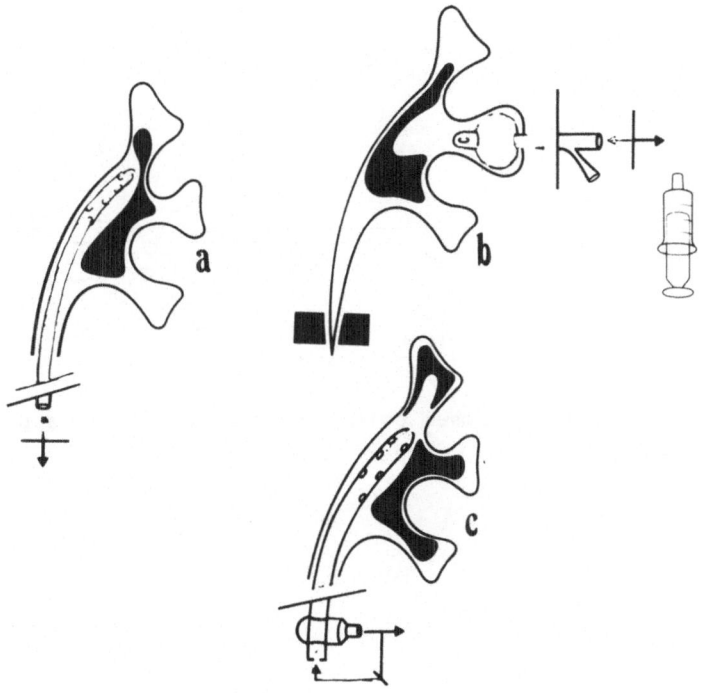

Fig. 2. Intermittent perfusion.

contains a ureteral catheter, which is employed for the introduction of the fluid.

The continuity of the flow accounts for the fact that this type
of perfusion only acts in close proximity to the catheters or within
the shortest route between them.

Its action is therefore usually slow and it requires the catheterization of the stone-containing calyces. As it does not allow
the cavities to fill up, the risks of distension and therefore of
parenchymal reflux are limited. Consequently this type of perfusion
must always be employed when the calculi occupy large parts of the
cavities.

Intermittent perfusion (Fig. 2)

This can be performed as follows:
a) Inlet and outlet through a simple ureteral catheter.

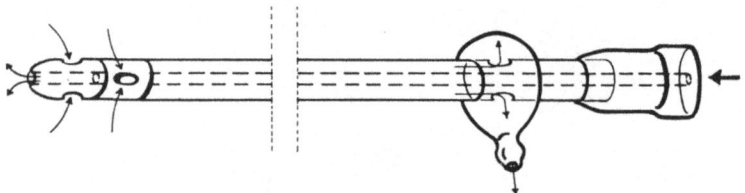

Fig. 3.   Concentric lumen catheter with automatic apparatus
          for intermittent perfusion.

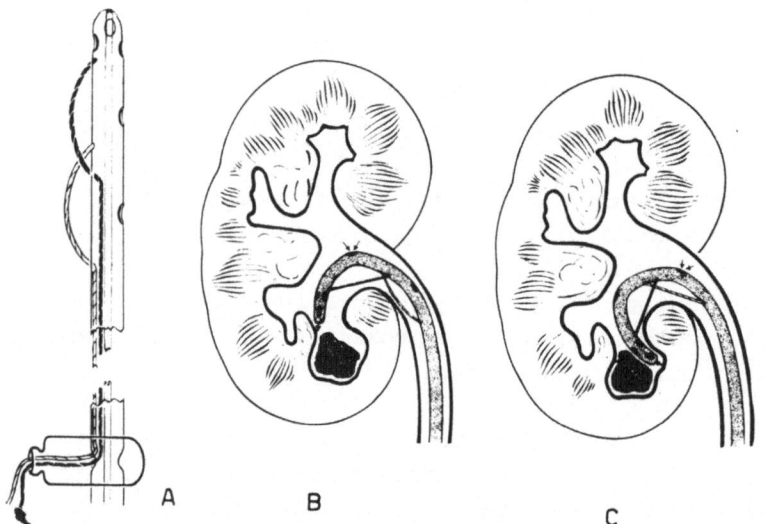

Fig. 4.   A wire, applied to the catheter, provides, in each
          case, the most appropriate curvature for the catheter
          tip.

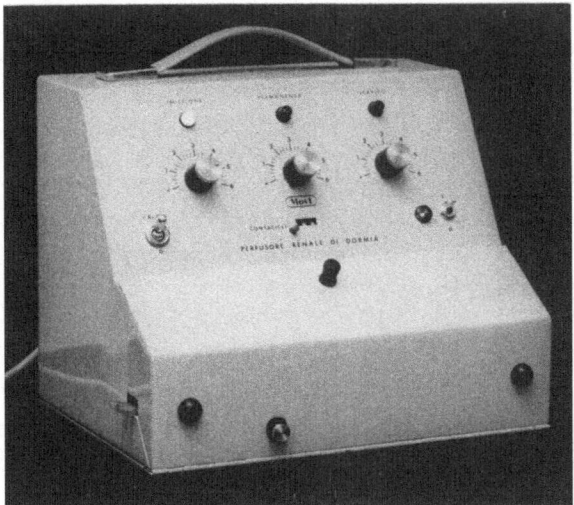

Fig. 5.   Automatic machine for intermittent perfusion.

b) Inlet and outlet through a nephrostomy tube. Only a syringe
   is necessary in this case.

c) Inlet and outlet through a two-way ureteral catheter.

This kind of perfusion, as opposed to the continuous one, allows
the diffusion of the solution throughout the kidney cavities, increasing their volume at each influx. For this reason it has a
faster action and does not normally require the catheterization of
the calyces; but for the same reason this can only be used in cases
where the calculus only partially occupies the renal cavities.

In order to avoid overdistension, perfect emptying must be
obtained in both techniques. In intermittent perfusion the same
volume of solution must be injected at each cycle.

These two fundamental requirements are fulfilled with the concentric lumen catheter and with the automatic apparatus for intermittent perfusion.

In above-mentioned catheter (Fig. 3) the holes giving exit to
the outflow solution are situated all around the contour of the
endorenal extremity. This greatly improves the drainage from the

renal cavities as compared with the parallel-channel catheters. When catheterization of the calyces is necessary, a wire can be applied to the catheter, providing, in each case, the most appropriate curvature for the catheter tip (Fig. 4).

The automatic machine (Fig. 5) for intermittent perfusion allows the inflow of a constant volume of fluid, regulates the time of retention in the kidney, and ensures free emptying of the cavities before successive injections.

Different kinds of solvent solutions are described in the literature (Tab. 1). To dissolve calcium and struvite calculi we use a mixture of chelating solutions of trisodium ethylendiamino-tetraacetate (EDTA) and of bisodium nitryltriacetate (NTA), having an alkaline pH (Dormia's solution).

In the treatment of phosphate calculi we use alternatively the above-mentioned mixture and Alboulker's acid solution modifying the pH of solution day by day according to the presence of infection.

In my experience, when oxalic calculi are to be dissolved, it is necessary to use a chelating solution to which an appropriate dose of alkaline solution is added, thereby increasing its effectiveness. If uric acid is also a component of the stone, its dissolution is favoured by increasing the alkalinity of the solution.

To dissolve cystine calculi a sodium hydroxide solution is the most active, whereas piperazine solution is more effective against uric acid calculi.

During the night, and at other intervals, we use a mucolytic wash with a solution of acetyl-cysteine, to which antibiotics are usually added. The latter solution has allowed us to dissolve four soft matrix stones.

The indications for instrumental litholysis remain controversial.

If one considers litholysis as an alternative method to a surgical operation, at least in selected cases, the choice depends not only on the surgeon's experience but also on the different evaluation that he gives to the risks of surgery versus those of

Tab. 1 Solutions for litholysis.

Uric acid - Urates	Alkaline solution (Aboulker)	Pyperazine g 16 Glycine g 30 H_2O g 1000	
	P 30 Solution (Timmermann)	Lithium EDTA salt	
Calcium-oxalate	Calsol (Gehres-Raymond-Timmermann)	Bisodium EDTA 15%	
	Litholys (Dormia-Zardini)	Trisodium EDTA Bisodium NTA equimol solution 3%	
Calcium - Ammonium Maghesium Phosphates	Renacidin (Mulvaney)	Citrates-gluconates malates-methylic and ethylic esters	
	Acid solution (Aboulker)	Citric acid, maleic acid and sodium gluconate in water	
	Calsol		
	Litholys		
Cystine	Sodium lye solution (Steg-Thomas)	Sodium hydrate ml 8 Glycine g 27 H_2O 1000	
Soft (matrix)	Fluimucil	N-acetyl-cysteine 4% in physiologic saline solution	

litholysis.

The following points are to be carefully evaluated:

a) Litholysis requires a longer hospitalization period, varying
   with the volume of the calculus.

b) It can cause primary infection or, more often, produce an
   acute flare-up of a chronic infection.

c) Rarely it can be the cause of stenosis of calyces, ureter
   or urethra.

d) Its results may be incomplete.

We have used litholysis therapy:

1) For calculi, in recurrent stone-forming patients, with no
   obstruction in the urinary tract, if there is a recurrence
   as well as a definite risk that further interventions may
   produce urinary tract stenosis.

2) For other calculi, usually staghorns (almost always relapsing), under three circumstances:

a) when the stone is of limited size, and therefore easy to
   dissolve, especially if contained in an intrarenal pelvis,
   so that a wide nephrotomy be the only alternative procedure.

b) When, in a kidney already submitted to several previous
   surgical procedures, the operation appears likely to be
   exceedingly difficult, unlikely to allow a complete removal
   of multiple recurrent stones and is associated with a
   very high surgical risk.

3) For calculi left after an operation.

4) For calculi in inoperable patients.

5) For uric acid stones in inoperable patients when alkalinisa-
   tion cannot be employed.

As an example, figure 6 shows the results obtained in a patient with a recurrent phosphatic stone in a single kidney.

We do not consider as candidates for litholysis:

- any patients that can be cured by medical or surgical treatment;

- patients affected by unilateral lithiasis with severe renal functional impairment, having a healthy contralateral kidney. Here nephrectomy is to be preferred.

- calculi which are secondary to urinary tract obstruction, requiring surgical correction with simultaneous stone removal.

Lastly, I wish to describe our policy in cases of calculi constantly recurring in single kidneys or in both kidneys when preventive therapy has been ineffective. In such cases we remove the stones surgically and perform a definitive nephrostomy at the same procedure. The patient himself can perform daily irrigations at home through the nephrostomy tube, using lytic solutions, in order to prevent further relapses. The nephrostomy is applied in a "polopolar" position using a U-loop tube. We have devised a special catheter, with concentric channels which allows:

1) quick and safe replacement;

2) effective cleansing of the intrarenal cavities;

3) continuous or intermittent perfusion with antibacterial or litholytic solutions.

In my opinion, litholysis must not be considered as an alternative therapy to a surgical operation if this is easily feasible, but either as a complement to the latter or as the treatment of choice in cases with a very high operative risk.

Table 2 shows our results obtained in the last 11 years. I would like to make it clear that only 11 failures were due to technical reasons: in three patients oxalic calculi remained insoluble while in the other eight patients affected by phosphatic calculi it was not possible to reach the fragments in the calyces.

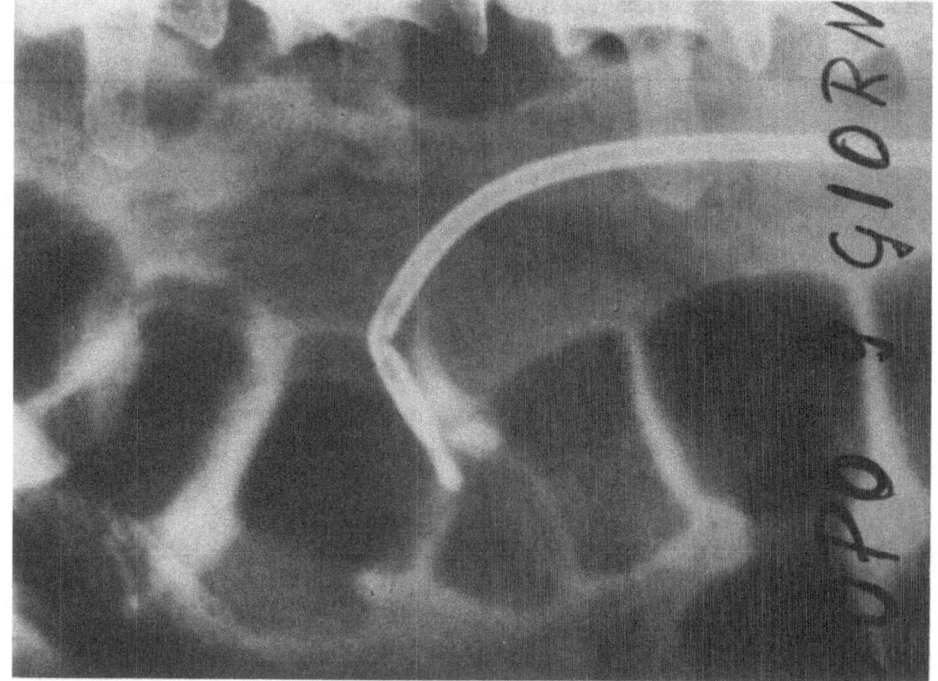

(B)

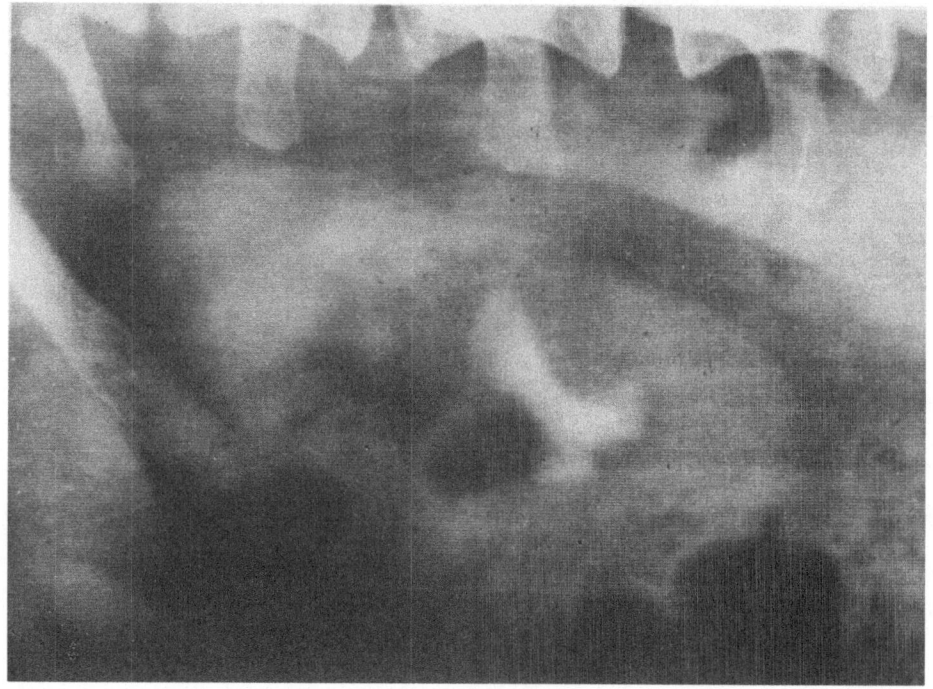

(A)

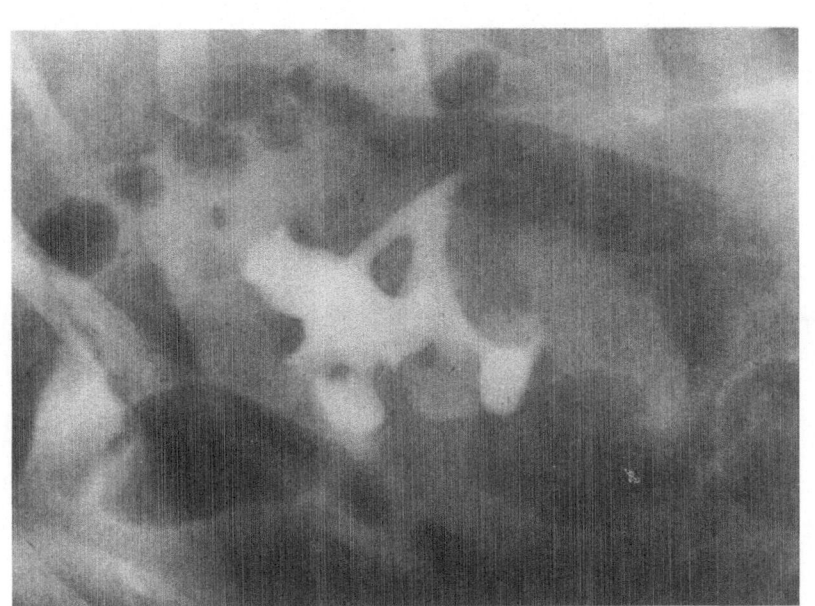

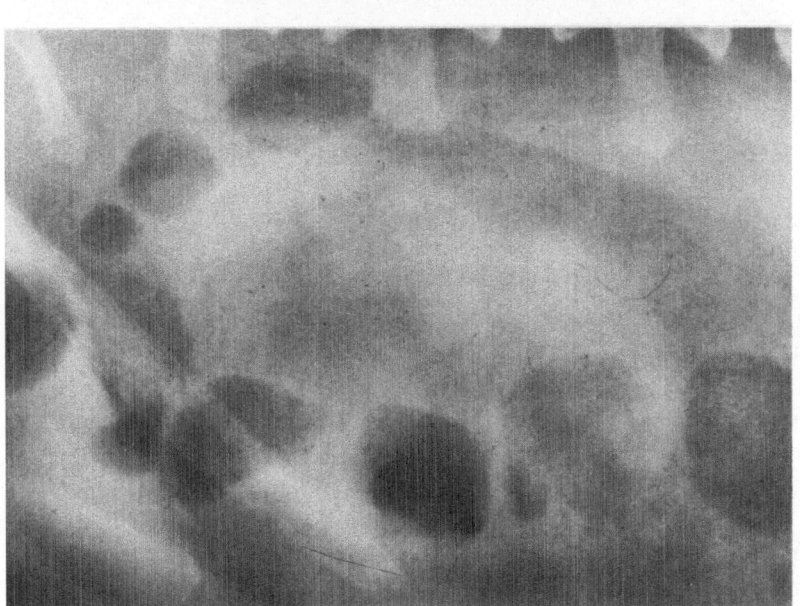

Fig. 6.   In situ dissolution of a phosphatic recurrent stone in a single kidney. A: X-ray examination before the treatment. B: After nine days of perfusion. C: After seventeen days of perfusion. D: Intravenous pyelography at the end of the treatment.

Tab. 2. Personal experience 1969–1979.

CALCULI		TOTAL DISSOLUTION	PARTIAL DISSOLUTION
Phosphates	174	127	47
Oxalates	71	67	4
Soft (matrix)	4	4	–
Urates (by perfusion)	5	5	–
Cystine	1	–	1
	255	203	52
Uric acid (per os)	73	69	4
	328	272	56

NEPHROSTOMY + PREVENTIVE LITHOLYSIS

Phosphates	15	No relapse and well functioning kidney (after 1 – 3 years): 12

The remaining cases are either waiting to be readmitted or have declined further treatment.

The long-term unfavourable results of instrumental litholysis are shown in table 3.

Among 169 patients who have obtained a complete dissolution of stones, relapses occurred in 34%.

We think that these results are of definite value. They can be very gratifying, for both the surgeon and patient, especially in the case of rapidly recurring staghorn calculi, particularly if bilateral, or located in solitary kidneys. In 8 out of 11 such cases, patients who underwent post-surgical litholysis have remained free from recurrences with a follow-up of 1-3 years.

Table 3. Side effects of and recurrence rate after litholysis.

UNTOWARD EFFECTS

- Ureteral stenosis   2   (0,78%)
- Urethral stenosis   1   (0,39%)
- Glomerular filtration
  rate reduction below   15   (5,88%)
  20 ml/min

RECURRENCES

- Out of 169 cases   57   (34 %)

REFERENCES

1. Aboulker, P., Thomas, J., Tanret, P. and Steg, A.: Dissolution des calculs pyélo-caliciels et urétéraux d'acide urique. J. d'Urologie 67: 224, (1962).

2. Dormia, E. and Zardini, O.: Esperienze di laboratorio su una nuova soluzione litica per i calcoli urinari. Arch. Ital. Urol. 36: 167, (1963).

3. Dormia, E.: Studio tossicologico sulla soluzione DZ. Arch. Ital. Urol. 35: 508, (1962).

4. Dormia, E.: La litolisi. In: M. Sorrentino - Aggiornamenti di Urologia - Minerva Medica, Torino, 1974, p. 315-335.

5. Mulvaney, W.P. and Henning, D.C.: Solvent treatment of urinary calculi. J. Urol. 88: 145, (1962).

6. Suby, H.: Dissolution of urinary calculi. J. Urol. 68: 96, (1952).

7. Timmermann, A.: Die Grundlagen der klinischen Nierenstein-chemolyse. Der Urologe 2: 243, (1963).

8. Timmermann, A.: Uber den Stand der Auflösung von Harnsteinen durch Nierenbecken-Dauerspulung. Die Therapiewoche 13: 105, (1963).

SPEAKERS

G. Andres, Buffalo, USA
J.F. Bach, Paris, France
F.C. Berthoux, Saint-Etienne, France
V. Bonomini, Bologna, Italy
M. Broyer, Paris, France
M. Camey, Suresnes, France
G. Camussi, Turin, Italy
R. Coppo, Turin, Italy
E. Dormia, Lecco, Italy
J. Egido, Madrid, Spain
M. Fini, Bologna, Italy
H. Fleisch, Berne, Switzerland
L. Giuliani, Genova, Italy
J. Hannappel, Aachen, West Germany
H. Kreis, Paris, France
F. Linari, Turin, Italy

Q. Maggiore, Reggio Calabria, Italy
R. Maiorca, Brescia, Italy
A. Martelli, Bologna, Italy
G. Maschio, Verona, Italy
L. Miano, Rome, Italy
P. Michielsen, Leuven, Belgium
R. Mongiorgi, Bologna, Italy
M. Pavone-Macaluso, Palermo, Italy
G. Piccoli, Turin, Italy
V. Pulini, Bologna, Italy
L. Riva di Sanseverino, Bologna, Italy
A. Rousaud, Barcelona, Spain
B. Van Damme, Leuven, Belgium
A. Vercellone, Turin, Italy

INDEX